People on the Move in a Changing Climate

Global Migration Issues

Volume 2

Series Editor:
Dr. Frank Laczko
Head of Research and Publications, International Organization for Migration (IOM), Geneva

This book series contributes to the global discussion about the future of migration policy through the publication of a series of books on emerging migration issues. Most reports on migration policy tend to focus on national or regional perspectives; books in this series will focus on global policy challenges, such as the impact of climate change or the global economic crisis, on migration.

This series is closely linked to the production of IOM's *World Migration Report*. Some of the books in this series will be based on research which has been prepared for the *World Migration Report*.

The series also includes a special focus on the linkages between migration and development, and the themes discussed each year at the Global Forum on Migration and Development (GFMD), given the growing policy interest in harnessing the benefits of migration for development.

For further volumes:
http://www.springer.com/series/8837

Etienne Piguet • Frank Laczko
Editors

People on the Move in a Changing Climate

The Regional Impact of Environmental Change on Migration

Springer

Editors
Etienne Piguet
Institute of Geography
University of Neuchâtel
Neuchâtel
Switzerland

Frank Laczko
Migration Research Division
International Organization for Migration
Geneva
Switzerland

ISSN 2213-2511 ISSN 2213-252X (electronic)
ISBN 978-94-007-6984-7 ISBN 978-94-007-6985-4 (eBook)
DOI 10.1007/978-94-007-6985-4
Springer Dordrecht Heidelberg New York London

Library of Congress Control Number: 2013951644

Printed on acid-free paper

Springer is part of Springer Science+Business Media (www.springer.com)

Book Presentation

Climate change is one of the major concerns for the international community. Even if precise estimates are unavailable, it is expected that more and more people will migrate due to changes in the environment in the future. Despite a growing body of case-studies, there was no volume which provides a synthesis and comparison of how each of the major regions of the world is likely to be affected by changes in the environment and migration in the coming years. This book therefore fills an important gap in the literature, and at the same time informs regional policy discussions. At present, most of the policy fora which exist to promote cooperation between States on migration issues at the regional level, such as regional consultative processes on migration, do not include policymakers dealing with environmental issues. The book will, therefore, serve as an important tool for policymakers interested in promoting regional cooperation on issues relating to migration, the environment and climate change. Each chapter of the book is written by leading experts of the region, on the basis of a unified framework, in order to produce an evidence-based and exhaustive assessment of the state of knowledge on the environment/migration nexus.

Contents

Regional Perspectives on Migration, the Environment and Climate Change

1

Frank Laczko and Etienne Piguet

Context

The perception that large numbers of people may be forced to migrate due to the effects of climate change has fuelled a renewed interest in the subject of migration and the environment. Recent estimates suggesting that between 200 million and 1 billion people could be displaced by climate change over the next 40 years have alarmed policymakers. Even though such estimates have been dismissed as, at best, "guesswork" by many experts (IPCC 2007; Foresight 2011), they have helped to focus policymakers' attention on the linkages between migration and climate change. Concerns about the migration-related consequences of climate change have encouraged policymakers around the world to focus more on how environmental change will affect people's lives and human security. The Chairman of the leading expert authority, the Intergovernmental Panel on Climate Change (IPCC), for example, has talked about the "human faces of climate change" (Piguet 2013). Unlike indicators of environmental health, such as carbon dioxide (CO_2) emissions or changes in rainfall or temperature, migration reflects the human dimension of climate change.

This book focuses on the likely effects of environmental change – particularly climate change – on migration. Its title (*People on the Move in a Changing Climate*)

Both authors are very grateful to Sieun Lee, Research Officer at IOM Geneva, for her valuable assistance.

F. Laczko (✉)
Migration Research Division, International Organization for Migration, Geneva, Switzerland
e-mail: flaczko@iom.int

E. Piguet
Institute of Geography, University of Neuchâtel, Neuchâtel, Switzerland
e-mail: etienne.piguet@unine.ch

E. Piguet and F. Laczko (eds.), *People on the Move in a Changing Climate*,
Global Migration Issues 2, DOI 10.1007/978-94-007-6985-4_1,

reflects the fact that many people who migrate due to environmental factors, like most migrants in the world today, do not cross international borders. Migrants who move for environmental reasons are also likely to fall into many different categories, and not all migration linked to environmental change can be described as forced displacement. Except in extreme cases, population movements tend to be the result of a multi-causal relationship between environmental, political, economic, social and cultural dimensions. Although media and policy attention tends to focus on disasters and displacement, it is important to keep in mind that environmental factors do not affect all individuals, households and communities equally, and information related to climate change is not perceived in the same way everywhere and by everyone (Piguet 2010a: 517). Even when confronted with severe environmental push factors, people and communities are resilient and have some degree of control over their decision to migrate or to choose other adaptation strategies. Some people, for example, may choose to migrate to avoid the impact of environmental change and therefore have time to plan their move. Others may decide to move temporarily, while the environmental situation is poor. In this sense, migration can become a way of adapting to climate change, for some people, rather than being viewed merely as a problem.

Research on Migration and the Environment

Migration and the environment is not a new area of research. As early as 1992, for example, IOM published a report – *Migration and the Environment* – that stated:

> Large numbers of people are moving as a result of environmental degradation that has increased dramatically in recent years. The number of such migrants could rise substantially as larger areas of the earth become uninhabitable as a result of climate change (IOM 1992).

For many years, however, the topic has been neglected by researchers. In the 2005 report of the Global Commission on International Migration, for example, there is barely a mention of migration and the environment. This may be partly due to the lack of consensus among researchers as to whether or not environmentally induced migration is a distinct form of migration worthy of special study. There has been considerable disagreement about how to conceptualize the relationship between migration and climate change and migration and the environment. Although many experts accept that climate change can be a factor in people's decision to migrate, the conceptualization of this factor as a primary cause of forced displacement has been questioned (Black 2001). While the environment can be a driver of migration, usually a complex combination of causes determines whether or not people move. Given the multiple causes of migration, therefore, drawing a clear line between voluntary and forced movements is not always straightforward.

This disagreement over the extent to which the environment induces migration is reflected in further disagreements over terminology. It is common to describe those who move for environmental reasons as 'climate change refugees' or as 'environmentally displaced persons' and to characterize such movements as 'forced migration'. Popular with the media, the term 'environmental refugees' has been used to describe the whole category of people who migrate for environmental reasons. This broad definition, while evoking an image that has brought public attention to the issue, is not sufficiently precise to describe the whole category of people who migrate due to environmental factors. As noted earlier, in some situations, such as natural disasters, people may have little choice about moving and may be forcibly displaced. In other situations where environmental change is gradual, movement is more likely to be voluntary as people have time to consider their options, and environmental change may be one of many factors inducing them to move.

The Purpose of This Book

Due to renewed policy and media interest in the subject of migration and the environment, there is a growing interest in improving data and research in this field. Policymakers are essentially interested in the following types of questions: how many people are likely to move, who is most likely to move, where will they move to, what will be the likely impact on the origin and destination areas, and how should decision-makers plan for such migration?

The main purpose of this book is to review and compare the existing evidence base in each major region of the world in order to inform policy responses, especially at the regional level. In most regions of the world, there are policy forums that deal with migration issues, such as regional consultative processes (Achieng 2012), but policymakers have yet to fully address questions linked to migration and the environment in these forums. Before launching new studies, it is important that policymakers learn from existing evidence. There are many studies on migration and its linkages to the environment, but the information is often scattered between countries and within regions. This book aims to conduct, for the first time, a systematic review of existing research on how environmental change affects migration across all the major regions of the world. Essentially, the book explores what lessons can be learned from the current body of research; what has been the main focus of this research at the regional level; in which regions of the world the evidence base is weakest; and whether studies that have been conducted in one region of the world could be replicated in another.

The aim of this book is not to elaborate upon, or to synthesize, conceptual and theoretical debates, as this has been done elsewhere (Laczko and Aghazarm 2009; Piguet 2013). Similarly, the issue of the environmental impact of migration is not considered herein, as it refers to a largely different – and significantly older – body of literature (Black 1998; Hugo 2008). The book focuses instead on mapping the existing research on migration and the environment at the regional level, and on pooling together the key findings and results from empirical research, field studies and surveys. To ensure coherence between chapters, the main theme of the book is the impact of environmental change on migration, while recognizing that the environment may not always be the sole factor driving migration.

Structure of the Book

Each chapter of the book is written by leading regional experts using a common framework. All authors were able to draw upon a common bibliographic database that has since been published as a separate reference document (*People on the Move in a Changing Climate: A bibliography*, IOM 2013). As the focus in this book is on regions, there is no specific chapter dealing with small island States. Discussion of the evidence base relating to these islands is included within the respective regional chapters – for example, the chapter on Oceania includes references to the Pacific Islands. In addition, given the paucity of data relating to the Middle East and North Africa, a special chapter was prepared for this region. A specific chapter on the Himalayas was also commissioned as this is one of the areas of the world most likely to be affected by climate change, and several new studies have been conducted in the region. Focusing on this region also allowed for the study of certain environmental factors specific to mountainous regions that have an impact on the movement of people.

In preparing their chapters for this book, authors used a common framework, covering the following key points:

- An historical overview of migration trends relating to environmental events
- A brief synthesis of the regional forecasts regarding climate change, with a special emphasis on those that are recognized as relevant to population movements
- An overview of the main current environmental issues in the region, whether connected or not with climate change
- An overview of the main current internal and international migration trends
- A synthesis of existing case studies on the links between migration and environmental stressors.

Research Database on Migration and the Environment, Created for This Project

The bibliographic database used in this book was prepared by the University of Neuchâtel, with additions from IOM. Keywords are used to enable the reader to identify empirical cases studies, and they are organized according to world regions and the methodology used in the study. Using the database, it is possible to obtain a detailed overview of the existing empirical literature on the environmental impact on migration. This analysis,[1] presented below, builds on earlier but less quantitative assessments completed by other authors (Erway Morinière 2009; Morrissey 2009; Piguet 2010a; Laczko and Aghazarm 2009). The database of research studies shows that, in recent years, there has been a significant linear increase in the number of research studies and publications focusing on migration and the environment (see Fig. 1.1). Indeed, since 2008, more reports and studies have been published on

[1] The quantitative analysis was conducted by Sieun Lee and Frank Laczko at IOM – Geneva.

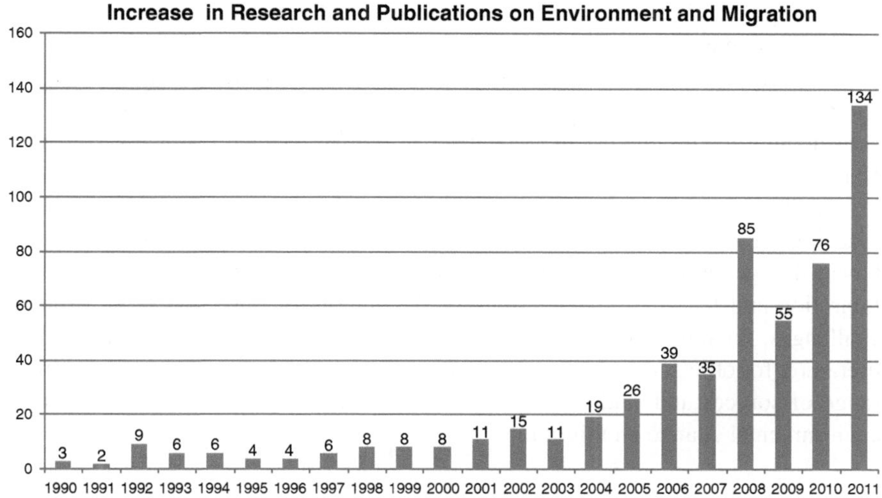

Fig. 1.1 Increase in the number of publications on migration and the environment, 1990–2011 (Source: IOM/University of Neuchâtel)

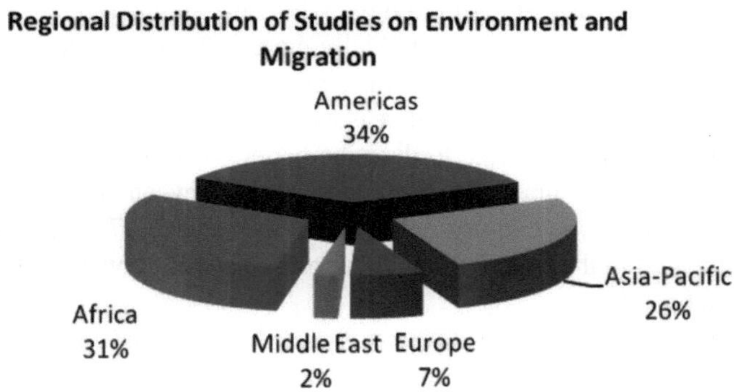

Fig. 1.2 Regional distribution of publications on migration and the environment (Source: IOM/University of Neuchâtel)

migration and the environment than during the whole period between 1990 and 2007. Between 2008 and 2011, 350 publications were produced, compared to 220 between 1990 and 2007. However, an increase in the number of publications does not necessarily mean that more empirical data on the subject are now available, as some reports may draw upon the findings of the same studies.

Most studies in the database focus on migration and the environment at the national level. In comparing regions, it can be seen that the evidence base is much stronger in some parts of the world than in others, with most publications focusing on the Americas (96), Africa (87) and the Asia-Pacific region (74). In Europe (18 publications) and the Middle East (6), the evidence base seems to be much weaker (see Fig. 1.2).

Intraregional differences are also striking, especially in Africa, where most of the studies conducted have been in Western Africa, Eastern Africa and Northern Africa, with very little research being done in Central and Southern Africa. In the Americas, studies focus more on South and Central America than on North America, despite considerable media interest in the impact of climate change following Hurricane Katrina.

Another interesting trend is the growing number of reports and publications focusing on small island States and coastal cities, with 25 publications, to date. Given this geographical imbalance in the number of studies and publications on migration and the environment, some of the authors of this book faced different challenges: for some, the challenge was to review a very large and growing body of literature; for others, the task was to produce a regional chapter using a more limited range of sources and studies that may be only indirectly related to the impact of environmental change on migration.

Typologies and Conceptual Approaches

The bibliographic database also enables the reader to identify another interesting trend concerning the use of terminology. While the number of publications on migration and the environment has gradually increased, the number of studies that use the term 'environmental refugees' (a term popularized in 1985 by El-Hinnawi (1985)) is decreasing. For a range of reasons, a growing number of studies have questioned whether it is accurate to describe people who move for environmental reasons as 'refugees' – noting, for example, that many environmental migrants remain within their own country. (Between 1985 and 2007, the term 'climate refugees' appeared in the title of 48 publications whereas, between 2008 and 2011, it appeared in only 13.)

On the other hand, the term *adaptation* and the view that migration can be a potential adaptation strategy have appeared more frequently in recent publications on migration and the environment. Many studies focus on migration as an adaptation strategy, with 33 of the 44 publications on migration and the environment produced in the period 2008–2011 referring to it as such. The links between migration and human security are also a growing area of research, with 44 publications on this topic having been produced in the same time frame. It is interesting to note that migration will be dealt with under the heading of human security in the forthcoming 2014 IPCC report (WG2-Report-chapter 12). Finally, more publications on migration and the environment focus on issues relating to governance, policy and legal frameworks, with nearly half (22 of the 45) of the publications dealing with this subject produced since 2010. Surprisingly, despite widespread calls for better data, publications addressing the challenges of data collection and research methods remain rare, with only 17 publications having been identified for inclusion in the bibliographic database.

The database also provides an indication of the extent to which publications have focused on particular environmental push factors. The categories and figures presented below show the number of publications relating to the various factors and suggest that researchers have studied a broad range of environmental factors.

Environmental Degradation

- Desertification: 12
- Drought: 17
- Sea-level rise: 17

Natural Disasters and Extreme Events

- Natural disasters: 10
- Extreme events: 5
- Hurricanes: 17
- Floods: 10
- Tsunamis: 5

Overview of the Regional Chapters

Each of the chapters in this volume begins with a general discussion of the expected regional impacts of climate change, which are summarized in Box. 1.1. Each chapter then goes on to review the evidence regarding the likely implications of these and other environmental changes on migration.

Box 1.1: Key Regional Impacts of Climate Change, from the IPCC *Climate Change* 2007: *Synthesis Report*[2]

Asia

- By the 2050s, freshwater availability in Central, South, East and South-East Asia, particularly in large river basins, is projected to decrease.
- Coastal areas, especially heavily populated mega-delta regions in South, East and South-East Asia, will be at greatest risk due to increased flooding from the sea and, in some mega-deltas, from the rivers.

Africa

- By 2020, between 75 and 250 million people are projected to be exposed to increased water stress due to climate change.

[2]See IPCC 2007.

(continued)

Box 1.1 (continued)

- By 2020, in some countries, yields from rain-fed agriculture could be reduced by up to 50 %. In many African countries, agricultural production, as well as access to food, is projected to be severely compromised. This would further adversely affect food security and exacerbate malnutrition.

Europe

- Climate change is expected to magnify regional differences in Europe's natural resources and assets. Negative impacts will include increased risk of inland flash floods, more frequent coastal flooding and increased erosion (due to storms and sea-level rise).
- In southern Europe, climate change is projected to worsen conditions (with high temperatures and drought) in a region already vulnerable to climate variability, and to reduce water availability, hydropower potential, summer tourism and, in general, crop productivity.

Latin America

- By mid-century, increases in temperature and associated decreases in soil water are projected to lead to gradual replacement of tropical forest by savanna in eastern Amazonia. Semi-arid vegetation will tend to be replaced by arid-land vegetation.
- Changes in precipitation patterns and the disappearance of glaciers are projected to significantly affect water availability for human consumption, agriculture and energy generation.

North America

- Warming in western mountains is projected to cause decreased snowpack, more winter flooding and reduced summer flows, exacerbating competition for over-allocated water resources.
- In the early decades of the century, moderate climate change is projected to increase aggregate yields of rain-fed agriculture by 5–20 %, but with important variability among regions. Major challenges are projected for crops that are near the warm end of their suitable range or that depend on highly utilized water resources.

Small Islands

- Sea-level rise is expected to exacerbate inundation, storm surges, erosion and other coastal hazards, thereby threatening vital infrastructure, settlements and facilities that support the livelihood of island communities.
- Deterioration in coastal conditions – for example, through erosion of beaches and coral bleaching – is expected to affect local resources.
- By mid-century, climate change is expected to reduce water resources on many small islands, such as those in the Caribbean and Pacific, to the point where they become insufficient to meet demand during low-rainfall periods.

Asia

Graeme Hugo and Douglas Bardsley start their chapter on Asia by giving a broad overview of current demographic and migration issues. They note that environmental change due to natural and anthropogenic causes has had enormous impacts on Asian societies and associated demographic processes, including human migration. They also point to the recent massive changes in the form and scale of human mobility on the continent. The chapter then reviews major environmental issues for Asia and suggests an important distinction between linear and non-linear impacts. Whereas the progressive increments of linear processes can, to a certain extent, be predicted, non-linear impacts are much more difficult to forecast, although potentially much more dramatic. Identifying thresholds is thus a major task for researchers. What, for example, would be the tipping point in any given country that would force large numbers of people to migrate, due to changes in the environment? With that question in mind, projections of future climatic changes are presented for the continent, and the interaction between such changes and current major environmental phenomena, as well as migration flows, is examined, with a focus on selected subregional countries (Bangladesh, China, Tajikistan and Thailand).

Hugo and Bardsley conclude their chapter by noting that climate change need not have catastrophic outcomes for migration in Asia. The outcomes of the last 40 years of human mobility in Asia suggest that the impact of environmental change can be evolutionary and beneficial, if managed effectively. Policymakers' capacities to respond effectively are, however, hampered by a lack of data and research on existing patterns, drivers and impacts of migration. The research undertaken in the region generally focuses on specific impacts of environmental events, but climate change also has the potential to globally alter the prevailing environments and resources that support livelihoods – an issue that should be addressed urgently, both scientifically and politically.

Europe

In their chapter, Mark Mulligan, Sophia Burke and Caitlin Douglas examine the interaction between environmental change and migration for countries throughout Europe, as well as the Mediterranean countries of North Africa. The authors examine the population, the gross domestic product (GDP), the infrastructure and the impact that environmental factors may have on the socioeconomic landscape, and how this may contribute to migration.

An increase in the frequency and intensity of hydro-climatic hazards, such as floods, droughts, soil erosion and landslides, is projected for Europe, with similar trends anticipated in land degradation, sea-level rise (combined with storm surges), and heat waves. All of these may have implications for patterns of migration, and the European region is potentially very sensitive to shifting climate, given the existence of its strong cultural, economic, political and demographic gradients in certain already climatically stressed conditions, especially in the south. At the same time, economic demand (as a pull factor) in affluent immigration countries has traditionally appeared to be the most important factor in migration in Europe and, historically, has had more impact than push factors in regions of origin. Although environmental drivers may exert an influence, their impact in Europe may thus be small, indirect and mediated through other social, political, cultural and, particularly, economic drivers. The authors also agree with the well-known forecast that migration induced by environmental changes is likely to be over shorter distances.

The authors conclude that, since GDP increases occur largely in the urban and industrial areas of the northern Mediterranean and northern Europe, cities in these regions may be particularly attractive to migrants from inside and outside Europe, generating increasing spatially concentrated pressures on ecosystems and exposure to hazards. These city regions are also where those who migrate for environmental reasons will be attracted. Meanwhile, and counterintuitively, the countries likely to suffer the most from environmental change over the next 100 years are not necessarily the poorer countries of the southern and eastern Mediterranean, but the wealthier countries around the Alps (due to projected changes in landslide frequency and snow-melt) and the low-lying Netherlands (due to sea-level rise and flooding). It is hard to tell if such countries will experience out-migration as a consequence of these evolutions, given that they have already developed mitigation and emergency-planning strategies.

Sub-Saharan Africa

In his chapter, James Morrissey reviews more than 30 empirical studies in sub-Saharan Africa. As he points out, the four types of studies identified in this particularly rich corpus appear to contradict the general claim that there is a dearth of empirical data on migration and the environment in this part of the world.

One type of study considers the impact on migration of the major droughts of the 1970s/1980s and of other environmental stresses such as rainfall and soil erosion. The findings largely confirm that such events do indeed influence human mobility but they also illustrate the complexities of that relationship and the fact that it is rarely the environmental stress alone that causes people to migrate. In fact, the author asserts that, under certain circumstances, established mobility strategies can be altered to the point of creating a paradoxical decrease in certain types of migration. This group of studies also illustrates the paramount importance of individual characteristics such as gender, class and ethnicity in determining who does and does not migrate.

A second group of studies took a longer-term view, assessing the impact of the progressive drying of the Sahel over half a century. The findings of three studies carried out among pastoralists are analysed in detail and found to confirm the main findings of the first group of studies – further clarifying the complex ways in which climate change can alter mobility patterns and the fact that trends are often characterized by a migratory drift over a long period, rather than by a sudden shift.

Under certain circumstances, migration constitutes an effective means of adaptation but, under others, migrants may end up worse off and risk being marginalized in their region of destination.

A third group of studies takes a much more static approach, evaluating to what extent environmentally better-endowed regions tend to experience more positive migration balances than less well-endowed ones. This approach is close to the subfield of amenity migration studies and to neo-classic migration theories. In this case, contrary to expectations, it was found that fewer people actually migrate out of areas with unfavourable climatic conditions – a paradox explained by the fact that, in the most depleted areas, people experience difficulty in accumulating enough money to migrate.

A fourth group of reviewed studies attempts to model migration behaviour at the individual level (agent-based modelling). Such studies underline the nonlinearity of migration processes and allow for some scenario-building, but the quality of the data on individual preferences for migration is still in need of substantial improvement.

Although not directly allowing for predictions regarding future migrations, this impressive review suggests that the dominant form of mobility in Africa will be cyclical, cross-border, rural–rural and rural–urban, possibly in a stepwise fashion, rather than long-term, long-distance and international. These features reflect the current situation of African migration, as a whole, which is presented at the end of the chapter.

Middle East and North Africa (MENA)

The chapter on the Middle East and North Africa (MENA) was prepared by a World Bank team, led by Quentin Wodon, which utilized a new data set collected in 2011 in Algeria, Egypt, Morocco, Syria and Yemen. The five countries were chosen because of their population size (to ensure that findings would be illustrative), vulnerability to climate change (drought), and levels of socioeconomic development (middle- and low-income countries). In all five countries, climatic and population data indicate that migration will increasingly be induced by water scarcity, aridity and droughts, soil infertility and, in some areas, sea-level rise. For MENA, however, unlike other regions of the world, there are very few empirical studies on past environmental shocks that might have affected migrations. Those that do exist nevertheless confirm the crucial importance of the household and its characteristics, such as socioeconomic status and political context, in the decision to migrate. Due to many intervening variables, the effect of environmental change is not uniform across households, and the decision to migrate is made by individual household members.

Evidence from new household surveys conducted by the World Bank yields new insights about the links between climate change and migration in the Arab world, especially regarding the way populations perceive and connect those two parameters: (1) In the combined sample of the five countries surveyed, more than 75 % of the households in areas susceptible to climate change said they believed that changes in climate patterns were taking place. (2) Regression results suggest that poor climate and extreme weather events lead to a higher probability of migration, but that the role of climate as a push factor remains smaller than that of socioeconomic characteristics and job prospects in cities. (3) It is estimated that a significant deterioration in climatic conditions would lead to an increase of about 1.5 percentage points in both temporary and permanent migration. Although still based on very few studies and small samples, these results point to the necessity of focusing more attention on MENA countries as possible hotspots where significant migration could take place in the future, due to climate change.

North America

In this chapter, Susana B. Adamo and Alexander M. de Sherbinin begin by discussing migration trends in North America, which are characterized by high levels of international immigration and substantial (but declining) internal population mobility, particularly from rural and small urban areas to large metropolitan areas. The principal climatic impacts that are likely to affect migration in this region are temperature rise, regional changes in precipitation, and decreased winter snow pack – all of which affect water supply. Rising sea levels and storm surges in coastal communities are also of concern, as is the likelihood of more frequent and intense droughts in Central North America. According to the authors, there are relatively few environmental issues in North America grave enough to cause massive out-migration, but the few existing studies on migration related to environmental events in the past nevertheless show numerous associations.

Although it remains difficult to document their effect beyond immediate displacement, hurricanes have, throughout history, impacted population mobility. A key finding is that the degree of housing damage, along with factors such as age, race, education and socioeconomic status, affect the ability to return. Studies of the droughts at the beginning of the twentieth century on the Great Plains of the United States and Canada highlight the fact that migration and displacement are multicausal. Recent hurricane events have led to an increase in the number of studies on the subject, which also confirms these findings.

These recent studies have also led to interesting advances in data development and alternative data sources for estimating displacement and population redistribution after hurricanes and other catastrophic environmental events. One study suggests combining, on an ongoing basis, census population data and administrative data with data collected after disasters. Louisiana, for example, currently has a negative migration balance, due to the displacement impacts of hurricanes Katrina and Rita. Other studies in the region have covered: population in low-elevation coastal zones; displacement and relocation of Arctic communities in Alaska and Canada; mitigation measures that could directly or indirectly affect mobility; and changes in water availability. In conclusion, the major driver of migration in North America is still clearly related to the economy. However, the examination of relatively new data on possible scenarios of climate change-induced migration suggests that the direct and indirect contribution of environmental factors at the local level is not insignificant, and that it could escalate and extend to other geographic areas through existing migration networks.

Latin America and the Caribbean (LAC)

After a brief history of the debate on the relationship between the environment and migration, the authors of this chapter – Raoul Kaenzig and Etienne Piguet – identify the main environmental consequences of climate change for Latin America. They then address more specifically the aspects related to tropical storms and hurricanes, floods, droughts, rising sea levels and melting glaciers, which are identified as the main potential migration drivers in the region. The chapter then proceeds with a summary of the past consequences for migration of these environmental changes, after which the authors identify the most important migration issues related to future climate changes. The relationship that can be observed in Latin America and the Caribbean between environmental change and migration confirms the main tendencies noted in other chapters: most displacements take place over short distances, with urban centres exerting a strong pull, while simultaneously being vulnerable to environmental hazards. In the case of sudden catastrophes, displacements are usually short term.

Migration also emerges as a coping strategy that includes leaving and returning, temporary stays, and multiple residences. In addition, the findings of these studies on Latin America confirm that environmentally related displacements are multi-causal and context-specific. The same kind of disturbance can have completely different consequences, depending on the economic, social and political context in which it occurs. Finally, environmentally driven displacements must be considered within their historical context and they occur most frequently between countries that have a pre-existing migration relationship.

Just as, globally, empirical research on migration and the environment is unevenly distributed, geographically, there is also an intra-continental imbalance. In Latin America, many more studies were found on Mexico, other Central American States and, to a lesser extent, Brazil, than on other countries in this region. The Andean countries, for example, along with the continent's north-eastern countries, have remained relatively unexplored. The authors suggest a series of hypotheses to explain this finding: the existence, or lack thereof, of local research centres on migration, pre-established relationships of migration with the United States and, finally, political motives relating to the politics of migration and security.

Oceania

Oceania is comprised of Pacific Island countries (PICs), Australia and New Zealand, but the chapter on this region, prepared by Richard Bedford and John Campbell, focuses mainly on the PICs, as the impacts of climate change are likely to be greater on these island countries than on Australia and New Zealand. The latter two countries are mainly discussed as destinations for Pacific Island migrants, which they have been, for increasing numbers of peoples, since the mid-1940s – especially Polynesians from the eastern Pacific. Both countries are heavily urbanized (85 % of the population live in towns and cities) and offer opportunities for wage employment and a host of social services that cannot be found on the islands.

The historical overview shows that environmental factors and resettlement have long been a major policy concern in the PICs, especially since the mid-1940s, when the colonial administration purchased an island in Fiji for the purpose of resettling people from an island in Kiribati where the environment was being progressively destroyed by large-scale phosphate mining. In 2012, the Government of Kiribati approved the purchase of land in Fiji to provide long-term security for the people of Kiribati, given the impact of climate change. Migration linked with environmental change long pre-dates the current concerns with global warming and associated rising sea levels. In the future, however, internal resettlement options are likely to be culturally unacceptable in the PICs, as 90 % of the land is held under forms of customary title. Over 500 cases of extreme events and migration have been documented but few allow for accurate estimates of the impact on mobility patterns.

Three key areas of concern have been identified for this region:

1. All but a few Pacific countries are located either fully or partially in areas of tropical cyclones, which are the most frequently reported triggers of disaster in the region. Torrential rains and flood events are associated with tropical cyclones. However very little information is available about mobility patterns after these events.
2. Droughts are relatively common and affect fresh water availability and quality, although there is little literature indicating that droughts cause migration.
3. Sea-level rise has been identified by numerous observers as likely to be the most disruptive of climate change effects, the greatest risk to PICs and the most likely to trigger migration. However, there are as yet no clear cases of climate change having caused the sea level to rise and render atolls uninhabitable.

There is a need for a more nuanced consideration of migration as a climate change response, rather than considering it in dualistic ways as either an inevitability or something that will not happen at all, and one that incorporates both migrating/relocating and host communities. In conclusion, few attempts have been made to predict the number of people who are likely to migrate (within and from the Pacific Island countries) as a result of climate change. For the islands of Oceania, there is a lack of detailed data on the projected effects of climate change, and limited information is available as to how much and whether such effects will reduce the land, livelihood and habitat security of these island communities. Consequently, despite the fact that Pacific islanders are often labelled in the media as the 'first climate refugees', the existing data for this region provided limited support for this notion.

Himalayas

This chapter, prepared by Soumyadeep Banerjee, Richard Black, Dominic Kniveton and Michael Kollmair, shows that, although there are insufficient data on the state of the environment in the Hindu Kush Himalayan (HKH) region, it is highly likely that heat waves, glacial retreats and permafrost degradation will lead to increasingly unstable slopes, mass movements, and glacial lake outflows. Heavy precipitation, which will affect landslides, is also anticipated. Currently, there is no coordinated mechanism in the region to collect and compile data on mobility due to natural disasters. The available evidence is based on post-disaster rapid assessments or sporadic case studies.

Despite this lack of data, the HKH region is believed to be one of the hotspots of future climatic impacts as it is highly sensitive to small changes in temperature and precipitation. One of the key factors affecting livelihoods is exposure to stresses and shocks relating to the availability and changing quality of water. It remains a major challenge simply to assess the migration flows to and from the region. It is clear, however, that most of the migrants from this region remain within their country of origin and that cities are the main destinations. In Bhutan, India, Nepal and Pakistan, for example, internal rural–urban migration has been a dominant trend. There are few disaggregated data and studies available on mountain-specific migration but one study nevertheless observed a direct relationship between the type of migration and the altitudinal profile of the native communities in Afghanistan and Pakistan: migrants from mid-altitude regions are mainly seasonal workers with relatively sophisticated professional skills involved in trade and commerce, whereas migrants from high-altitude areas are mainly younger semi-skilled or unskilled individuals. Although not connected to the issue of environmental change, such findings underline the specificities of different mountain areas and the need to adjust data and studies accordingly.

The subsection on environmental drivers of migration reviewed all available empirical evidence from the region and, more specifically, four case studies focused on the relationship between environmental change and migration.

Another important question addressed in the review is the extent to which migrants' remittances help to build resilience against environmental change. The two reviewed studies show that financial remittances are a significant source of cash income for many mountain households and that remittances have been used to procure food, to meet other basic needs during/after a disaster, to re-establish livelihoods, and to rebuild lost assets. In some cases, remittances can also contribute to disaster preparedness, such as strengthening of housing quality or procurement of boats when there are floods. The findings also illustrate the potential role that migration can play as an adaptation strategy in times of environmental variability and change. In conclusion, the evidence from across this region (albeit sporadic) indicates once again that environmental stressors do influence people's migration decisions but not in isolation from non-environmental drivers and other challenges.

Regional Policy Perspectives

This chapter by Karoline Popp presents, for the first time, an overview of selected policy and cooperation efforts among governments on the issue of migration and the environment at the regional level. It aims to complement existing accounts of normative frameworks and policy on this topic, principally from the perspective of international law, international institutions and national policy. Three main conclusions emerge:

- The issue of migration and the environment has arisen in the context of other discourses and policy priorities such as migration, environment, security and human rights. While implementation of explicit migration and environment policies remains rare, the issue has nonetheless begun to permeate policy awareness. Regional policy and cooperation have remained at the level of informal, non-binding dialogue that has, in some instances, translated into 'soft regional policies' and, in a few cases, may acquire legal force.
- Limited interest, limited capacity and limited participation may explain why regional consultative processes on migration (RCPs) have been less prominent as a forum for raising awareness of migration and the environment. By contrast, more formal regional institutions have made comparatively greater strides, particularly where the institutional architecture is sufficiently elaborate to prompt different policy sectors within the same regional body to take an interest in migration and the environment.
- In sum, regional policy action on migration and the environment remains incipient, indirect, informal and often incoherent.

Conclusion

A diverse picture emerges from the regional reviews presented in this book. On the one hand, many common features regarding the relationship between migration and the environment worldwide can be observed. On the other hand, many regional specificities point towards differentiated challenges regarding future migration flows, which will necessitate specific policy approaches.

In addition to presenting these observations, the detailed regional overviews further validate many of the insights already reported in recent studies on environmental change and migration:

- *Multi-causality*: Migration is rarely explained by one single driver, be it environmental or not. It is the product of a set of livelihood options, framed by a mix of biophysical conditions and social, political and economic contexts.
- *Migration as a response to environmental stress*: Migration is often seen as a last-resort option, once a certain threshold of stress is reached. Although this may be true for some, migration also represents a potential means of reducing the burden of the environment on the sending household. Migration may therefore be a coping mechanism as well as a consequence of environmental change.

- *Affected populations as victims or agents*: Some studies consider environmentally vulnerable populations to be passive victims, whereas it is important to bear in mind that such populations develop strategies to cope with – and adapt to – environmental change.
- *Remittances*: The transfer of money or goods by migrants to their family members in the place of origin is a proven coping strategy, but the impact of remittances is not yet factored into climate change adaptation plans.
- *Those left behind*: Although migrants are often the focus of media attention as the "human faces of climate change", most populations likely to be affected in the future will not be in a position to move and may be trapped in their place of residence. These people, rather than the migrants, may be the worst affected by climate change.
- *Short distance*: Where mobility occurs in response to environmental change, it is likely to be over shorter distances, in which case the term *displacement* might be more appropriate than *migration*.
- *Short time*: In the past, when hazardous environmental events have occurred around the world, displacement has usually been temporary, with most people wishing to go back to their place of residence as soon as possible.
- *Threshold*: The literature on migration often implicitly posits certain linearities – or, at least, stepwise processes – in the evolution of migration flows. Yet the patterns of migration that will be induced by environmental change might well be non-linear and (once certain thresholds are reached) characterized by very quick changes – for example, from low to high levels of migration. This underlines the need for forecasting scenarios so that adequate policies can be developed.
- *Urban zones*: In many parts of the world, towns are the main destinations for internal and international migrants, who are driven by, among other factors, environmental change. However, urban settings are often themselves vulnerable to environmental changes, such as sea-level rise, landslides and hurricanes, with the result that migrants may find themselves in a vicious circle of vulnerability.
- *Selectivity*: As in the case of other migration processes, migration linked to environmental change is highly selective and different population groups are affected in different ways, depending on their gender, class, ethnicity, livelihood, social capital, networks, etc.

The book's regional chapters also reveal some interesting new local specificities regarding the connection between migration and environmental change:

- Different parts of the world will be confronted with very different environmental challenges (see Box. 1.1), which may result in different regional migration scenarios.
- Some regions are more likely than others to be affected in the near future. Whereas certain regions of the world may be affected within decades (for example, by an increased number of hurricanes), others may not be affected for a century or more but may then face huge challenges linked, for example, to sea-level rise.

- Pre-existing migration patterns differ greatly between regions. Even if they had little to do with environmental changes in the past, migration channels and networks will have a significant impact on the development of future environmentally induced migration flows.
- Although it is often suggested that poorer countries are more likely to be affected by climate change-induced migration, the chapters on Europe and North America show that, although rich countries are in a much better position to develop policies for coping with environmental change, they are also likely to see migration linked to such change in the future.
- Physical geography appears to be a very important factor in population movements. Elevation, soil type and erosion are three examples of characteristics that will interact with climatic processes in specific ways in each region of the world.
- Social factors will also be of paramount importance. The chapter on Oceania points, for example, to the issue of the land tenure system as a key factor affecting opportunities for relocation. Regional political blocs and their agreements regarding freedom of movement will also play an important role. Whereas countries of the Northern Mediterranean basin will, for example, benefit from European mechanisms of solidarity and, if necessary, free circulation within the European Union (EU), Southern countries confronted by the same challenges will be in a much less favourable position, in terms of developing migration-related coping strategies.

This book also suggests a new agenda for policy-oriented research on the linkages between migration and the environment, focusing on three key areas:

1. The evidence base in a number of key regions of the world needs to be strengthened to promote new forms of research. In some regions, such as the MENA countries, North America and Europe, relatively little research has been done. In other regions, such as Oceania, South America, Africa and Asia, numerous specific case studies have been carried out but few comparative studies have been conducted. A much more cross-national research, using a common research design, is needed to facilitate meaningful comparisons between countries and regions.
2. The methodological coherence of empirical research should be enhanced. In 2009, a six-group typology of methods used to assess the weight of environmental drivers of migration was identified (Piguet 2010b). This typology encompassed: ecological inferences, individual sample surveys, time series, multilevel analysis, agent-based modelling and ethnographic methods. Although qualitative studies are vitally important, there is a need for large-scale survey studies too – ideally including time series data. More research using quantitative methods is needed, to provide a potentially more representative picture of the linkages between environmental change and migration (Laczko and Aghazarm 2009). The value of this kind of research is reinforced by Piguet, who reports: "The most illuminating and original studies that we have referred to make use of data specially developed through time-consuming collection processes involving qualitative as well as quantitative methods. Comparable efforts shall hopefully be intensified" (Piguet 2010b: 522).

3. There is a need for more studies on specific aspects of environmental change and the likely implications for migration, focusing on mountainous regions such as the Himalayas, the Andes and the Alps, which are experiencing glacial retreat; coastal areas affected by sea-level rise; urban areas affected simultaneously by population growth, infrastructure challenges and environmental disruptions; and drylands experiencing declining water availability and decreasing food production.

Policymakers and practitioners also need assistance in identifying and interpreting the findings from the growing number of reports and studies on migration and climate change. The establishment of a clearing-house for research on migration and the environment would be useful, as existing studies are currently scattered across different countries, disciplines and journals. The publication of the analytical bibliography that accompanies this book is a first step in that direction. The database will be regularly updated and a series of keywords developed, enabling researchers worldwide to quickly access all relevant available data. The authors count on the community of researchers to help make publications available and to keep a critical eye on this database. The IOM will also be a key actor in supporting this project, which will help spread knowledge and promote sound and innovative research practices for what is likely to be a key global issue in the years ahead.

References

Achieng, M. (2012). Regional and inter-regional processes: Advancing the discourse and action on migration and development. In I. Omelaniuk (Ed.), *Global perspectives on migration and development*. Dordrecht: Springer.

Black, R. (1998). *Refugees, environment and development*. London: Longman.

Black, R. (2001). *Environmental refugees: Myth or reality?* New Issues in Refugee Research Series (Working Paper #34). Geneva: United Nations High Commissioner for Refugees (UNHCR).

El-Hinnawi, E. (1985). *Environmental refugees*. Nairobi: United Nations Environment Programme (UNEP).

Erway Morinière, L. C. (2009). Tracing the footprint of 'environmental migrants' through 50 years of literature. In A. Oliver-Smith & X. Shen (Eds.), *Linking environmental change, migration & social vulnerability* (pp. 22–29). Tokyo: United Nations University.

Foresight. (2011). *Migration and global environmental change* (Final Project Report). London: The Government Office for Science.

Hugo, G. (2008). *Migration, development and environment*. Geneva: International Organization for Migration (IOM).

Intergovernmental Panel on Climate Change (IPCC). (2007). Climate change 2007: Synthesis report. In: Core writing team: R. K. Pachauri & A. Reisinger (Eds.), *Contribution of working groups I, II and III to the fourth assessment report of the Intergovernmental Panel on Climate Change*. Geneva: IPCC.

International Organization for Migration (IOM). (1992). *Migration and the environment*. Geneva: International Organization for Migration (IOM).

International Organization for Migration (IOM). (2013). *People on the move in a changing climate – A bibliography*. Geneva: International Organization for Migration (IOM).

Laczko, F., & Aghazarm, C. (Eds.). (2009). *Migration, environment and climate change: Assessing the evidence*. Geneva: International Organization for Migration (IOM).

Morrissey, J. (2009). *Environmental change and forced migration: A state of the art review.* Oxford: Refugee Studies Center, Oxford Department of International Development (ODID), University of Oxford.

Piguet, E. (2010a). Climate and migration: A synthesis. In T. Afifi & J. Jäger (Eds.), *Environment, forced migration and social vulnerability* (pp. 73–86). Bonn: Springer.

Piguet, E. (2010b). Linking climate change, environmental degradation and migration: A methodological overview. *Wiley Interdisciplinary Reviews: Climate Change, 1*(4), 517–524.

Piguet, E. (2013). From "primitive migration" to "climate refugees": The curious fate of the natural environment in migration studies. *Annals of the Association of American Geographers, 103*(1), 148–162.

Migration and Environmental Change in Asia 2

Graeme Hugo and Douglas K. Bardsley

Introduction

There are major challenges emerging for Asia both in relation to future human mobility and the impacts of climate change. The interaction between these two processes is little understood and yet the implications of future environmental change on societies, and the consequent potential for fundamental changes in the migration of people both within countries and across international borders could be enormous. For example, in recent times there has been no environmental event that has had as great an impact as the Asian Tsunami of December 2004, which killed approximately 300 thousand people in 12 Asian and African countries surrounding the Indian Ocean and left some five million people in immediate need of assistance (UNHCR 2006: 21). Estimates of the numbers of people displaced vary between more than one million (UNHCR 2006: 21) to over two million (AidWatch 2006). In Sri Lanka 450,000 people were forced to move, and in Indonesia, in the province of Aceh, there were 533,000 Internally Displaced Persons by the end of 2004 (Yin 2006). Overwhelmingly the people forced to move by the Tsunami moved to other locations *within* their regions, although some travelled longer distances due to effective social networks. Yet, based on the experience of the Asian Tsunami of 2004, Naik et al. (2007) have pointed out that the nexus between migration, development and environmental hazards remains relatively uncharted territory among researchers and policy makers. The massive scale of displacement associated with the Asian Tsunami, not to mention the tragedy of loss of life and prosperity, "sparked an extraordinary mobilisation of resources. Governments, private citizens and corporations, NGOs in the effected countries and beyond were quick to respond

G. Hugo • D.K. Bardsley (✉)
Geography, Environment and Population, University of Adelaide, Adelaide, SA, Australia
e-mail: graeme.hugo@adelaide.edu.au; douglas.bardsley@adelaide.edu.au

E. Piguet and F. Laczko (eds.), *People on the Move in a Changing Climate*,
Global Migration Issues 2, DOI 10.1007/978-94-007-6985-4_2,

with offers of money, supplies and manpower" (UNHCR 2006: 21). The disaster has also bought into sharp relief the need to better understand the challenges of environmental migration in Asia.

The chapter aims to synthesise the current academic debate on environmental change and migration for Asia. It introduces some of the key theoretical elements for both climate change adaptation and human migration, and provides a historical overview of the relationship between environmental change and migration in Asia. Projections of future climate change are presented for the continent, and the manner in which such climatic changes interact both with current major environmental issues and migration flows are examined. The chapter continues on to draw from a report undertaken by Hugo et al. (2009), which was published in part by the Asian Development Bank (ADB 2012a). However, it emphasises the value of conceptualising the relationship between environmental change and migration according to likely linear and non-linear change, with a focus on the sub-regional example countries of China, Thailand, Bangladesh and Tajikistan.

Over the last five decades, there have been massive changes in the form and scale of human mobility in Asia. One of the major features of contemporary Asia which is shaping migration levels and patterns is its rapid economic growth. Asian modern development has been hugely successful over the last three decades from a perspective of economic growth and social opportunity, such that the region is now perceived as the engine for the global economy (Yeung 2012). In 2011, average Asian GDP grew at 7.2 %, the highest of any global region (ADB 2012b). In the same year, Asia for the first time accounted for more than half of global economic activity. Increased population mobility between and within Asian countries has been an important cause and consequence of rapid economic growth. Nevertheless, Asian countries vary greatly, in both their rate of economic growth and in the socio-economic wellbeing of their inhabitants, with per capita GDP in 2011 ranging from US$584 and US$678 in Afghanistan and Bangladesh compared with US$49,271 and $36,584 in Singapore and Brunei (ADB 2012b; IMF 2012). Rates of growth of GDP in 2011 ranged from 17.3 to 9.9 % in Mongolia and Turkmenistan compared with 0.1 % in Azerbaijan and Thailand (IMF 2012).

While economic drivers of mobility have been dominant in the modern-era, there is a growing appreciation that environmental change plays a significant role. In particular, projections of future climate change suggest that environmental drivers of migration will become an increasing proximate or contributory factor to human mobility throughout the Twenty-First Century. A number of factors indicate that climate change impacts could be a particularly important driver of migration in Asia, and these will be outlined at some length below, briefly these are:

- The population and environmental management pressures in the region are already enormous;
- Climate change impacts are projected to be large in the region, particularly as monsoonal rainfall patterns could become less reliable and cyclonic activity intensifies; and,
- A large percentage of the population is poor and their well-being is already highly sensitive to environmental hazards and the degradation of natural resources.

In any discussion of global trends the Asia-Pacific must loom large. In 2011 it was home to 60.5 % of the world's population (4.21 billion) and the number living in the region will increase to 4.81 billion in 2030 (United Nations ESCAP 2011). The Asia-Pacific region is enormously diverse with three of the four largest nations in the world in 2011 (China 1,348 million, India 1,241 million and Indonesia 242 million) but also some of the world's smallest nations, especially among the Pacific Island countries (Niue and Tokelau 1,000, Nauru 10,000, Tuvalu 11,000). The report of the Global Commission on International Migration (2005) in discussing the reasons for increasing migration between countries points to widening international differentials in what it refers to as the three Ds – development, demography and democracy. This refers to increasing gaps between nations, firstly in income and job opportunities, the rate of growth (or, in the case of more high income nations, decline) in the workforce ages as a result of decreasing fertility and the extent to which citizens enjoy freedom and human rights. It is the first two Ds which are driving an increase in international labour migration within, into and out of the Asian region. As Martin (2008) points out there is a 21–1 gap in per capita income between the 30 richer and 165 poorer countries in the world. The Asia-Pacific region has some of the world's poorest nations such as Afghanistan (Gross National Income PPP Per Capita in 2009 of US$860), Nepal (US$1,180), Bangladesh (US$1,550) and Cambodia (US$1,820) as well as wealthy nations such as Japan (US$33,440), Macao-China (US$57,390), Hong Kong-China (US$44,540) and Singapore (US$49,780). Similarly with respect to demography, the region has experienced substantial change as both a cause and consequence of the social and economic transformation of the region. Table 2.1 reflects the scale of these changes. Since 1970 the population of the region has doubled but the annual growth rate of the population has halved over that period. This has been due to a remarkable decline in fertility which has seen the average number of children borne by Asian women more than halve from 5.4 in 1970 to 2.1 in 2010. The average life expectancy of Asian men and women has increased by around 16 years over the same period.

Levels of poverty and inequality suggest that the livelihoods of many millions of people across the region are already highly vulnerable due to their socio-economic situation (ADB 2012b). Future vulnerability to environmental change will be also associated with the sites where, and situations within which, people live. Asia contains some of the most densely settled areas in the world such as Java in Indonesia, the deltas of Southern Asia, coastal China and the river valleys of India and Bangladesh (Hugo 2011). As will be detailed later, these include some areas most vulnerable to the impact of climate change. The Asian region is experiencing rapid urbanisation, with the percent of the population living in urban areas increasing from 24 % in 1970 to 42 % in 2007, and within the next decade it is likely half of the population in the region will be living in cities, many of which will be strongly concentrated in coastal areas (McGranahan et al. 2007; World Bank 2008).

Responses to climate change either focus on mitigation of the rate of change or societal adaptation to the impacts. Human migration can be conceptualised as a component of adaptation, but it is likely that *in situ* (in place) adaptations will

Table 2.1 ESCAP region: major demographic changes, 1970–2011

Demographic variable	1970[a]	2011	Percent change 1970–2011
Total population (m)	2,041.2	3,998	+95.9
Percent of world population	55.2	57.2	+3.6
Annual growth rate[b]	2.2	0.9	−59.1
Percent urban[b]	24	43	+79.2
Percent aged 0–14[b]	40	25	−37.5
Percent aged 65+[b]	4	7	+75.0
Dependency ratio[b]	80	47	−41.3
Total fertility rate[b,c]	5.4	2.1	−99.6
Expectancy of life at birth – males[b]	52	68	+30.8
Expectancy of life at birth – females[b]	54	72	+33.3

Source: UNESCAP (1984, 2011)

[a]The data exclude the countries of Central Asia which were not part of the ESCAP region in 1970

[b]Includes Central Asia in 2011

[c]TFR and Life Expectancies refer to the average of the 5 years prior to 1970

dominate at least in the early stages of environmental change. People are strongly tied to place and local systems of production and exchange, and the evidence of human adaptation suggests that people will try to adjust their local systems to adapt to environmental change, rather than predominantly seek out opportunities elsewhere (Lewicka 2011). In fact, individuals, communities and societies are constantly adjusting, or adapting, their socio-ecosystems *in situ* to improve their adaptation to constantly changing local conditions (Adger et al. 2005). The levels of adjustment required to cope with future climate change infer that local responses will become increasingly inadequate for many people across Asia.

Given the potential importance of future climate change impacts on migration, there has been relatively little empirical research on the topic for Asia and it remains very difficult to provide precise guidance on how many people may move or provide any details on their sources or destinations. Indeed there is still a strong contention between those who would argue that environmental change is already a significant, although complex driver of migration and will be more important in the future (McLeman and Smit 2006; Reuveny 2007) and those who suggest that the evidence does not support a strong association between environmental degradation or hazards and human migration (Bohra-Mishra and Massey 2011; Findlay and Geddes 2011; Gray and Mueller 2012). Nevertheless, it is possible, and we would argue essential, to review the complex implications of climate change for migration in Asia as we have here, to better inform policy and practice that will govern the complex interactions between environmental and demographic change in a strategic manner. If climate change migration could be governed and managed effectively, humanitarian crises will be minimised, conflicts avoided, and opportunities will be provided for countries to benefit from the processes of environmental and social change. However, the development of effective policy responses will require a re-conceptualisation of the role of migration in relation to environmental risk.

A History of Environmental Change and Migration

Environmental change due to natural and anthropogenic causes has had enormous impacts on Asian societies and associated demographic processes, including human migration. The strongest influence of the environment is not due to change itself, but rather to the state of local and regional conditions that leads to the creation of unique sites and situations to provide socio-ecological opportunities for people to exploit and thrive. In other words, the environmental combination of: fertile river valleys and plains, including major river deltas; mountain ranges that collect and provide water sources for some of the world's major river systems; coastal and marine resources; a dominant summer monsoon rainfall pattern; and more marginal drylands to the north and west, together have partly pre-determined how the societies of Asia have developed. People have moved historically to exploit the resources the environment provides, especially for agriculture, fisheries and forestry, and associated trade, and populations have grown locally in association with those primary and secondary industries. The environment, in the most comprehensive sense of the term, has therefore been a vital influence on Asian population distribution (for a Chinese example see Chen et al. 2007). As a result, any major changes in the condition of environmental resources or hazards would suggest a proportional shift in population distributions. However, in the colonial and modern eras, processes of human migration are less directly rationalised by environmental conditions, and are driven more by colonial processes, especially associated with major port cities, and in recent times, by the massive economic and political forces of modern global development. Therefore, it could be argued that recent eras have seen a weakening in the direct links between environmental conditions in particular locations and patterns of population growth in Asia.

Environmental change is not a new phenomenon in Asia, although global climate change has the potential to significantly enhance local change in some places and fundamentally change locations in others. Away from the impacts of individual natural disasters, rarely does the evidence suggest that environmental change has directly and strongly influenced broader contemporary migration patterns in Asia (Perch-Nielsen et al. 2008; Seto 2011). Therefore, while increases in forced displacement of people due to climate change impacts are likely, the majority of climatic influence on mobility is likely to be associated with the complex forces and pre-conditions for migration, especially associated with precautionary adaptation to experienced and perceived climatic risk. That said, non-linear forced migratory responses to climate risk are also likely to increase. A large percentage of the population in the region lives in poverty and their well-being is highly exposed to environmental hazards and degradation of natural resources. In fact, analyses of the implications of historical events suggest that a more rapidly changing environment could be a major driver of future migration in Asia.

Previous experiences with environmental hazards such as flooding have caused significant displacements of large populations across Asia (Dilley et al. 2005; Perch-Nielsen et al. 2008). While the Asian Tsunami of December 2004 displaced several million people, most moved to nearby locations or at least within their own countries. In fact, coastal and low-lying delta areas are places where large populations are exposed to major flood events that leads to altered migratory activities (Ericson et al. 2006). Flooding in China has had a large impact on human migration historically (Ye et al. 2012). In recent years, floods in Thailand, Pakistan, Vietnam and Nepal have inundated vast areas, forcing many thousands of people to move (Bardsley and Hugo 2010; Dun 2011; Mirza 2011). The impacts of sea-level rise on Small Island States is another major area of debate, because unlike larger countries there is often limited terrestrial space in countries for people to migrate to internally (Barnett and Campbell 2010).

Bangladesh is often cited as being one of the most vulnerable countries to environmental change because, being situated at the Ganges-Brahmaputra river delta, it is both one of the most densely populated nations, and highly vulnerable to a range of environmental impacts, including flooding and river bank erosion, storm surges, cyclones and droughts (Shamim 2008; Findlay and Geddes 2011). People are forced into highly exposed situations such as living on chars (river islands) or along coastlines through socio-economic processes of poverty and landlessness (Lein 2000; Nasreen et al. 2006; IOM 2010). McGranahan et al. (2007: 32) note that the coastal population in Bangladesh grew at about twice the national population growth rate between 1990 and 2000. The experiences of migration into and out of spaces of vulnerability to flooding or drought in Bangladesh are not isolated from broader socio-economic impacts on migration. Although individual events do not necessarily lead to significant out-migration (see Paul 2005), the migration of people both internally and internationally in and from Bangladesh can be partly attributed to environmental change. Bangladesh experiences regular flooding to about one-quarter of its landmass, but major floods also occur in decadal timeframes and are associated with considerable negative social and economic impacts (Karim and Mimura 2008; von Storch and Woth 2008). For example, a number of studies indicate that in-migration to the slums of Dhaka from rural areas are due in part to the impacts of environmental hazards and physical displacement (Begum 1999; Haque 2005; Hossain 2005). Similarly the large flow of migrants from Bangladesh to India is linked to environmental degradation and other socio-ecological risks (Alam 2003; Reuveny 2007).

Major developments provide other examples of forced migration that suggest that environmental change can have huge migratory impacts. Hydro projects have led to the displacement of 100s of thousands of people in Bangladesh and China (Cernea 1990; Cernea and McDowell 2000). The Three Gorges Project in China has produced the world's largest planned human displacement and resettlement, involving the relocation of over 1.33 million people across a 16-year period to 2008 (Tan 2008). Displacement produced by dam construction is characterised by compulsory mass removal, involving complex and risky processes of socio-economic and cultural change (CEDEM 2008a).

Environmental degradation has been used to support state-led migration initiatives and these initiatives provide important lessons for the future. In China, environmental migration is being used as an explicit state policy to facilitate the movement of people defined as impoverished away from rural areas classified as ecologically fragile to resettle elsewhere (Mailisha 2004). Already agricultural production and water use in the defined ecologically fragile zones are seen as insufficient to sustain local populations, but the local environments will also be exposed to future climate change and impacts of ongoing social deprivation. Inhabitants of the zones are mostly poor and from ethnic minorities, and state-led migration is largely involuntary based on a perceived need to remove people from the vulnerable areas (Shi et al. 2007; Yeh 2009). From 1983 to 2006 around two million people in west China were displaced, of which 1.02 million people were displaced from fragile environments during the 2000–2005 period (West China Development Leadership Office of the State Council 2005; Tan and Guo 2007).

Although often not considered explicitly, and significantly under-researched considering their importance, environmental hazards and degradation of natural resource condition, major developments and state-led environmental migration clearly have played a role in human mobility across the region, and will continue to do so. Climate change will exacerbate these forms of environmental drivers of migration, occasionally through current patterns of environmental degradation and human mobility, but also through new forms of environmental impacts on human populations.

Projected Impacts of Climate Change in Asia

Climate change is projected to have a significant impact on the environments and societies of Asia, particularly because of coastal impacts and because monsoonal rainfall patterns could change. It is not the intention here to repeat the extensive work reviewing the likely impacts of climate change on the region by the IPCC and others (Preston et al. 2006; Cruz et al. 2007; Meehl et al. 2007; IPCC 2012). Rather, it is important to note the growing scientific consensus suggests that human-induced climate change is already occurring, and impacts are projected to increase in the future. For example, Preston et al. (2006: 2) summarize the major projected primary climate change impacts on Asia as:

- Temperature increases: 0.5–2 °C by 2030; 1–7 °C by 2070, greater in arid areas of Northern Pakistan, India and Western China;
- Increasing rainfall during the summer monsoon, but greater monsoon variability;
- Decline in winter rainfall in South and Southeast Asia;
- Rise in global sea level 3–16 centimeters (cm) by 2030 and 7–50 cm by 2070;
- More intense tropical cyclones; and,
- Changes in important modes of climate variability such as the El Niño/La Niña Southern Oscillation.

Primary impacts of climate change will have direct impacts on societies, but there will also be flow-on secondary impacts that due to altered environmental systems and tertiary impacts of change to societal systems. Together, there are a number of ways in which climate change will impact on migration:

1. Primary impacts, which are resulting immediately from changes to the climatic patterns themselves.
2. Secondary impacts, which are changes to environmental systems resulting from the primary impacts of climate change.
3. Tertiary impacts, which include the broader impacts on societal systems, such as the implications for migration patterns and the effects of policy responses by governments.

Warmer temperatures would, for example: have direct and indirect health impacts; lead to increasing evapotranspiration rates, alter crop production and water management systems; and, increase rates of snow pack and glacier melt. Glacial melting has received a lot of attention due to contentious projections of the rate of glacial retreat in the Himalayas. Any melting of glaciers, in association with changes to the Asian monsoon, would increase the variability in river flows in many important rivers, including the Yellow, Yangtze, Mekong, Ganges-Brahmaputra, Indus, and Amu-Darya (Vörösmarty et al. 2000; Xu et al. 2009; IPCC 2012).

Monsoonal activity is projected to increase in intensity, leading to increased average precipitation across much of the region including South, East and Southeast Asia (Schewe et al. 2011). Average annual runoff is projected to increase significantly in areas such as Northeast India-Bangladesh, eastern China and Southeast Asia. Flooding is already a major concern, but in areas where monsoonal activity is projected to intensify in South and East Asia, this hazard could become more problematic (Meehl et al. 2007). There is projected to be a net drying trend for much of Western and Central Asia, and runoff is likely to decline in association. It may also be possible that intra-annual rainfall variability will increase in many areas, because large parts of South and Southeast Asia are projected to experience decreased winter precipitation, even as monsoons bring more summer rain.

At a global scale, the number of intense cyclones is projected to increase in association with warmer seas and this shift may already be occurring. Webster et al. (2005) suggest that the number of category 4 and 5 cyclones have increased over the last 30 years. Although the relationship is not clear, there is a significant regional concern that the northern Indian Ocean has warmed more than other oceans, leading to rates of increase in cyclonic intensity in South Asia that exceed global averages (Vecchi and Soden 2007). Cyclonic activity by itself will potentially have an increasing impact on populations and economies in those areas that already experience cyclones or typhoons, including Japan, North and South Korea, coastal China, Vietnam, the Philippines, Burma, Bangladesh, northeast and northwest coastal India, and the south coast of Pakistan (Dilley et al. 2005).

Upper level projections for sea level rise to the end of the Twenty-First Century are in the order of 1 m above current levels (Meehl et al. 2007). Given the significant uncertainty and the possibility of non-linear melting of terrestrial ice, the possibility of much greater global mean sea level rise is real. Some areas will simultaneously be experiencing sinking landscapes, particularly deltas where flood control measures are in place upriver. Of particular concern could be the deltas of the Indus, Ganges-Brahmaputra, Irrawaddy, Mekong, Red, and Yangtze Rivers, which already experience cyclones (Hugo et al. 2009). In Bangladesh, storm surges over 10 m have been recorded (Mikhailov and Dotsenko 2006; Karim and Mimura 2008). In the Red River delta in Viet Nam, 3 m storm surges are experienced during intense cyclones (Pilarczyk and Nuoi 2005). As storm-surge results from an association of high wind speeds, wave action and low air pressure, often in association with high tides, stronger cyclones are likely to impact on many low-lying areas in the form of more regular, destructive storm surges and greater salt-water intrusion (von Storch and Woth 2008).

While there a major environmental changes projected for Asia, there are already important environmental issues that influence socio-ecosystems across the region. Future climate change impacts need to be discussed in the context of experienced threats and trends to identify likely future impacts on human livelihood, wellbeing and mobility.

Major Environmental Issues in Asia

There are enormous environmental issues for Asia irrespective of future climate change. Thresholds of local capacity to adapt to environmental change *in situ* are being passed as natural hazards such as coastal and river flooding, landslides and typhoons become more regular and/or intense in the region, and as natural resources, and especially access to water, decline in quantity and quality. The management of water resources is a priority irrespective of the potential implications of an intensification of the Asian monsoon, the drying of Western and Central Asia, or coastal and island inundation and erosion.

Together with major hazards including sea-level rise, typhoon impact and flooding, it is possible to examine the risk of drying to identify a number of hotspot areas in Asia are at relatively high risk of adverse impact from one or more natural hazards which result from climate change (Dilley et al. 2005). For example, in South Asia, the deltas of the Ganges-Brahmaputra, the Mahanadi, Godavari, Krishna and the Indus Rivers already experience regular cyclonic activity and these rivers are likely to flood more often with increased monsoonal rainfall. Coastal flooding is likely to significantly increase in Bangladesh, West Bengal and south to Chennai, and from Karachi, Pakistan to Mumbai, India (Fig. 2.1).

Particular environments and sub-regions across Asia will be highly sensitive and exposed to the projected changes, including:

(i) low-lying coastal areas (the location of many of the world's major urban areas);
(ii) deltaic regions (two of the world's three largest deltas are in the region—the Mekong and the Ganges/Brahmaputra);
(iii) low-lying small island states (where climate change induced sea level rise puts the long-term habitability of atolls at risk);
(iv) and semi-arid or low-humidity regions (mostly in Central Asia or western China, where drought and availability of water are already problematic and climate change is likely to exacerbate water shortage problems).

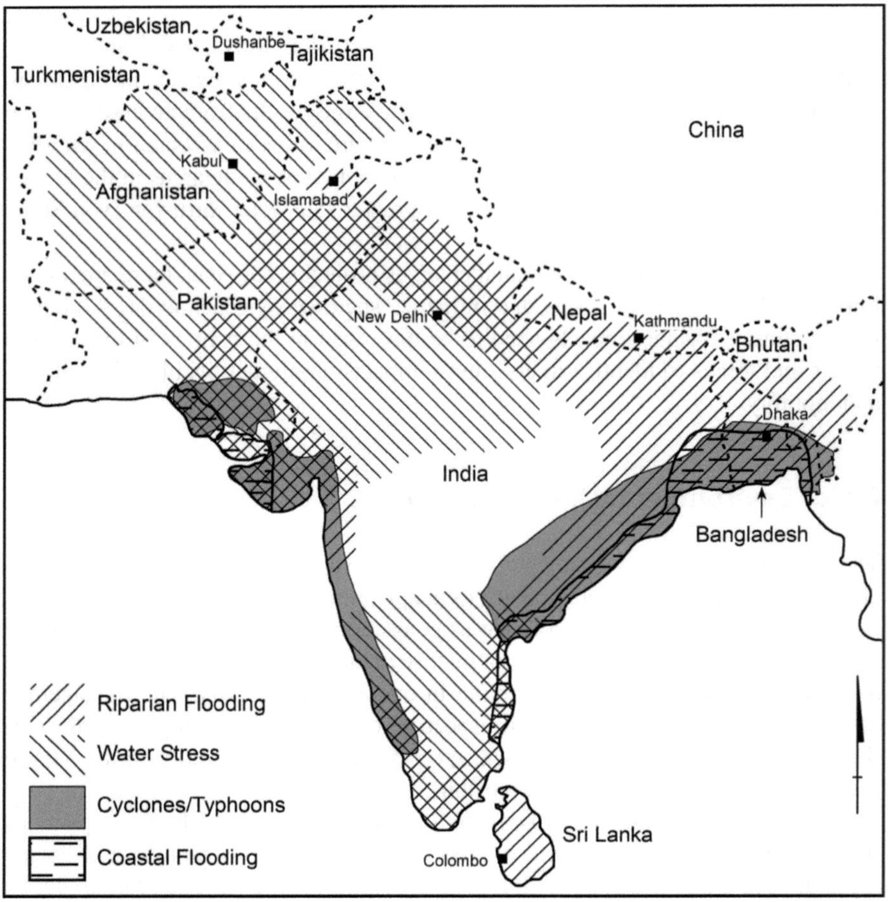

Fig. 2.1 Location of areas in South Asia likely to be most impacted by coastal flooding associated with sea level rise riparian flooding, cyclones/typhoons and water stress as a result of climate change (Source: Hugo et al. 2009)

By mapping key impacts of climate change on the sub-region, it is possible to analyse how many people would be exposed within these areas. Table 2.2 presents projections of populations living in areas exposed to climate change impact areas in the main South Asian countries over the 2000–2050 period.

The complexity of impacts and their causes extends beyond relatively simplistic models of spatial exposure of populations to increased hazard. For example, in east Asia, the deltas of the Yellow River, the Yangtze River and the Pearl River are identified as highly vulnerable (Hai-Lun and Kang 2001; Cruz et al. 2007). In these areas, socio-economic and ecological systems could be disrupted through sea-level rise, salt water intrusion, marshland inundation and frequent floods, and storm surges. Yet, in these deltas, as elsewhere in Asia, the geomorphology of coastal areas is continually changing (Bardsley and Hugo 2010). There has been a failure to

Table 2.2 South Asia: projected population at multiple risk from climate change ('000)

	2000	2020	2030	2050	% Change 2000–2050
Afghanistan	21,524	40,380	52,455	78,234	263.5
Bangladesh	136,946	193,332	217,930	254,082	85.5
India	860,451	1,175,611	1,283,481	1,413,489	64.3
Nepal	22,142	34,310	39,929	49,638	124.2
Pakistan	137,163	202,702	233,802	284,332	107.3
Sri Lanka	18,924	20,229	20,249	18,715	−1.1

Source: Hugo et al. (2009)

create and maintain sustainable coastal management systems on many islands and landmasses already, such that future climate change impacts of sea-level rise, storm surge or cyclonic activity will only be enhancing deteriorating coastal conditions (Harvey 2006; Mimura 2008). Thus, inappropriate development of coasts will have a major influence on the future exposure of large populations to future climate change. Similarly, the over-withdrawal from freshwater aquifers; the disturbance of sand drift patterns; the mining of sediments; the destabilisation of natural protection such as reefs, dunes or vegetation along coasts or in oceans; the changing sedimentary loads from rivers; the changing erosive capacities of rivers; and the development of engineered structures can and do degrade coastal systems. Clearly, as climate change increases the vulnerability of low-lying parts of Asia, current decisions and previous mismanagement will also play vital roles in the vulnerability of populations.

While the risk of environmental change having direct effects on human migration is increasing, the indirect effects of environmental change on society via agricultural or health systems, and the consequent implications for mobility, are highly complex and not fully understood. Environmental change could lead to rapid and dramatic change in human mobility via such indirect drivers this century, especially through the impacts of climate change on food security across Asia (for a Mongolian example see Janes 2010). Already, food security issues are enormous across Asia, with most of the malnourished people of the globe living in the region (WFP/FAO 2009). For example, groundwater resources in India are severely depleted in vital agricultural regions such as Punjab, Haryana and Rajasthan (Rodell et al. 2009). Similarly, drought in China is a major problem and there is increasing competition between rural and urban uses. Further water resources for the development of agricultural irrigation is scarce or contested already, and climate change will worsen shortages in many places (Hanjra and Qureshi 2010; Ringler et al. 2010). Not only are there limited opportunities for further expansion of agricultural production areas around the globe to meet the growing demand for food in Asia, in many places climate change will significantly deplete resources, as well as increase environmental hazards (Nelson et al. 2009; Gornall et al. 2010). There are many major environmental issues that cannot be considered in any depth here, but apart from the focus on water resource, coastal, flood and food security issues discussed above, a brief list of relevant socio-ecological issues that will be accentuated by climate change at local regional scales in Asia would also include urban air pollution and sprawl; depleted biodiversity, including agricultural, wetland, forest and marine resources; mining impacts; and depleted fossil fuel and other minerals. Together, all of these issues will have an increasing influence over the decisions of people to migrate within, from and to Asia.

Major Asian Migration Flows

Climate change needs to be conceptualised as an additional driver of migration, which will add to an array of existing drivers. It is important to establish the existing patterns of population mobility in the Asia as a preliminary to projecting the future impact of climate change on migration (Barnett and Webber 2009; Hugo et al. 2009). While it is not possible in the space available to provide a comprehensive account of contemporary regional population mobility, a summary of the main features is presented here.

One of the most striking recent demographic trends in Asia having major implications for economic and social change has been an increase in the level and complexity of migration (Lin and Yeoh 2010). Changing one's place of residence on a permanent or temporary basis has drawn within the calculus of choice to a majority of people as they contemplate their opportunities. As a result, levels of mobility *within* and *between* nations has increased exponentially (United Nations 2006). In 1950, only 16.8 % of the region's population lived in urban areas (United Nations 2008). By 2005 this had increased to 38.6 % and by 2025 half of the population will live in urban areas (Table 2.3). Rural–urban migration is one of the prominent causes for rapid urban growth in Asia (for a Chinese example see Wang et al. 2011). In 2002, the average annual urban growth rates in Bangladesh India, China, and Vietnam were 4, 2.8, 2.4 and 2.2 % respectively (Skeldon 2003).

Most environmentally induced migration in Asia has involved *internal* movement within countries on a temporary or permanent basis (Hugo 1996). Similarly, most forced migrations due to developments or government intervention have involved rural migrations. As development has increased, mobility is being used by more Asian residents as a strategy to adjust to changing circumstances or improve their socio-economic position. For example, in East Asia, rapid economic growth, especially in the 'demographic giant' China, over the last two decades has seen a massive increase in individual mobility, both internally and internationally. At least 10 % of Chinese (130 million people) are internal migrants in the sense that they are living away from the place in which they are registered (*hukou*) (World Bank 2008). Most have moved from rural to urban areas, especially to the coastal cities of the east. Just as in China, many of the very large cities of Asia are in locations that suggest they will be vulnerable to climate change impacts. In particular, the massive growth in megacities in coastal areas significantly increases the risk of climate change for large populations (McGranahan et al. 2007) (Fig. 2.2).

There are particular components of the complex rural out-migration process which could be significantly affected by climate change, both for the voluntary

Table 2.3 Asia: growth and projected growth of megacities, 1950–2025 (millions)

1950		2000		2005		2020		2025	
City	Population	City	Population	City	Population	City	Population	City	Population
Tokyo	11.275	Tokyo	34.45	Tokyo	35.327	Tokyo	36.399	Tokyo	36.4
Shanghai	6.066	Mumbai	16.086	Mumbai	18.202	Mumbai	24.051	Mumbai	26.385
		Shanghai	13.243	Delhi	15.053	Delhi	20.484	Delhi	22.498
		Calcutta	13.058	Shanghai	14.503	Dhaka	19.422	Dhaka	22.015
				Calcutta	14.282	Calcutta	18.707	Calcutta	20.56
				Dhaka	12.576	Shanghai	18.466	Shanghai	19.412
				Karachi	11.553	Karachi	16.922	Karachi	19.095
				Osaka-Kobe	11.258	Manila	13.892	Manila	14.808
				Manila	10.761	Beijing	13.807	Beijing	14.545
				Beijing	10.717	Jakarta	11.689	Jakarta	12.363
						Osaka-Kobe	11.368	Guangzhou	11.835
						Guangzhou	11.218	Osaka-Kobe	11.368
								Lahore	10.512
								Shenzhen	10.196
								Chennai (Madras)	10.129

Source: United Nations (2008)

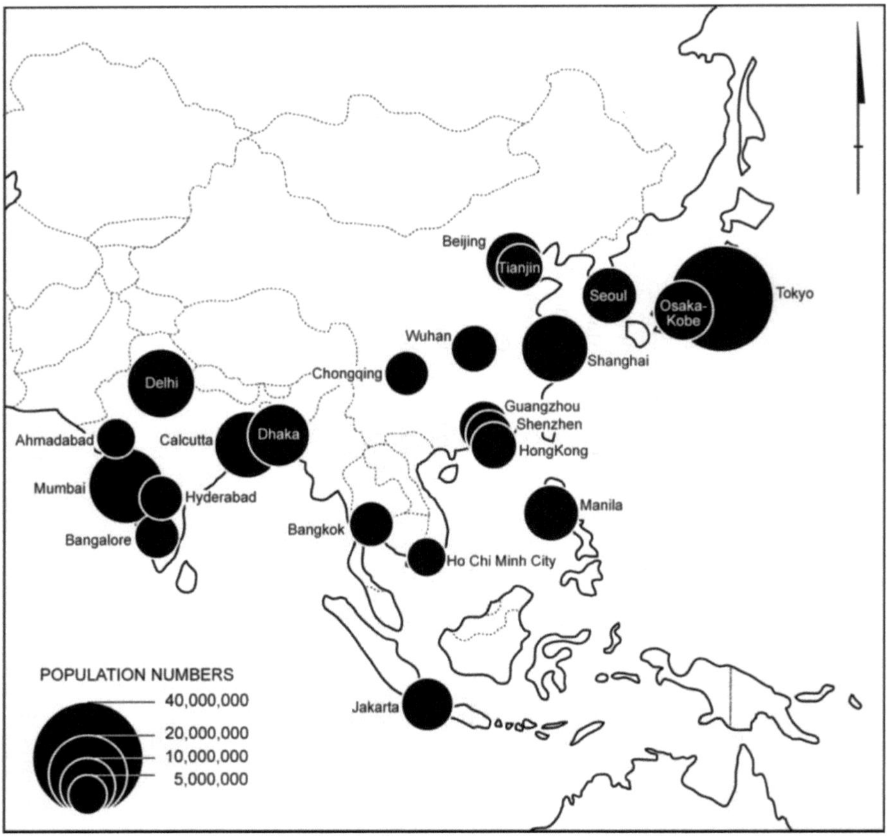

Fig. 2.2 Asia: location of cities with a population of five million and over, 2005 (Source: United Nations 2008)

migration of relatively well informed and wealthy individuals, and those who use migration as mechanism for survival (Warner et al. 2008). Rural–urban flows on temporary, cyclical or permanent bases are creating strong linkages between rural and urban areas. For that reason, the dominant processes of urbanisation and circular migration are likely to frame future migration in a changing climate. For example, during the 2011 flooding in Bangkok, Thailand situated on the delta of the Chao Phraya river, significant numbers of people were forced or chose to leave the city, but many of those people still had links to villages in the regions, so they had places to go on a temporary basis until the floods receded.

While people do have coping strategies to manage environmental hazards and change, it is the large numbers of rural poor across large parts of Asia, that will be most exposed to climate change impacts (McGranahan et al. 2007), because:

- They are often forced to exploit natural resources on the environmental margins such as coasts, flood plains or semi-arid areas;

- They are more directly dependent on natural resource management and environmental services for their livelihoods; and,
- They have poor accommodation, no means to flee and are often living in groups that are comparatively poorly informed and/or educated about environmental hazards.

Simultaneously, the rural poor are often most sensitive to environmental change because:

- They have the fewest personal or community economic resources to support autonomous adaptation, including migration;
- They often have the least access to formal institutional assistance or collective adaptation responses; and,
- They are often distanced from national and international decision-makers and therefore, their plight may go un-noticed or responses may be comparatively weak or late.

Together, we could conclude that while the cities will experience climate change impacts, their adaptive capacities are relatively significant, due to a combination of their relative wealth, their social and human capitals, and their attractiveness for political, economic and research interests to ensure effective adaptation (Betsill and Bulkeley 2007). In contrast, the large poor rural populations in Asian countries that remain heavily reliant on natural resource exploitation or wage labour associated with primary industries may be particularly vulnerable to climate change impacts. For that reason, the rural–urban, and in fact rural-rural migration networks will continue to provide a vital adaptation mechanism for rural populations in distress (for South Asian examples see Revi 2008; Roy 2009). Rural to urban migration across the region is generated both by the real and perceived inequalities in opportunities between rural and urban areas, and increasing impoverishment in rural areas (Seto 2011). In contrast to East and Southeast Asia, there is still a relatively low level of urbanisation in South and Central Asia and there is considerable scope in the future for rural to urban migration in these sub-regions (World Bank 2008). On the other hand, new opportunities to settle in productive rural areas in Asia are relatively limited, with people increasingly settling in areas exposed to significant environmental risks such as storm surges, floods and droughts.

There has also been important international migration to and from Asia. In the first instance, it is necessary to conclude from the general lack of good data on international migration from and to Asian countries that current information weakens opportunities to make realistic projections of future international mobility levels. Although Asia is a net region of emigration, numerous countries have important levels of labour immigrants. Southeast Asia, for example, is highly diverse from a migration perspective with several net immigration countries (Singapore, Brunei, Malaysia, Thailand), but also some of the world's major emigration nations (Philippines, Indonesia, Vietnam) (Haque 2005; Lin and Yeoh 2010). There are very substantial flows *between* the countries of the region, which includes many of the largest migration corridors in the world identified by the World Bank (2008), a third involving a South or Central Asian country. As well as differentials in income and

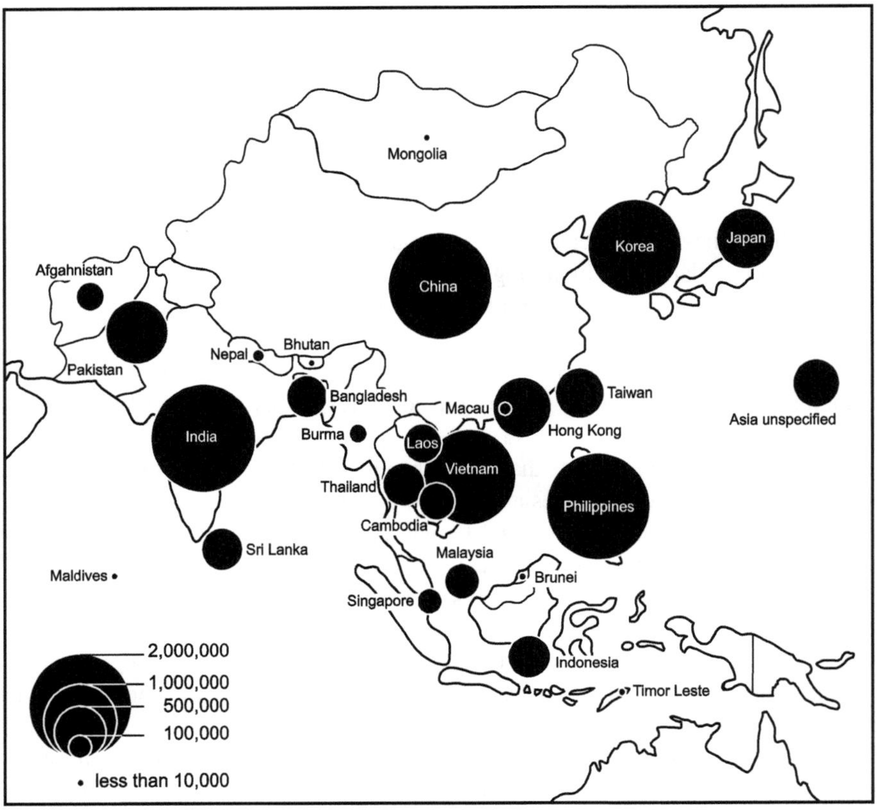

Fig. 2.3 Asian countries: number of persons born in Asian countries living in OECD countries in 2000 (Source: OECD)

population growth, there are a number of elements which are encouraging economic migration within and from Asia. These include:

- Labour market segmentation due to structural needs of the economies, which create a demand for labour migrants on one hand, and remittances on the other;
- A highly developed and self perpetuating immigration industry; and,
- Social networks that link origin and destination become key elements in sustaining and enhancing populations flows.

There are very large numbers of permanent and temporary immigrants from Asian countries in OECD countries around the world (Fig. 2.3). The large international diasporas provide established networks and contacts for future mobility across the globe for many populations. Those networks also assist in facilitating business and trade links (for a Chinese example see Buckley et al. 2007). Such diasporas could be expected to continue to provide resources for adaptation, as well as influencing directions of mobility for people making decisions about how to adapt to experienced or future change.

Previous forced displacements provide important guides for future climate change migration. Conflict has periodically been an important cause of population movement, or in some cases migration has led to conflicts, but movements have been primarily contained within the countries of Asia. In Indonesia, for example, there were significant forced migrations caused by post-independence conflicts, while in the late 1990s unrest caused the displacement of over a million people within the country (Hugo 2002). India's Farakka barrage has been implicated in driving environmental change in Bangladesh that has led in part to the displacement of several hundred thousand Bangladeshis to India, that in turn has occasionally led to violent conflicts between native Indians and Bengali Muslim migrants (Datta 2004). Large numbers of Afghani refugees have sort asylum in Iran and Pakistan. In that case, as in previous large refugee migrations have been established, people primarily move along corridors established by original refugee flows. In fact, most migration occurs along social networks established by earlier generations of migrants (Massey et al. 1993).

The majority of movement of people due to climate change impacts, even in extreme cases, will likely follow or reinforce established channels of movement or at least depend upon established networks and relationships (Hugo 2010). However, adaptations within locations may be making a clear analysis of the future relationships between migration and environmental change difficult – in other words, the easy gains to support people to adapt to environmental change *in situ* may be already reaching thresholds of effectiveness in many places, and more and more people may be forced to resort to *ex situ* adaptation involving the movement of themselves, their communities, and their infrastructure and assets (Bardsley and Hugo 2010). In fact, it is becoming increasingly clear that environmental pressures resulting from climate change across large parts of Asia could so significantly alter local socio-ecosystems that they will also drive the development of new corridors and new scales of migration. Furthermore, local adaptation policies could in themselves lead to displaced populations and climate change will impinge upon some important destination areas and hence act to reduce the 'pull' of potential destinations. Thus, as we have here, it is possible to conclude that there will be linear changes to Asian migration due to climate change, but also that there will be non-linear changes which could surprise policy and practice throughout the region (Bardsley and Hugo 2010).

Linking Climate Change and Migration in Asia

It is impossible to comprehensively examine the relationship between migration and climate change in Asia, due to the enormity of the issues, especially the considerable uncertainty surrounding climate change impacts. Knowledge of existing patterns, drivers and impacts of population mobility in Asia also remains constrained by the lack of comprehensive and accurate data (ADB 2012a; McAdam 2010; Foresight 2011). What research has been undertaken examining the link between climate change and migration generally focuses on specific impacts of environmental events, but climate change has the potential to not only change the impacts of environmental hazards but also alter the prevailing environments and resources that support livelihoods (Hugo et al. 2009). Irrespective of this complexity, it is possible to deduce some major themes from amongst the complexity. It is essential that:

- Environmentally induced population mobility be seen as a wide array of mobility types and not just displacement; and,
- That it be seen as only one of the responses among an array of potential mitigation and adaptation strategies that populations will implement or undertake.

Changes to existing migration patterns in Asia as a result of climate change will take many forms, but will principally involve an increase in the 'push' forces which lead to emigration and outmigration. There are three main processes through which climate change impacts are likely to manifest themselves on migration patterns:

- increasing experienced risk of environmental hazards and associated socio-ecological events;
- changing resource condition in linear and non-linear trends through time which alter access and effective utilisation of natural resources; and
- altering perceptions of risk of impacts of climate change, irrespective of real experiences.

It is neither possible nor responsible to infer that precise predictions of future migration numbers due to changes in climate can be created, nevertheless recent work by Hugo et al. (2009) has reviewed 'populations at risk' due to climate change in Asia. The discussion below draws from that review and acts to integrate many of the issues raised to date in this chapter within particular sub-regional and national contexts. Such projections can help guide future policies and practices to respond to the imminent threat of future change. It is important to emphasise that it is poor populations in both urban and rural contexts which will be particularly disadvantaged by climate change impacts and are therefore, likely to require specific, targeted policy responses.

Climate Change Impacts on Migration in East Asia

Considerations of climate change in this region focus particularly on China partly because of its large population, but also due to the relatively high degree of vulnerability of many of the Chinese population living in regions likely to be exposed to significant climate change (Hugo et al. 2009). Major movements are already underway in China due to government policies resettling people from defined "Ecological Fragile Zones" or from developments such as the Three Gorges Dam. Several of China's megacities, such as Tianjin, Shanghai, Shenzhen, Guangzhou and Haikou are located in coastal and inner valley areas that are vulnerable to coastal and river flooding. Furthermore, urban populations at risk in China are projected to increase one and a half times over the next half century. In contrast, in Japan the numbers in 'at risk' urban areas will decrease over the period and this reflects the fact that the Japanese population has already begun to decline. Similarly in South Korea only limited growth in urban populations is anticipated. In contrast, for each country in the region except Mongolia, the rural populations living in areas most exposed to environmental change will decline substantially. Thus, in the East Asian context most population growth in vulnerable places over the next five decades will be in China and most of it will be in urban areas, and particularly Chinese megacities. The potential impacts of environmental change on Chinese migration can be summarised according to likely linear and non-linear changes:

Linear change

- Rural-to-urban migration has been the dominant population movement in China since the mid-1980s. Ongoing environmental change will increase the likelihood of strong push forces from rural areas, as well as maintain the reasons why government supported resettlements are being undertaken.

Non-Linear change

- Many of the vulnerable areas in coastal and river valley areas in China are currently important destination areas for rural–urban migrants. Hence, any migration response to climate change impacts in China would involve reversing or limiting existing migration trends which are channelling migrants into cities located in areas projected to be at greater risk of flooding.
- 41 % of China's population of 1.32 billion (in 2007) are located in coastal regions, mostly in large cities. Many of these cities could be vulnerable to a large-scale forced relocation of people, permanently or temporarily, due to greater frequency and severity of climatic hazards such as floods and cyclones, along with rising sea levels. The incorporation of environmental factors into settlement planning in these cities would reshape urban development along the coast and human mobility would change in China as a result.

Climate Change Impacts on Migration in Southeast Asia

Coastal flooding poses the greatest climate change risk in Southeast Asia, with around a third of the population living in areas considered to be at risk of coastal flooding and its associated impacts (Hugo et al. 2009). Populations at risk are especially concentrated in Indonesia, the Philippines, Vietnam, Myanmar and Thailand, mainly in the megacities of Jakarta, Manila, Ho Chi Minh City and Bangkok which are located in low-lying coastal areas. Migration patterns in and from the Mekong delta would also be particularly impacted. Already, anecdotal evidence from the Mekong delta suggests that 'successive flooding events leading to destruction of crops on more than one occasion can drive people to migrate elsewhere in search of an alternative livelihood' (Warner et al. 2008: 39). A large number of Southeast Asian people live in areas projected to be further affected by cyclones, especially in Vietnam, the Philippines, Myanmar and Thailand. Considering that much of the local population remains socio-economically dependent on agriculture, lack of alternative livelihoods combined with increasing risks of sea level rise and flooding may eventually lead to large-scale population displacement in coastal areas across the sub-region. The potential impacts of environmental change on Thai migration can be summarised according to likely linear and non-linear changes:

Linear change

- A number of specific impacts are likely to reduce local opportunities for development, which in turn will enhance migration linearly along established corridors of migration. The variability in the timing and amount of rainfall associated with the monsoon is a significant factor in the viability of dryland rice farming, particularly in Northeast and North Thailand. Flood risk is also significant in the North and Central areas of Thailand.
- If resources decline or more extreme rainfall events occur more regularly, more people way wish to leave hazardous regional areas, enhancing rural–urban migration.

Non-Linear change

- The interaction of sea-level rise, storm surges and river flood risk associated with climate change could lead to the displacement of people from the low-lying sectors of Bangkok, leading to a fundamental change in the role of the city: as a site for Thai residents; as a sink destination for internal immigrants; and, as a stop-over point for future international migrants.
- The existing corridor across Thailand could be extremely sensitive to additional migration arising due to climatic change. This is a point that is rarely considered in relation to climate change impacts on migration – important destination areas for migrants may be seriously affected by impacts and thereby the livelihoods of many migrants and their communities in origin countries could be threatened.

Climate Change Impact on Migration in South Asia

Projected climate change would have significant impacts on South Asia. It is of concern that by 2050 it is anticipated that 1.4 billion Indians will be living in areas experiencing negative impacts from climate change (Hugo et al. 2009). Clearly while most will strive to adapt *in situ*, the potential for a considerable redistribution of population through migration is substantial. India could face increasing intensity and frequency of floods along important rivers such as the Ganges and Brahmaputra in the north, Mahanadi, Godavari, Krishna in the south, Indus in the west and along the north eastern and eastern regions from Assam and West Bengal to Andhra Pradesh and Tamil Nadu. Given local communities' dependence on agriculture for daily subsistence and livelihood, climatic changes in the form of droughts, floods, landslides, and cyclones may increase their vulnerability and potentially lead to both internal and international displacement. Significant internal migration – both permanent and temporary – from rural to urban areas is already attributed to an increasing frequency of droughts and floods on agricultural production. Over time, coastal regions within India could witness a particularly non-linear change in their vulnerability to climate change impacts.

Climate change impacts such as more extreme monsoonal rainfall and associated landslides and flood events, would exacerbate the impoverishment of many rural Nepalis in the Terai, hill and mountain valley regions, potentially leading to increased landlessness and outmigration. There will be over 250 million people in highly vulnerable areas in both Bangladesh and Pakistan by the middle of this century (Hugo et al. 2009). The potential impacts of environmental change on Bangladeshi migration can be summarised according to likely linear and non-linear changes:

Linear change

- migration is already utilised within Bangladesh as a coping mechanism in the face of significant environmental challenges such as droughts, floods and cyclones, and so mechanisms and networks are established. Already a clear majority of temporary labour migrants from Bangladesh in India originate from 5 of the country's 64 districts that are especially prone to flooding and other environmental hazards.
- projected climate change would strengthen such environmental forces so that pressure for permanent and temporary migration out of large areas of southern Bangladesh areas would increase.

Non-Linear change

- sea level rise in association with increasing cyclone intensity and runoff variability could, over extended periods of time result in non-linear changes to the local and regional environment, and render large areas unsuitable for inhabitation. Internal mobility within Bangladesh, and international migration from the country, could increase significantly as a result.

Climate Change Impact on Migration in Central and Western Asia

Drought has been a recurrent hazard in various parts of Central and Western Asia and as Stringer (2008: 14) notes "desertification and land degradation are not new environmental problems in Central Asia.". In the last two decades, many communities have had to relocate to explore new livelihood opportunities. For example, the degradation of the Aral Sea environment has led to a significant regional out-migration within Kazakhstan and Uzbekistan since the 1960s (Glantz 2004; McLeman 2011). Poverty and ineffective governance further adds to the vulnerability of local populations to environmental disasters such as drought, loss of agricultural productivity and resulting food insecurity. Projected climate change would significantly increase economic, social and political pressures in the sub-region, thereby resulting in a likely increase in migration (Hugo et al. 2009). For example, in Kazakhstan the combined effect of high temperature, limited water availability and changing precipitation patterns could also result in severe impacts on agricultural production. Increasing aridity would also make access to freshwater more difficult and potentially exacerbate desertification in both semi-arid and steppe sub-humid areas.

Tajikistan is vulnerable to socio-economic and environmental crisis due to changes in climatic conditions in and around the region. The country's mountains are considered the main glacial area for Central Asia, while its rivers are the main water sources that have replenished the Aral Sea for several centuries. According to one estimate, 'Up to 80 % of stream flow of the Amudarya River that flows into the Aral Sea is formed in Tajikistan' (Ministry for Nature Protection 2002: 61). Increasing warming and climatic changes have led to constant retreating of small glaciers and also 'in the mid- and long-term, a catastrophic reduction of water flow in many rivers is expected' (Ministry for Nature Protection 2002: 15). The glaciers have retreated – both in area and volume – primarily due to temperature increases and changes in precipitation, and that trend is likely to continue. Furthermore, the intensity and frequency of floods in Tajikistan has also increased over the last few decades. National records suggest that increased rainfall combined with melting of the glaciers has led to disastrous floods in 1969, 1981, 1993, 1998 and 2002 (Ministry for Nature Protection 2002; CEDEM 2008b; Khakimov and Mahmadbekov 2009). The potential impacts of environmental change on Tajikistan migration can be summarised according to likely linear and non-linear changes:

Linear change

- Tajikistan is already highly vulnerable to environmental degradation of varying kinds – floods, mudslides, salinity, water and soil erosion and desertification. The entire country's population was identified as living in a region where multiple impacts of climate change may be experienced. These environmental issues are further compounded by the presence of low-paid jobs, widespread poverty and high unemployment, thereby enhancing rural–urban and international labour migration.

Non-Linear change

- Reduced agricultural outputs, along with increased flood frequency and lack of water resources, could increase the vulnerability of particularly large rural populations and increasing pressures leading to fundamental changes in rural to urban and international migration if thresholds of resource depletion are met.

Conclusion

Climate change projections for Asia suggest that there are very real new risks emerging for societies throughout the region. Of great concern, is that current projections for climate change impacts are mostly not limited, in the sense that changes are only projected to increase throughout the Twenty-First Century. While there is still enormous uncertainty about the scale and scope of those changes, this chapter indicates that some important steps can be made to review locations of high risk of impacts and vulnerabilities to future impacts, and begin to respond accordingly. It is likely to be those who are already most vulnerable due to their socio-economic situation who are also likely to be most negatively impacted by future climate change. However, the growing importance of Asia for the global economy suggests that if significant impacts of climate change are experienced, there would also be important implications for global systems of migration. *In situ* adaptation to reduce the exposure and sensitivity to climate change impacts will form the bulk of societal responses in Asia as elsewhere, but also there needs to be re-conceptualisation of migration as part of a broader societal adaptation strategy. Researchers must engage strongly with decision-makers to assist them to establish iterative governance structures; to develop and enact plans; and, to ensure resources are provided for a timely and strategic response (Bardsley and Rogers 2011).

There can be no doubt that the impacts of climate change will add to already high levels of migration complexity in Asia. We cannot be specific about the scale and scope of those impacts, but they will be significant. However, you only need to look at the last 40 years of human mobility in Asia to understand that major change in population distributions need not have catastrophic outcomes, but could be evolutionary and benefit most people over the century to come. It is our conclusion that it is possible to not only cope with environmental change and migration in Asia, but if sound policy could be implemented in a timely manner, new forms and scales of population mobility could be harnessed to reduce poverty and enhance economic and social development in the region. In order to support migration as a component of effective, largely autonomous, adaptation to environmental change, however, there will need to be major policy improvements in many areas: more effective migration management; stronger governance; more appropriate funding and remittance mechanisms to facilitate adaption; enhanced international cooperation on climate change issues; expanded and improved development assistance mechanisms; and, the implementation of sound economic development policy and practice throughout the region. Moreover there is an *urgency* in making these changes to accommodate the effects of climate change on mobility. This urgency derives from two things. Firstly some of the impacts are already in evidence, and secondly because the changes which are required involve substantial institutional, structural and cultural change, they will take time to achieve. Many of the impacts of climate change are likely to be felt hardest at least several decades into the future. However, for both human mobility, and societal adaptation more broadly, responses must be developed and implemented now to ensure that the environmental risks of the future will be managed effectively.

References

ADB (Asian Development Bank). (2012a). *Addressing climate change and migration in Asia and the Pacific*. Manila: ADB.

ADB (Asian Development Bank). (2012b). *Asian development outlook 2012: Confronting rising inequality in Asia*. Manila: ADB.

Adger, W. N., Arnell, N. W., & Tompkins, E. L. (2005). Successful adaptation to climate change across scales. *Global Environmental Change, 15*, 77–86.

AidWatch. (2006). *UN Office of the special envoy for tsunami recovery*. http://reliefweb.int/organization/un-special-envoy-tsunami-recovery. Accessed 24 Apr 2012.

Alam, S. (2003). Environmentally induced migration from Bangladesh to India. *Strategic Analysis, 27*, 422–438.

Bardsley, D. K., & Hugo, G. J. (2010). Migration and climate change: Examining thresholds of change to guide effective adaptation decision-making. *Population and Environment, 32*, 238–262.

Bardsley, D. K., & Rogers, G. P. (2011). Prioritizing engagement for sustainable adaptation to climate change: An example from natural resource management in South Australia. *Society and Natural Resources, 24*, 1–17.

Barnett, J., & Campbell, J. (2010). *Climate change and small island states: Power, knowledge and the South Pacific*. London: Routledge.

Barnett, J., & Webber, M. (2009). *Accommodating migration to promote adaptation to climate change*. A policy brief prepared for the Secretariat of the Swedish Commission on Climate Change and Development and the World Bank Report 2010 Team.

Begum, A. (1999). *Destination Dhaka: Urban migration: Expectations and reality*. Dhaka: The University Press.

Betsill, M., & Bulkeley, H. (2007). Looking back and thinking ahead: A decade of cities and climate change research. *Local Environment, 12*, 447–456.

Bohra-Mishra, P., & Massey, D. S. (2011). Environmental degradation and out-migration: New evidence from Nepal. In E. Piguet, A. Pécoud, & P. de Guchteneire (Eds.), *Migration and climate change* (pp. 74–101). Paris: UNESCO.

Buckley, P. J., Clegg, L. J., Cross, A. R., Liu, X., Voss, H., & Zheng, P. (2007). The determinants of Chinese outward foreign direct investment. *Journal of International Business Studies, 38*, 499–518.

CEDEM (Center for Ethnic and Migration Studies). (2008a). *China "Forced Migration and the Three-Gorges Dam"*. Case study report for the EACH-FOR project. Bonn: United Nations University Institute for Environment and Human Security (UNU-EHS).

CEDEM (Center for Ethnic and Migration Studies). (2008b). *Environmental Change and Forced Migration (EACH-FOR) scenarios – General overview study NIS and Central Asia*. Brussels: Center for Ethnic and Migration Studies (CEDEM) for European Commission.

Cernea, M. M. (1990). Internal refugee flows and development-induced population displacement. *Journal of Refugee Studies, 3*, 4.

Cernea, M., & McDowell, C. (2000). *Risks and reconstruction, experiences of resettlers and refugees*. Oxford: Berghahn Books.

Chen, M., Xu, C., & Wang, R. (2007). Key natural impacting factors of China's human population distribution. *Population and Environment, 28*, 187–200.

Cruz, R. V., Harasawa, H., Lal, M., Wu, S., Anokhin, Y., Punsalmaa, B., et al. (2007). Asia climate change 2007: Impacts, adaptation and vulnerability. In M. L. Parry, O. F. Canziani, J. P. Palutikof, P. J. van der Linden, & C. E. Hanson (Eds.), *Contribution of Working Group II to the fourth assessment report of the Intergovernmental Panel on Climate Change* (pp. 469–506). Cambridge: Cambridge University Press.

Datta, P. (2004). Push-pull factors of undocumented migration from Bangladesh to West Bengal: A perception study. *The Qualitative Report, 9*, 335–358.

Dilley, M., Chen, R. S., Deichmann, U., Lerner-Lam, A. L., & Arnold, M. (2005). *Natural disaster hotspots: A global risk analysis*. Washington, DC: The World Bank.
Dun, O. (2011). Migration and displacement triggered by floods in the Mekong delta. *International Migration, 49*, e200–e223.
Ericson, J. P., Vörösmarty, C. J., Dingman, S. L., Ward, L. G., & Meybeck, M. (2006). Effective sea-level rise and deltas: Causes of change and human dimension implications. *Global and Planetary Change, 50*, 63–82.
Findlay, A., & Geddes, A. (2011). Critical views on the relationship between climate change and migration: Some insights from the experience of Bangladesh. In E. Piguet, A. Pécoud, & P. de Guchteneire (Eds.), *Migration and climate change* (pp. 138–159). Paris: UNESCO.
Foresight. (2011). *Foresight: Migration and global environmental change, future challenges and opportunities* (Final project report). London: The Government Office for Science.
Glantz, M. H. (2004). *Creeping environmental problems and sustainable development in the Aral Sea basin*. Cambridge: Cambridge University Press.
Gornall, J., Betts, R., Burke, E., Clark, R., Camp, J., Willett, K., et al. (2010). Implications of climate change for agricultural productivity in the early twenty-first century. *Philosophical Transactions of the Royal Society B, 365*, 2973–2989.
Gray, C. L., & Mueller, V. (2012). Natural disasters and population mobility in Bangladesh. *Proceedings of the National Academy of Sciences of the United States of America, 109*, 6000–6005.
Hai-Lun, Z., & Kang, W. (2001). *Flood control and management for large rivers in China: A case study from the Huaihe River Basin*. http://www.unescap.org/esd/water/disaster/2001/china.doc.. Accessed 22 Apr 2009.
Hanjra, M. A., & Qureshi, M. E. (2010). Global water crisis and future food security in an era of climate change. *Food Policy, 35*, 365–377.
Haque, S. (2005). Migration trends and patterns in South Asia and management approaches and initiatives. *Asia-Pacific Population Journal, 20*, 39–60.
Harvey, N. (2006). *Global change and integrated coastal management: The Asia-Pacific region*. Dordrecht: Springer.
Hossain, S. (2005). Poverty, household strategies and coping with urban life: Examining 'livelihood framework' in Dhaka city, Bangladesh. *Bangladesh e-Journal of Sociology, 2*, 45–52.
Hugo, G. J. (1996). Environmental concerns and international migration. *International Migration Review, 30*, 105–131.
Hugo, G. J. (2002). Pengungsi – Indonesia's internally displaced persons. *Asian and Pacific Migration Journal, 11*, 297–331.
Hugo, G. J. (2010). Climate change induced mobility and the existing migration regime in Asia and the Pacific. In J. McAdam (Ed.), *Climate change and displacement: Multidisciplinary perspectives* (pp. 9–36). Oxford: Hart Publishing.
Hugo, G. (2011). Future demographic change and its interactions with migration and climate change. *Global Environmental Change, 21*, S21–S33.
Hugo, G., Bardsley, D. K., Tan, Y., Sharma, V., Williams, M., Bedford, R. (2009). *Climate change and migration in the Asia-Pacific region* (Final report to the Asian Development Bank). Manila: Asian Development Bank.
IMF (International Monetary Fund). (2012). *World economic outlook: Growth resuming, dangers remain*. Washington, DC: International Monetary Fund.
IOM (International Organization for Migration). (2010). *Assessing the evidence: Environment, climate change and migration in Bangladesh*. Bangladesh: IOM.
IPCC (Intergovernmental Panel on Climate Change), Field, C. B., Barros, V., Stocker, T. F., Qin, D., Dokken, D. J., et al. (Eds.). (2012). *Managing the risks of extreme events and disasters to advance climate change adaptation. A special report of Working Groups I and II of the IPCC*. Cambridge/New York: Cambridge University Press.
Janes, C. R. (2010). Failed development and vulnerability to climate change in central Asia: Implications for food security and health. *Asia-Pacific Journal of Public Health, 22*, 236S–245S.

Karim, M. F., & Mimura, N. (2008). Impacts of climate change and sea-level rise on cyclonic storm surge floods in Bangladesh. *Global Environmental Change, 18*, 490–500.

Khakimov, P., & Mahmadbekov, M. (2009). *Republic of Tajikistan: EACH-FOR case study report*. Brussels: Center for Ethnic and Migration Studies (CEDEM) for European Commission.

Lein, H. (2000). Hazards and 'forced' migration in Bangladesh. *Norwegian Journal of Geography, 54*, 122–127.

Lewicka, M. (2011). Place attachment: How far have we come in the last 40 years? *Journal of Environmental Psychology, 31*, 207–230.

Lin, W., & Yeoh, B. S. A. (2010). Questioning the 'field in motion': Emerging concepts, research practices and the geographical imagination in Asian migration studies. *Cultural Geographies, 18*, 125–131.

Mailisha. (2004). Ethnic minority immigrants under the Western Region Development: A report from the Sunan Yugur autonomous region. *Inner Asia, 6*, 111–117.

Martin, P. (2008). Another Miracle? Managing Labour Migration in Asia. Paper presented at United Nations Expert Group Meeting on International Migration and Development in Asia and the Pacific, United Nations Economic and Social Commission for Asis and the Pacific, Population Division, Department of Economic and Social Affairs, Bangkok, Thiland, 20–21 September.

Massey, D. S., Arango, J., Hugo, G., Kouaouci, A., Pellegrino, A., & Taylor, J. E. (1993). Theories of international migration: A review and appraisal. *Population and Development Review, 19*, 431–466.

McAdam, J. (2010). *Climate change induced mobility and the existing migration regime in Asia and the Pacific*. Oxford: Hart Publishing.

McGranahan, G., Balk, D., & Anderson, B. (2007). The rising tide: Assessing the risks of climate change and human settlements in low elevation coastal zones. *Environment and Urbanization, 19*, 17–37.

McLeman, R. A. (2011). Settlement abandonment in the context of global environmental change. *Global Environmental Change, 21S*, S108–S120.

McLeman, R., & Smit, B. (2006). Migration as an adaptation to climate change. *Climatic Change, 76*, 31–53.

Meehl, G. A., Stocker, T. F., Collins, W. D., Friedlingstein, P., Gaye, A. T., Gregory, J. M., et al. (2007). Global climate projections. In S. Solomon, D. Qin, M. Manning, Z. Chen, M. Marquis, K. B. Averyt, M. Tignor, & H. L. Miller (Eds.), *Climate change 2007: The physical science basis. Contribution of Working Group I to the fourth assessment report of the Intergovernmental Panel on Climate Change*. Cambridge/New York: Cambridge University Press.

Mikhailov, V. N., & Dotsenko, M. A. (2006). Peculiarities of the hydrological regime of the Ganges and Brahmaputra river mouth area. *Water Resources, 33*, 353–373.

Mimura, N. (2008). *Asia-Pacific coasts and their management: States of environment*. Dordrecht: Springer.

Ministry for Nature Protection. (2002). *The first national Communication of the Republic of Tajikistan to the United Nations Framework Convention on Climate Change*. Dushanbe: Ministry for Nature Protection.

Mirza, M. M. Q. (2011). Climate change, flooding in South Asia and implications. *Regional Environmental Change, 11*, 95–107.

Naik, A., Stigter, E., & Laczko, F. (2007). *Migration, development and natural disasters: Insights from the Indian Ocean tsunami*. Geneva: International Organisation for Migration.

Nasreen, M., Hossain, K. M., & Kundu, D. K. (2006). The interrelationship between poverty, environment and sustainable development. *Bangladesh e-Journal of Sociology, 3*, 59–79.

Nelson, G. C., Rosegrant, M. W., Koo, J., Robertson, R., Sulser, T., Zhu, T., et al. (2009). *Climate change: Impact on agriculture and costs of adaptation*. Washington, DC: International Food Policy Research Institute.

Paul, B. K. (2005). Evidence against disaster-induced migration: The 2004 tornado in north-central Bangladesh. *Disasters, 4*, 370–385.

Perch-Nielsen, S. L., Bättig, M. B., & Imboden, D. (2008). Exploring the link between climate change and migration. *Climatic Change, 91*, 375–393.

Pilarczyk, K. W., & Nuoi, N. S. (2005). Experience and practices on flood control in Vietnam. *Water International, 30*, 114–122.

Preston, B. L., Suppiah, R., Macadam, I., & Bathols, J. (2006). *Climate change in the Asia/Pacific region: A consultancy report prepared for the climate change and development roundtable*. Canberra: Commonwealth Science and Industry Research Organisation (CSIRO).

Reuveny, R. (2007). Climate change-induced migration and violent conflict. *Political Geography, 26*, 656–673.

Revi, A. (2008). Climate change risk: An adaptation and mitigation agenda for Indian cities. *Environment and Urbanization, 20*, 207–229.

Ringler, C., Bryan, E., Biswas, A. K., & Cline, S. A. (2010). Water and food security under global change. In C. Ringler, A. K. Biswas, & S. A. Cline (Eds.), *Global change: Impacts on water and food security* (pp. 3–15). Berlin: Springer.

Rodell, M., Velicogna, I., & Famiglietti, J. (2009). Satellite-based estimates of groundwater depletion in India. *Nature, 460*, 999–1002.

Roy, M. (2009). Planning for sustainable urbanisation in fast growing cities: Mitigation and adaptation issues addressed in Dhaka, Bangladesh. *Habitat International, 33*, 276–286.

Schewe, J., Levermann, A., & Meinshausen, M. (2011). Climate change under a scenario near 1.5°C of global warming: Monsoon intensification, ocean warming and steric sea level rise. *Earth Systems Dynamics, 2*, 25–35.

Seto, K. C. (2011). Exploring the dynamics of migration to mega-delta cities in Asia and Africa: Contemporary drivers and future scenarios. *Global Environmental Change, 21*, S94–S107.

Shamim, C. (2008). Alternative views of environmental security in a less developed country: The case of Bangladesh. *Journal of Third World Studies, 25*, 253–272.

Shi, G. Q., Zhou, J., & Li, J. Y. (2007). Protection of rights and interests and government's responsibilities: A case study of the Tarim River ecological migration as an example. *Jilin University Journal Social Sciences Edition, 47*, 78–86.

Skeldon, R. (2003). Migration and Migration Policy in Asia: A Synthesis of Selected Cases. Regionla Conference on Migration, Development and Pro-poor policy Choices in Asis, Dhaka.

Stringer, L. C. (2008). From global environmental discourse to local adaptations and responses: A desertification research agenda for Central Asia. In R. Behnke (Ed.), *The socio-economic causes and consequences of desertification in Central Asia* (pp. 13–31). Dordrecht: Springer.

Tan, Y. (2008). *Resettlement in the Three Gorges Project*. Hong Kong: Hong Kong University Press.

Tan, Y., & Guo, F. (2007, May 26–29). *Environmental concerns and population displacement in west China*. Paper presented at the 8th APMRN conference, Fuzhou, China.

UNHCR (United Nations High Commission for Refugees). (2006). *The state of the world's refugees 2006: Human displacement in the new millennium*. Geneva: UNHCR.

United Nations. (2006, May 18). *International migration and development, report of the Secretary-General*. Sixtieth session: Globalisation and interdependence: International migration and development, United Nations.

United Nations. (2008). *World urbanization prospects: The 2007 revision*. New York: United Nations.

United Nations Economic and Social Commission for Asia and the Pacific (UNESCAP). (1984). *Third Asian and pacific population conference: Selected papers*. Bangkok: ESCAP.

United Nations Economic and Social Commission for Asia and the Pacific (UNESCAP). (2011). *2011 ESCAP population data sheet*. Bangkok: UNESCAP.

Vecchi, G. A., & Soden, B. J. (2007). Effect of remote sea surface temperature change on tropical cyclone potential intensity. *Nature, 450*, 1066–1070.

von Storch, H., & Woth, K. (2008). Storm surges: Perspectives and options. *Sustainability Science, 3*, 33–43.

Vörösmarty, C. J., Green, P., Salibury, J., & Lammers, R. B. (2000). Global water resources: Vulnerability from climate change and population growth. *Science, 289*, 284–288.

Wang, C., Yang, Y., & Zhang, Y. (2011). Economic development, rural livelihoods, and ecological restoration: Evidence from China. *Ambio, 40*, 78–87.

Warner, K., Dun, O., & Stal, M. (2008). Field observations and empirical research. *Forced Migration Review, 31*, 13–15.

Webster, P. J., Holland, G. J., Curry, J. A., & Chang, H. R. (2005). Changes in tropical cyclone number, duration, and intensity in a warming environment. *Science, 309*, 1844–1846.

West China Development Leadership Office of the State Council (2005). Unordinary five years. Available on http://www.chinawest.gov.cn/web/NewsInfo.asp?NewsId=28366 (in Chinese).

WFP (World Food Programme)/FAO (Food and Agriculture Organisation). (2009). *The state of food insecurity in the world economic crises – Impacts and lessons learned.* Rome: FAO.

World Bank. (2008). *Migration and remittances factbook 2008*. Washington, DC: World Bank.

Xu, J., Grumbine, R. E., Shrestha, A., Eriksson, M., Yang, X., Wang, Y., et al. (2009). The melting Himalayas: Cascading effects of climate change on water, biodiversity, and livelihoods. *Conservation Biology, 23*, 520–530.

Ye, Y., Fang, X., Aftab, U., & Khan, M. (2012). Migration and reclamation in Northeast China in response to climatic disasters in North China over the past 300 years. *Regional Environmental Change, 12*, 193–206.

Yeh, E. T. (2009). Greening western China: A critical view. *Geoforum, 40*, 884–894.

Yeung, H. W. (2012). East Asian capitalisms and economic geographies. In T. J. Barnes, J. Pecj, & E. Sheppard (Eds.), *The Wiley-Blackwell companion to economic geography* (pp. 118–131). Chichester: Blackwell.

Yin, S. (2006). *The plight of internally displaced persons*. Washington, DC: Population Reference Bureau.

Environmental Change and Migration Between Europe and Its Neighbours

3

Mark Mulligan, Sophia Burke, and Caitlin Douglas

Introduction

Study Area

We focus particularly on migration within Europe and between Europe and its neighbours rather than global patterns of migration. The geographical area of study includes the Mediterranean countries as defined below, only some of which are European, as well as the potential migratory "pull" countries of northern Europe. The Mediterranean as defined here comprises the following Mediterranean Basin countries (each followed by its UN country code): Morocco (169), Algeria (4), Libya (145), Tunisia (248), Egypt (40765), Israel (121), West Bank (267), Lebanon (141), Syria (238), Turkey (249), Greece (97), Albania (3), Montenegro (2647), Bosnia and Herzegovina (34), Croatia (62), Slovenia (224), Italy (122), France (85), Sardinia (122), Malta (156), Cyprus (97), Sicily (122), Spain (229) and Portugal (199). Whilst environmental changes in these countries might act to increase the 'push' factors for migration out of them, we should understand that the pull factors for migration into specific countries are also subject to the impacts of environmental change on those countries. We thus also analyse changes in 'pull' factors as a result of environmental change by replicating our analysis in those countries that we consider as being potential European destination countries for migrants from the Mediterranean (within and outside of Europe). We are considering the Northern European 'pull' countries outside of the Mediterranean as UK (256), Ireland (119), Belgium (27), Netherlands (177), Germany (93), Switzerland (237),

M. Mulligan (✉) • C. Douglas
Department of Geography, King's College, London, UK
e-mail: mark.mulligan@kcl.ac.uk

S. Burke
Ambiotek Community Interest Company, Leigh-on-Sea, UK

E. Piguet and F. Laczko (eds.), *People on the Move in a Changing Climate*, Global Migration Issues 2, DOI 10.1007/978-94-007-6985-4_3,

Austria (18), Denmark (69), Sweden (236), Norway (186) and Finland (84). Given legal restrictions on migration from outside of Europe, most future migration in likely to be within and between these countries. However, through an analysis of historical migrations we place Europe with the broader context of global migration.

Environmental Drivers of Migration

There are many drivers of migration in Europe including economics, politics and security and environment. For economic migration push factors include poverty, lack of employment, high cost of living and pull factors include perceived opportunity or lower cost of living. For politically or security driven migration, push factors include internal or international conflict, lack of democracy, marginalisation. For environmentally driven migration push factors include perceived quality of life, amenable climate (driving seasonal migrations) and hazard or land degradation leading either to economic migration or -rarely in the European context – to environmentally forced migration. Clearly, many of these drivers are not environmental or biophysical but are, nonetheless, affected by the environmental state and will be affected by environmental change.

The History of European Migration

Pre – 1945

Until 1945, Europe's migration history was characterised by emigration (Fassmann and Munz 1992). Between 1850 and 1914, over 40 million people emigrated, and, as a result, Europe lost more than one-tenth of the population and one-quarter of its labour force (Easterlin 1961). The target for this migration was the 'New World', with two-thirds going to the United States of America, and other common destinations included Canada, Australasia, and Latin America (especially Argentina, Brazil and Cuba) (Hatton and Williamson 1994). The early emigrants were from the Northwest of Europe, particularly the United Kingdom and Scandinavia. However, after 1890 the vast majority of emigrants came from southern Europe – Italy, Portugal and Spain, and eastern Europe – Austria-Hungary and Poland. Migrants were typically young, unskilled males who left in search of employment and an increased living standard. An important influence on their decision to migrate was the presence of friends and/or relatives in the host country who could provide assistance travelling to, and within, the country (Hatton and Williamson 1994).

The exodus of people from Ireland in the late Nineteenth Century was prompted by a combination of economic and environmental drivers (Guinnane 1994; Fitzpatrick 1980). Emigration would have occurred as a result of the country's economic stagnation, but the potato famine served to hasten the emigration. Migrants went to the Britain, North America and Australasia (Fitzpatrick 1980).

The period of emigration from Europe prior to World War II represented an unprecedented population transfer which is said to have profoundly influenced the demography and economy of countries globally.

Post – 1945

The pattern of emigration was reversed shortly after World War II as western European countries accepted refugees, displaced people and citizens returned from former colonies (Fassmann and Munz 1992; Heisbourg 1991). By the mid-1950s European countries – particularly, France, Germany, Netherlands, Belgium, Switzerland, Austria and the UK – had a high demand for labour which could not be satisfied domestically (Hansen 2003; de Haas 2011; Fassmann and Munz 1992). As a result these countries looked first to southern Europe – Italy, Greece, Spain and the former Yugoslavia to satisfy their labour shortages. They then looked to the southern Mediterranean/north African countries of Turkey, Morocco, Algeria and Tunisia to supply workers. Until this point most migration within the southern Mediterranean had been internal (de Haas 2011). Countries with a colonial history, particularly the United Kingdom and France but also Belgium and the Netherlands, also looked to their colonies to find workers (Hansen 2003).

The 1970s hailed an important landmark in European migration patterns as increased restrictions on immigration marked the end of the European immigration boom (de Haas 2011). Despite the tight controls, immigration continued but at a slower rate. The continuance was largely due to family re-unification of immigrants already settled within Europe (de Haas 2011; Hansen 2003; Bade 2003). As birth rates of immigrants were higher than national averages this led to significant changes in the composition of the migrant populations and of the host countries (Fassmann and Munz 1992). Originally migrant populations had consisted of men of employable age, but as a result of family reunification and new births, the proportion of women, children and adolescents increased. This dramatically altered the labour picture of the host countries as migrant females entered the work force.

Several events occurred during the 1990s which resulted in increased migration to and within Europe. The end of the Cold War and the fall of the "Iron Curtain" in 1991 reconnected Europe and this new 'era' brought considerable migration from Eastern Europe to the West (Bade 2003; Heisbourg 1991). This period also coincided with a rapid increase in asylum application which was spurred by the opening of travel between east and west Europe, falling transportation costs and violent conflicts (Fassmann and Munz 1992; Hansen 2003; Koser and Lutz 1998). The integration of countries into the European Union, and the consequent free movement of labour within Europe, also had a significant impact on migration within Europe (Leitner 1997). A new migration route towards the northern-Mediterranean countries emerged in response to their inclusion into the EU. Southern-Mediterranean nationals travelled to Spain and Italy; South Americans with Spanish, Portuguese or Italian heritage moved to their respective countries; and, Spanish and Portuguese nationals working in northern Europe returned to their home countries (Fassmann and Munz 1992; Cornelius 1994). These migrations were primarily driven by employment opportunities due to economic growth.

Whilst migration within Europe has been facilitated by the European Economic Union, migration from outside Europe has been strictly regulated (Bade 2003). One consequence of the immigration restrictions has been an increase in the number of illegal immigrants (Fassmann and Munz 1992). Differences in immigration regulations among countries has resulted in certain countries serving as an entry point into other parts of Europe (Cornelius 1994; Huntoon 1998; Jokisch and Pribilsky 2002). In the 1990s mass emigration to Europe from Ecuador was triggered by a combination of drivers in Ecuador including: conflict; economic downturn triggered by El Nino floods destroying banana exports and infrastructure; very low oil prices; and, increased border controls and surveillance in the USA which had previously been the favoured destination for emigration (Jokisch and Pribilsky 2002). Most Ecuadorians entered through Spain due to that country's amenable immigration regulations and whilst the majority have stayed Spain, many have subsequently settled in France, Italy and the Netherlands. Specific countries such as Norway now offer temporary protection for environmentally forced migrants and this may now act as a pull factor under specific circumstances.

A relatively recent group of migrants is that of 'elites' such as business people, employees of multinational companies/ organizations, artists, research personnel, students and retirees (Fassmann and Munz 1992). In particular, skilled workers are in high demand and are able and willing to move internationally to secure employment (Mahroum 2001).

Drivers of Historic and Current Migration

Non Environmental Drivers

Foresight (2011) grouped the main drivers of global migration into five categories: social, political, economic, environmental and demographic, and conceptualised these as vertices on a pentagon. Each of these five drivers cannot be considered in isolation from the other five dimensions, and they interact or overlap in different ways in different places and at different times. Environmental change will affect each of these drivers, and their interactions with other drivers. As a result it is very difficult to predict the impacts of environmental change on migration. This section will briefly outline these main drivers within a European context, and the next section will analyse how global environmental change might influence these drivers into the future.

Economic demand is the most important factor in the decision to migrate in Europe (Hansen 2011). This was a significant driver for migration from the Mediterranean region to northern European countries during the 1950s and 1960s, initiated by the post-war development and economic boom. This continued until an improvement in economic conditions in the European Mediterranean led to an influx of migrant workers from the Southern Mediterranean countries into Northern Mediterranean countries (de Haas 2011). Political drivers in determining migration in Europe have included the expansion of the European Union. This Union has steadily grown a core of wealthy of European countries, altered the labour market structure and has drawn in increasing numbers of migrants to these countries from elsewhere within Europe (Hansen 2011) and also from neighbouring southern Mediterranean countries (de Haas 2011). The expansion of the EU in May 2004 to include ten new member states (NMS) made it possible for workers in some central and eastern European countries to take up work in the EU-15.

According to Krings (2012), immigrants from NMS to the EU-15 increased to two million between 2003 and 2007. In particular, Denmark, Sweden, Ireland and the United Kingdom had less restrictive immigration controls which allowed a relatively free movement of workers compared to the other EU- 15 countries. Recent studies have shown many of these migrants have since returned to their home countries (Hansen 2011). Demographic trends have changed only modestly in Europe in recent decades relative to migration trends and so demography has had only an indirect effect on migration (Hansen 2011; de Haas 2011) whereas social factors (such as language, family ties, historical and colonial links) can have huge influence on the country of choice for the migrants (Hansen 2011) in such a diverse suite of countries, especially those with a colonial history and thus strong cultural and language connections around the world.

Environmental Drivers

The complex linkages between environmental change and migration have been subject to recent interest (Laczko and Aghazarm 2009; Narusova-Schmitz and Lilleorg 2009). There are two types of environmental migration: first,: gradual and sudden environmental disruptions that can result in substantial human movement and displacement; and second, the impact of migration on the environment in the communities of migrant's origin and destination, in particular the interaction of migration with existing environmental problems o The developing countries are usually considered the most vulnerable to environmental degradation, but even Europe has its vulnerabilities, which have been the subject of recent research. The European Commission co-sponsored the Environmental Change and Forced Scenarios (EACH-FOR, www.each-for.eu) project to assess the impact of environmental change on migration at the local, national, regional and international level (Laczko and Aghazarm 2009). De Haas (2011) argues that the role of environmental drivers within Europe may be small, indirect and mediated through other drivers. So far, environmental hazards around the world have been unlikely to increase migration to Europe, despite the often strong social links between vulnerable countries (such as India and Bangladesh) and European countries (such as the UK). Where migration occurs as a response to environmental drivers, it is much more likely to be over shorter distances (Findlay 2011). Extreme events may force migrants to be displaced far afield such as to Europe, although cases where this has occurred are rare. Rather, refugees tend to be displaced by war and extreme political events, such as the 30,000 Ugandan Asians that fled to Britain in 1972–1973 after being expelled by President Idi Amin, or the 18,000 Vietnamese 'Boat People' that arrived in 1979–1982 at the end of the Vietnam war (Crawley 2011). The most significant short term migration to the UK was the 250,000 Belgians displaced by war in 1914. Crawley (2011) reports that the only recent displacement of people to Europe by purely environmental causes was from Montserrat in the West Indies, after the volcanic eruption that started 18 July 1995 and that devastated much of the island. This is however a very unusual situation. Nevertheless, into the future, environmental drivers both inside and outside Europe such as drought, crop failures and desertification, flooding and inundation may play a more significant role on European migration, as we saw with the recent increase in migration from North Africa to southern Mediterranean countries (Goria 1999).

The European Environment

We begin by analysing the current environmental status of Europe with a specific emphasis on differences both within and between countries. This analysis sets the context for environmental change and for the potential biophysical drivers for migration.

Climate and Recent Climate Variability

European countries have strong latitudinal precipitation (P) and temperature (T) gradients. Precipitation (Fig. 3.1) decreases from north to south and from west to east and temperature (Fig. 3.2) increases along the same gradients. These strong gradients mean that some regions of Europe are very sensitive to small shifts in climate. There are also significant montane and coastal effects on temperature and precipitation with mountain and coastal areas generally being cooler and wetter. Precipitation for large southerly areas of the southern Mediterranean national territories is close to zero, rising steeply towards the southern Mediterranean coast and further increasing in the northern Mediterranean and northern Europe.

Land Use and Productivity

According to the best available satellite assessments, dryland agriculture dominates in Europe and its environs (Fig. 3.3). Significant areas of dryland agriculture are present especially in the northern Mediterranean and in coastal and deltaic areas of Morocco, Algeria, Tunisia and Egypt. These dry-cropping areas may be more vulnerable to the effects of drought than those areas supporting sustainably irrigated cropland.

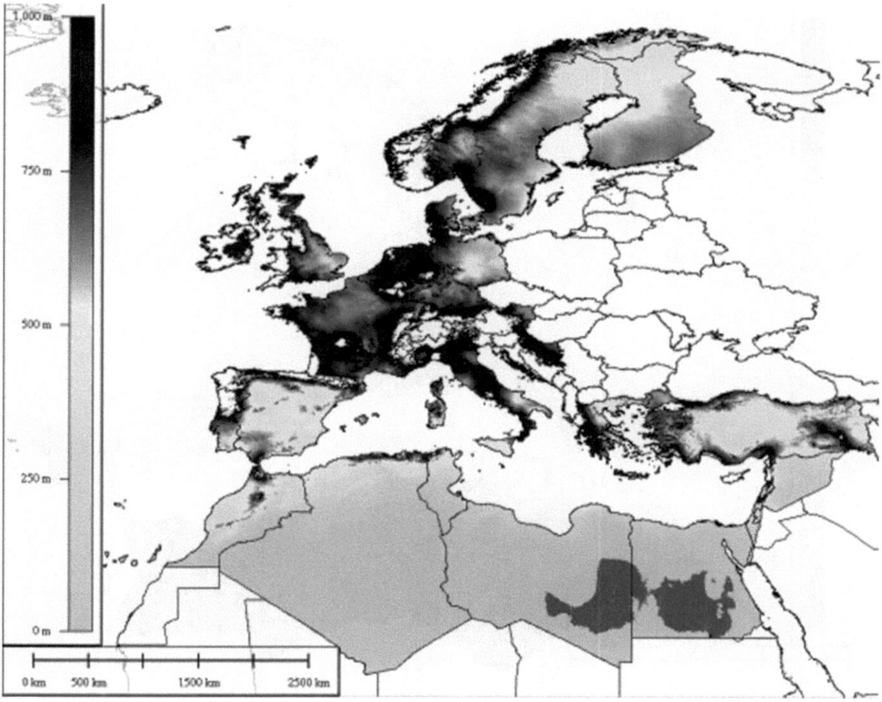

Fig. 3.1 Mean annual precipitation (mm) (Source: Hijmans et al. 2005). Areas shown in *dark* are zero and in *white* are greater than the legend maximum or outside of the region of interest

Figure 3.4 shows satellite-derived annual mean dry matter productivity (as a mean for the period 2000–2010) for the Mediterranean and northern Europe. This measure is for all plants, not only for crops and shows strong gradients with the highest observed values in the northern Mediterranean (Italy, Sardinia, northern Spain, Greece) and a strong decline towards the southern Mediterranean and far northern European land masses. Towards the south and east, water becomes limiting because of the observed gradient in precipitation (Fig. 3.1), towards the north, solar radiation and temperature become limiting (Fig. 3.2), especially in the winter.

Figure 3.5 indicates the extent of agricultural irrigation in the study area. Irrigation is clearly highly localised with areas of intensive irrigation in the Nile Delta, northern Italy, parts of Greece, France, Spain, Morocco and Turkey. Extensive (rather than intensive) irrigation occurs throughout the Northern Mediterranean and parts of lowland Northern Europe. Heavily irrigated areas are clearly less vulnerable to short term drought (if irrigation water is available) and can sustain higher productivity but are more vulnerable to degradation of soil and water resources if evapo-transpiration rates are maintained at values consistently greater than the groundwater recharge rates or fluvial water supply capacity.

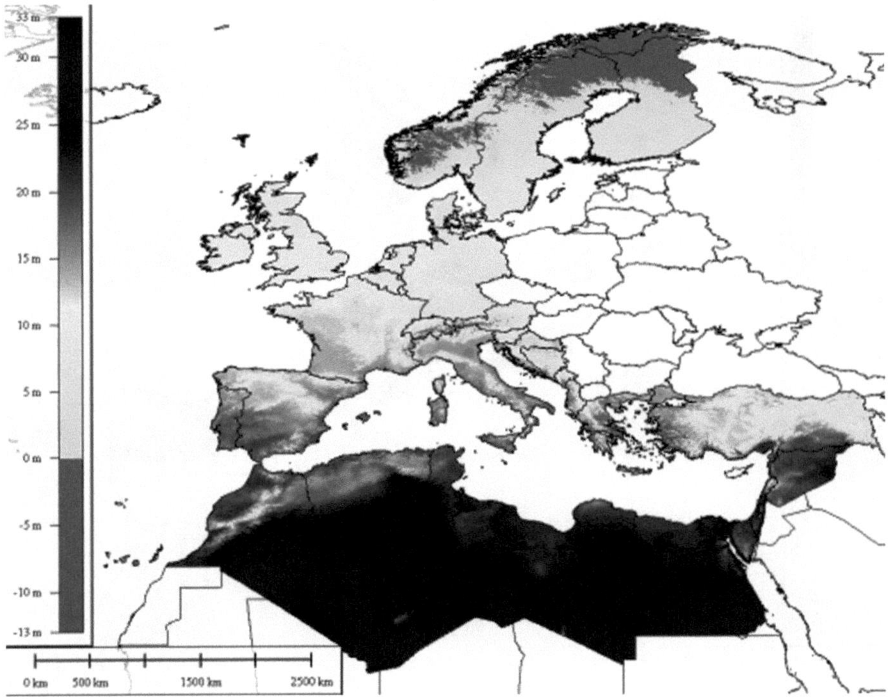

Fig. 3.2 Mean annual temperature (°C) (Source Hijmans et al. 2005). Areas shown in *dark shades* are less than zero and in *white* are greater than the legend maximum or outside of the region of interest

Agriculture in Europe is highly productive and highly mechanised but prone to the effects of drought and desertification in the south and to harsh winters or summer waterlogging in the north. The European environment is generally benign compared with many parts of the world so environmental risks are generally low. However, high population densities and significant concentration of populations and infrastructure tend to mean that exposure to extant risks is high.

Socio-economic Context

Europe is a highly human-influenced region and thus the distribution of population and infrastructure is an important consideration in understanding environment, environmental change and migration. Figure 3.6 shows the distribution of population density for Europe. There is a clear spatial association between high populations, amenable climate and high agricultural productivity. In the southern Mediterranean

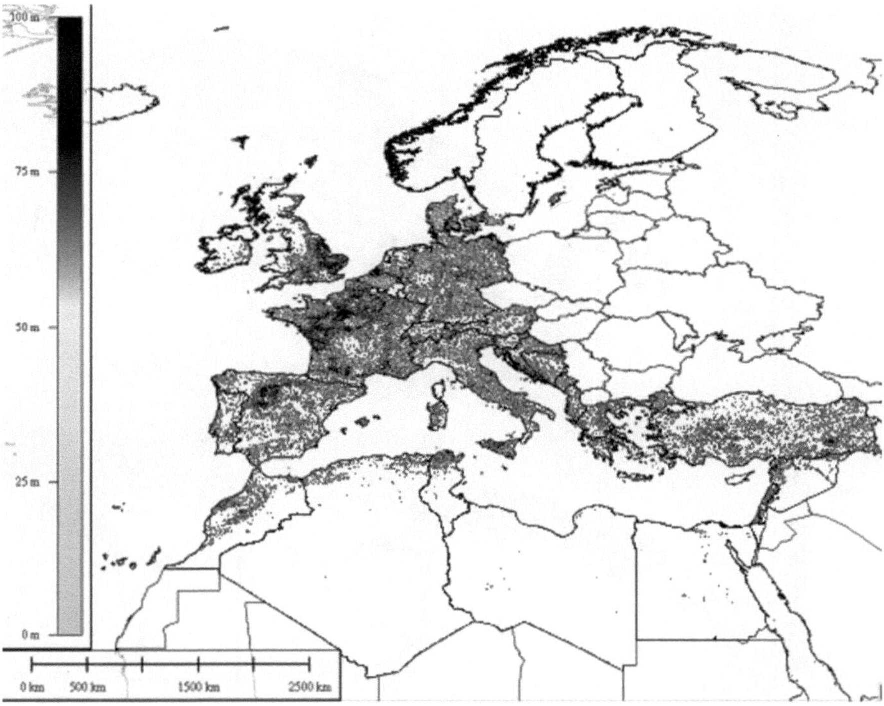

Fig. 3.3 Dry crops in Mediterranean and northern European countries according to GlobCover continuous fields (Mulligan 2009)

there are large areas of low rainfall and agricultural production and thus very low or no human population. Populations in the southern Mediterranean are concentrated in the highlands, coastal fringes and in the major river valleys which provide more amenable climates and access to markets, water and more fertile soils.

The northern Mediterranean is highly urbanised (Fig. 3.7) with these urbanisations being densely populated. Population densities are highest in non-Scandinavian north of Europe with contiguous high density urbanisation throughout UK, Netherlands, Belgium and Germany. Lower populations in the Scandinavian countries correspond with the very low temperatures and challenging environments observed there.

There are few quantitative measures of socio-economic context available at the sub-national scale and in a comparable form for all the countries of this analysis. One such dataset produced by the Center for International Earth Science Information Network (CIESIN) represents Gross Domestic Product GDP (Fig. 3.8). GDP is highly concentrated in urban areas, especially the financial and industrial centres of northern Europe and the northern Mediterranean and is low in parts of North Africa and the eastern Mediterranean. Low rural GDP may indicate vulnerability and a low adaptive capacity in the face of environmental change.

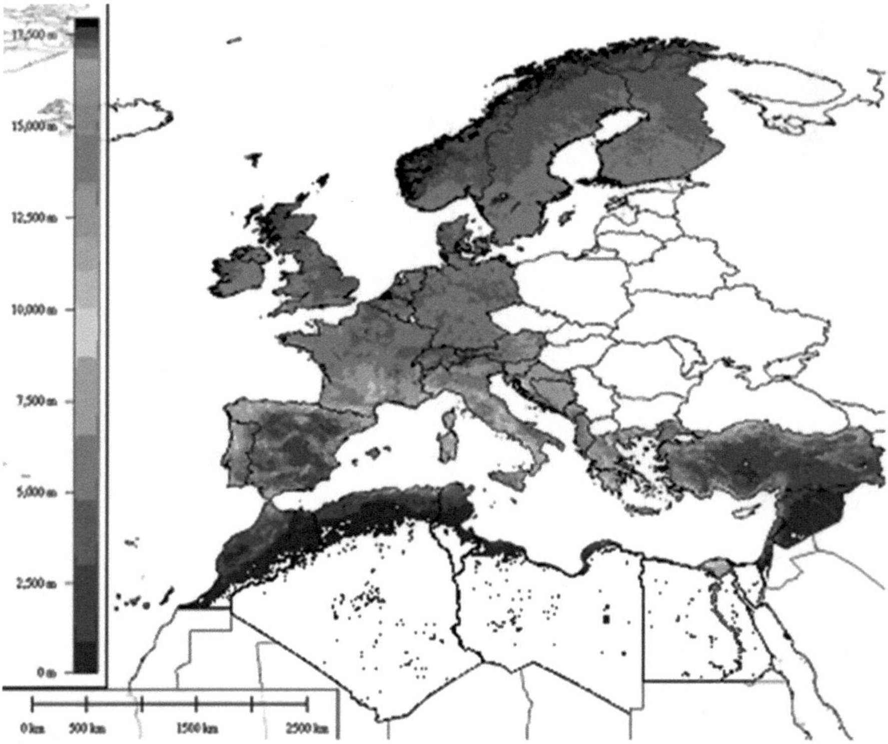

Fig. 3.4 Dry matter productivity (Dg/ha/day) for the Mediterranean and northern Europe (Source: Mulligan 2009). Areas in *white* are zero or are outside of the area of interest

Current Environmental Hazards in Europe

Stressors: Water

Water resources are limited and highly unevenly distributed in Europe in line with the pattern of rainfall (Fig. 3.1). The availability of sustainable water resources is perhaps the main environmental driver for population growth and development in the drier areas in and around Europe. This is exemplified by the distribution of percentage of land area flowing into dams for the region (Mulligan et al. 2009): Morocco (29.1 %), Algeria (12 %), Libya (0.1 %), Tunisia (10.3 %), Egypt (15.8 %), Israel (1.1 %), Lebanon (33.4 %), Syria (83.3 %), Turkey (68.2 %), Greece (15.4 %), Albania (29.0 %), Bosnia and Herzegovina (24.2 %), Croatia (13.1 %), Slovenia (20.8 %), Italy (20 %), France (18.3 %), Malta (0), Cyprus (9.8 %), Spain (73.4 %) and Portugal (28.9 %). Clearly, most of the southern and eastern Mediterranean countries capture much of their rainfall in reservoirs. Water quality in Mediterranean Europe is also considered to be decreasing due to chronic pollution from untreated agricultural, domestic and industrial discharges especially

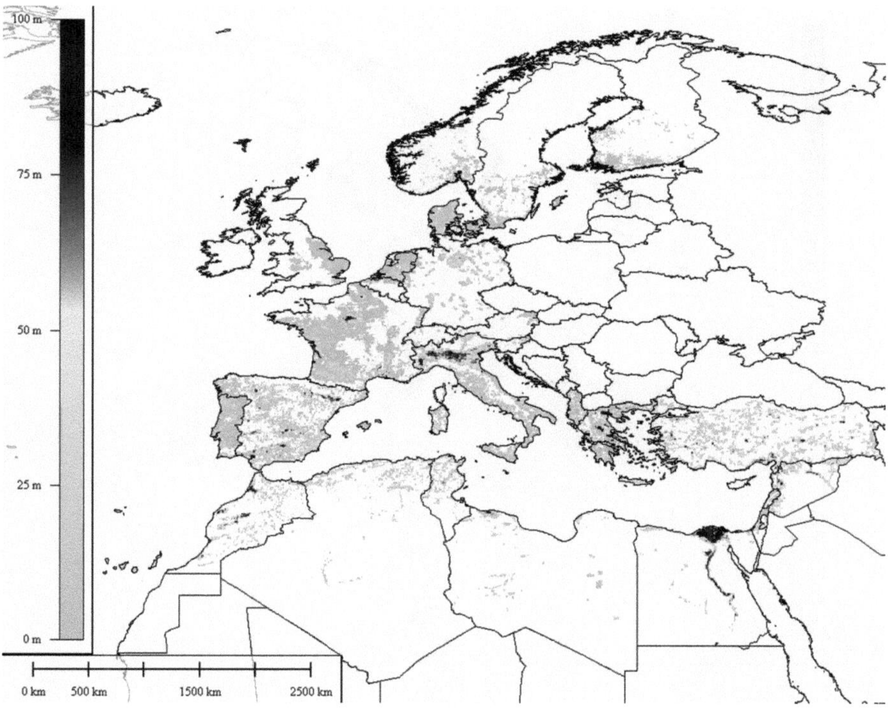

Fig. 3.5 Percentage of land irrigated (Source: Siebert et al. 2007). Areas in *white* are zero or are outside of the area of interest

from pesticides and nitrates (Plan Bleu 2009). Water also constrains agricultural productivity (Fig. 3.4), especially in the south and the east, leading to the adoption of intensive and extensive irrigation in the region (Fig. 3.5) which is sometimes unsustainable, drawing more water from fossil groundwater than is replenished by groundwater recharge on an annual basis.

Figure 3.9 shows the catchment average water balance per person for Europe, which indicates the capacity of renewable water resources to support the population and agriculture. We have calculated water balance as rainfall (*Rf*, Hijmans et al. 2005) minus potential evapotranspiration (*PET*, Zomer et al. 2008) on a 1 km spatial resolution basis with negative water balances ($PET > Rf$) set to zero. Note that this analysis does not include inflows of water from rivers (which sustain some areas such as the Nile) nor stores of fossil water but focuses on the locally produced *renewable* water resources since these will be critical for sustainable agricultural production in water scarce areas into the future. The water balance has been combined with a dataset for watersheds (Lehner et al. 2008) and population (Landscan 2007). Areas in black have plenty of water per person on a catchment average basis, whereas lighter shades show progressively lower *per-capita* water resources. The most water stressed catchments are in the southern Mediterranean (because of lack of water), in the eastern Mediterranean (because of lack of water

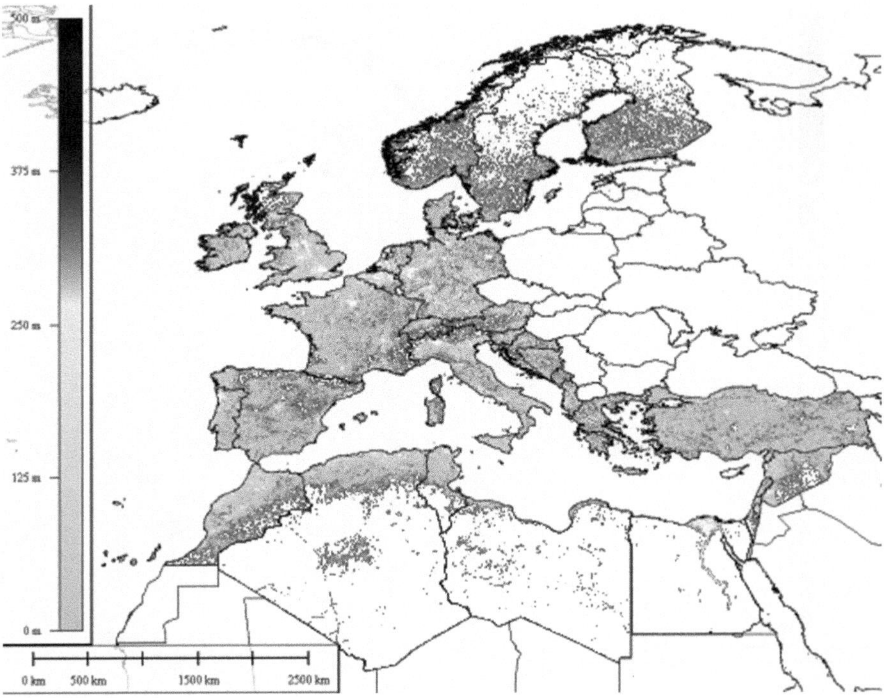

Fig. 3.6 Population density (persons/km^2) (Source: Landscan 2007). Areas in *white* are zero or are outside of the area of interest. or with population greater than the legend maximum

combined with high population) and in coastal northern Mediterranean (small catchments with high populations living near the coast and with a low water balance). This map is an indicator of environmental stress and it is interesting to note that some densely populated parts of northern Europe are also potentially water stressed on this measure.

Hazards: Flooding

In addition to water stress, the European populations and landscapes are subject to a range of natural hazards. As well as earthquakes, volcanoes and other geological hazards that occur in isolated areas of southern Europe, the key migration-relevant hazards influenced by environment and environmental change in Europe are river flooding, wildfires and sea-level rise.

In northern Europe flooding is frequent in the Netherlands, Belgium, parts of Germany and the UK. According to Llasat et al. (2010), flood frequency in southern Europe is highest in Spain and Italy, although there are also more available data there. The number of cases that affected Spain, Italy and France together amount to 59 % of the 185 Mediterranean-wide events considered. The distribution of floods

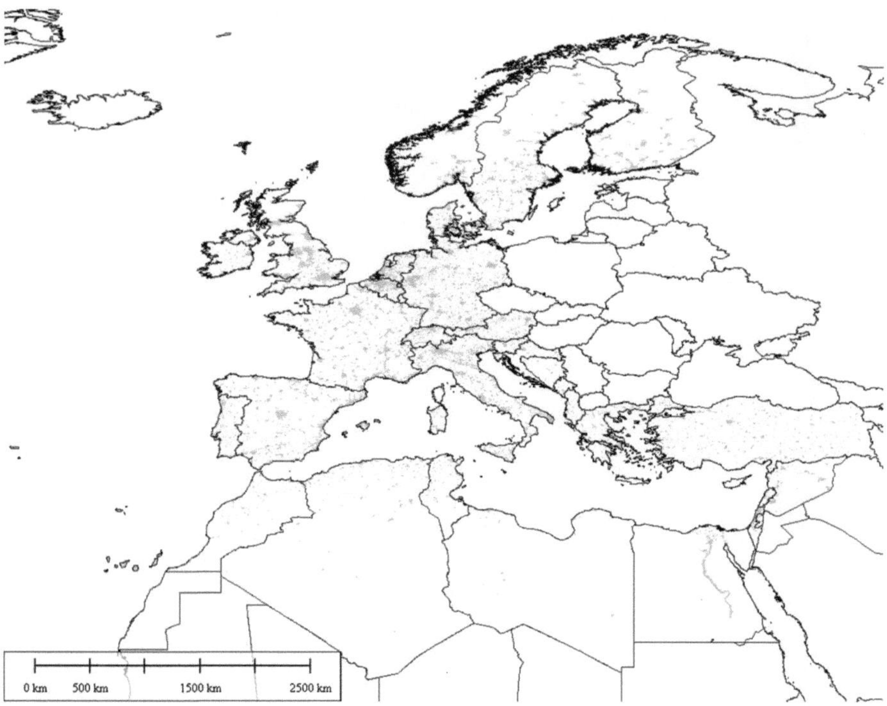

Fig. 3.7 Urban areas (Source: CIESIN et al. 2004)

is neither homogeneous across the region nor stationary over time, although most floods seem to occur in the western half of the Mediterranean. Floods are less frequent in the southern Mediterranean countries (northern Africa), but are usually catastrophic with a very high number of casualties.

There is also serious flood risk to urban areas worldwide owing to inadequate storm drainage infrastructure. The sensitivity to increased flooding will depend upon the proximity of infrastructure to major rivers, climate change impacts, and their land and water management contexts. (World Bank 2007). Coastal flooding can also have significant direct socio-economic impacts on coastal infrastructure and agriculture, and can cause displacement of coastal populations The risk of flooding and erosion from storm surges in coastal areas may increase with sea level rise, due to the subsequent reduction of the return period of high water levels. As sea level rises, the flood risk zone moves inland, so increasing the population and land area vulnerable to flooding, and increasing the risk of flooding to the population presently living in the coastal flood plain. Nicholls et al. (1999) document a reduction of the return period of high water levels due to storm surges in the Netherlands and in Egypt. Though flood defences and drainage systems have been constructed to manage the risks, particularly around the North Sea, significant risk of flooding still remains (Nicholls 2010). Flood risks may impact upon migration from outside

Fig. 3.8 Gross domestic product (GDP millions of US $) (Source: CIESIN 2002) Areas in *white* are zero or are outside of the area of interest or greater than the legend maximum value

Europe, as major flood disasters remain possible in any coastal region. Nicholls et al. (1999) suggest that, by the 2080s, the number of people flooded by storm surges will be more than five times higher due to sea-level rise, and many of these people will experience annual or more frequent flooding, suggesting that a response (such as migration or increased protection) will be required.

Areas most vulnerable to flooding are the southern Mediterranean (especially the Nile Delta), parts of Africa, and south and south-east Asia where there is a concentration of low-lying populated deltas. However, due to the smaller land mass, the Caribbean, the Indian Ocean islands and the Pacific Ocean small islands may experience the greatest relative increase in flood risk. These locations already have historical migration links to Europe and so migration to Europe may result. Myers (2002) postulated that tens of millions may be forced from their homes, especially given that coastal populations are increasing. However, if we can adapt successfully to these challenges by implementing coastal management policies, flood defences and land use planning in the flood plains, we can significantly reduce the populations affected (Nicholls 2010).

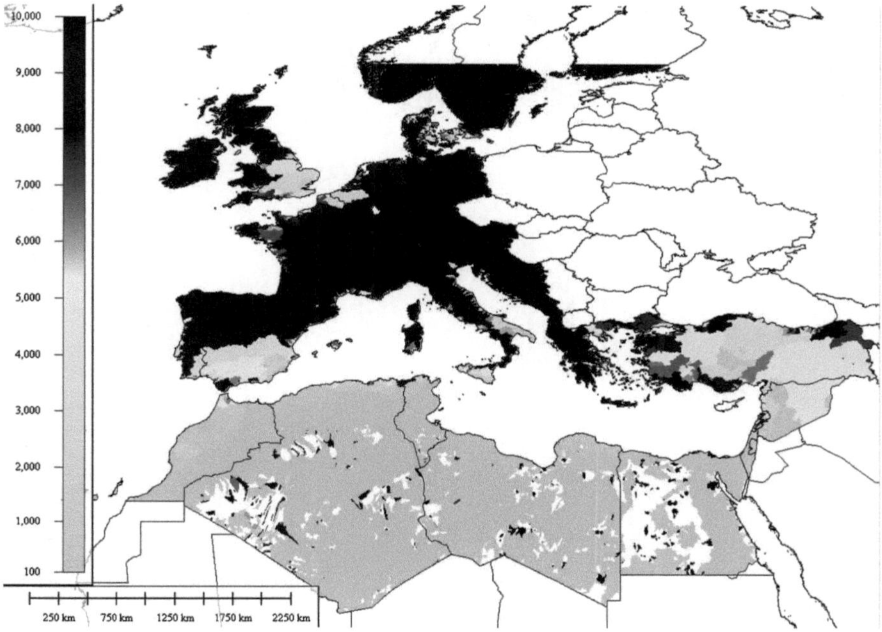

Fig. 3.9 Catchment average water balance (rainfall-ET) per person (m^3/person/year)

Hazards: Wildfires

Mediterranean forests in particular are regularly subjected to a large number of fires; 600,000 ha burn every year according to Cemagref (2009), some of which are started deliberately. Figure 3.10 shows the mean observed frequency of fire burn events between 2001 and 2010, compiled by Mulligan (2010) from the MODIS burned area product (Roy et al. 2008). Clearly the north and east Mediterranean are currently subject to significant fire risks which lead to significant burn areas and, in places, repeat burns. This hazard may be exacerbated by hotter, drier summers and by changes in population, land management and infrastructure. Burns in the Southern Mediterranean and northern Europe are much less frequent and extensive.

Environmental and Socio-economic Change in Europe

Environmental Change

Figures 3.11 and 3.12 show the difference in mean temperature and mean precipitation between 1900–2050 and 2041–2060 for the HADCM3 General Circulation Model (GCM), scenario SRES A1B scenario averaged by watershed. The figure shows that warming may occur throughout the region, with a strong north – south

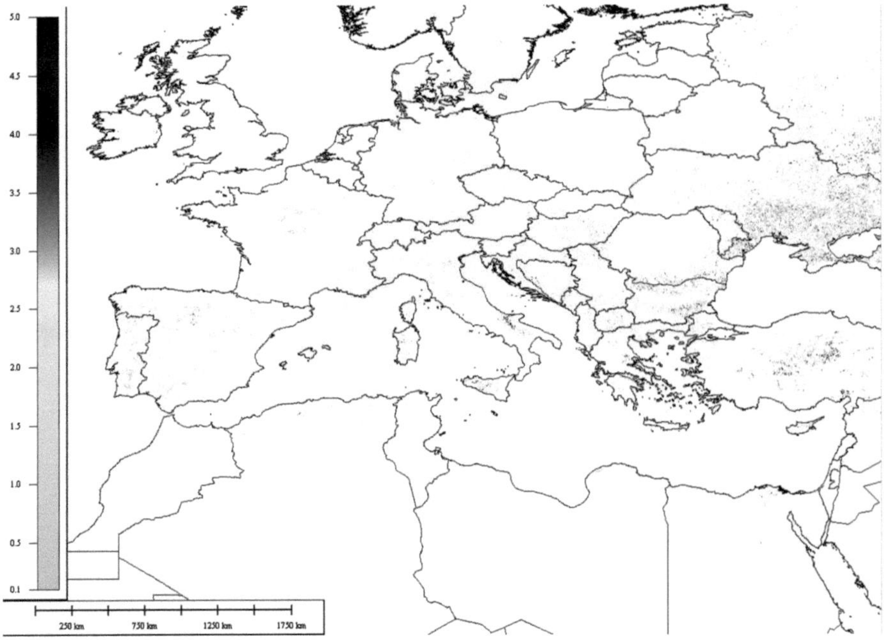

Fig. 3.10 Mean observed frequency of fire burn events (2001–2010) (% of MODIS observations) After Mulligan (2010)

gradient, the warming being greatest in coastal areas and to the south. Figure 3.12 shows that precipitation decreases may occur throughout Europe except the extreme south and the north, which may become wetter.

There are, however, significant uncertainties in the climate projections, both between models for a given scenario and between scenarios for a given model. These differences are particularly high for precipitation projections for which different GCMs can give different directions of precipitation change as well as significant differences in projected magnitudes.

Socio-economic Change

Spatial projections for population and GDP are publicly available only for the Special Report on Emissions Scenarios (SRES) B2 scenario. Under SRES B2, population increases by 2025 are expected to be greatest in the south and east of the Mediterranean, concentrated in coastal areas (Fig. 3.13). De-population is expected to occur in parts of the north Mediterranean (especially Spain, Italy and Greece). Urbanisation in Northern Europe is expected to continue.

It is important to indicate that these projections are highly dependent on the socio-economic scenario used. The B2 storyline represents a world with an

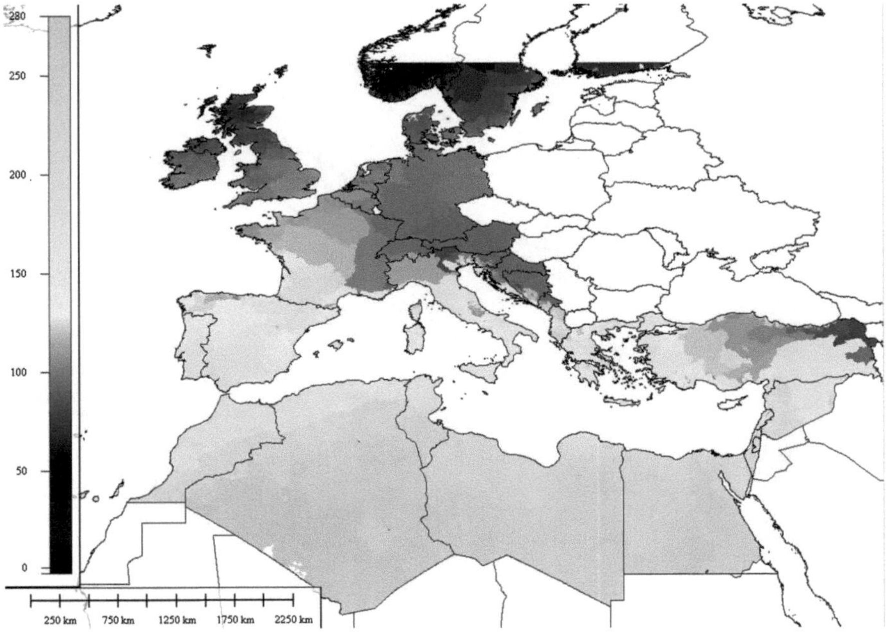

Fig. 3.11 Mean temperature change by basin (1900–2050 to 2041–2060) for HADCM3 SRES A1B (°C * 100)

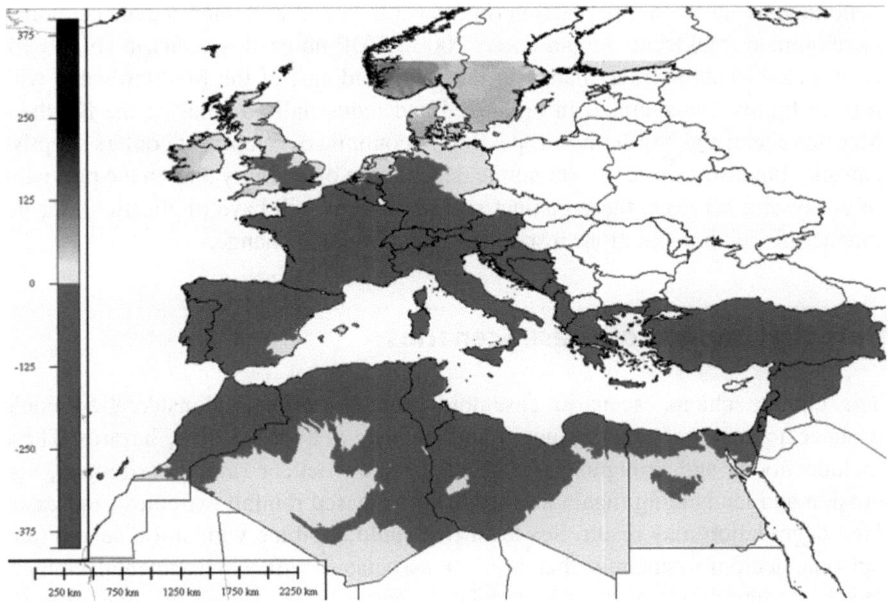

Fig. 3.12 Mean precipitation change by basin (1900–2050 to 2041–2060) for HADCM3 SRES A1B (mm/year)

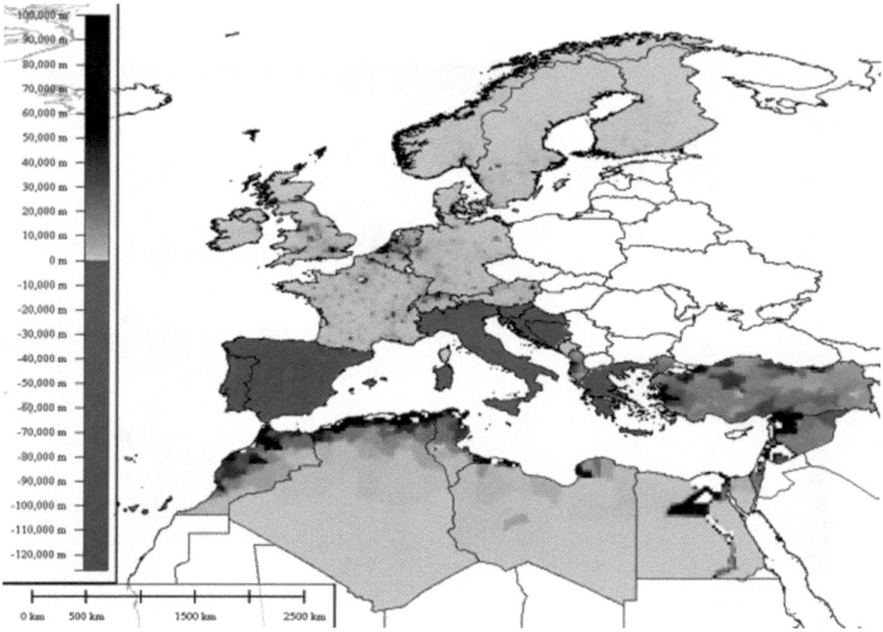

Fig. 3.13 Population projection for 2025 (persons/km^2, SRES B2) (Source CIESIN (2002))

emphasis on local solutions to economic, social, and environmental sustainability, a continuously increasing population (but lower than A2) and intermediate economic development (Nakicenovic and Swart 2000). GDP under this scenario (Fig. 3.14) is expected to show little change in the south and east of the Mediterranean with growth highly concentrated in urbanized and industrialised areas of the Northern Mediterranean and Northern Europe. Since economic development confers adaptive capacity but also exposes an economic dependence on stability and on the provision of ecosystem services, these distinct spatial patterns will have implications for the manner in which populations respond to environmental change.

Potential Impacts of These Scenarios

The climate change scenario ensemble projections under consideration could produce an increase in the frequency and intensity of hydro-climatic hazards. These include floods and droughts (as a result of snowmelt or rainfall extremes), soil erosion and landsliding (again as a result of increased rainfall extremes). Increased land degradation may result. Sea level rise could combine with storm surges (and episodic tectonic tsunamis that are not associated with environmental change) leading to inundation of densely populated coastal areas. Heatwaves, may increase in frequency and severity, affecting ageing and urban populations in particular. All of these may have implications for patterns of migration and the European region

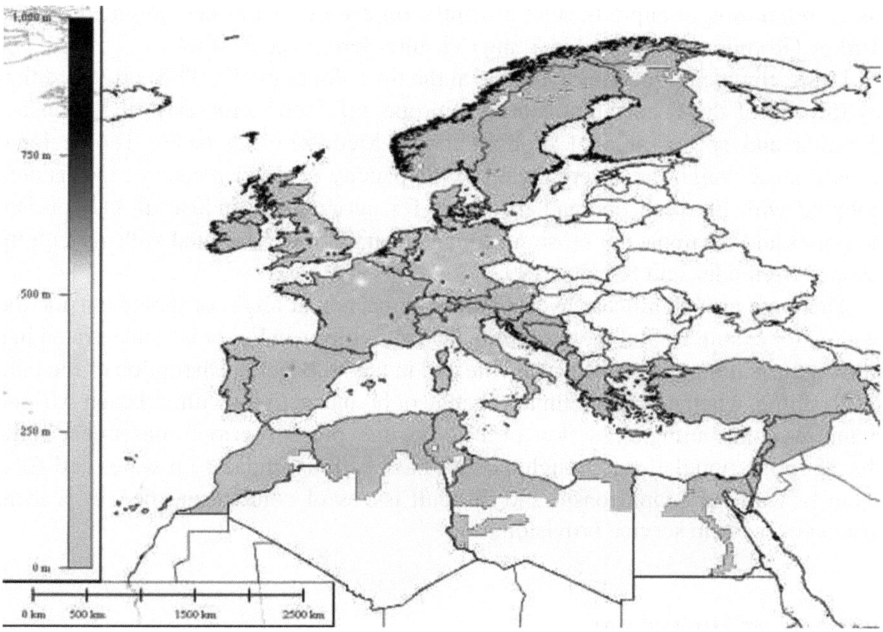

Fig. 3.14 GDP change projection for 2025 (SRES b2, GDP millions of US $) (Source: CIESIN (2002))

is potentially very sensitive to shifting climate, because Europe is characterized by strong cultural, economic, political and demographic gradients sometimes in already climatically stressed conditions, especially in the south.

Impacts on Droughts

The strong rainfall seasonality of the Mediterranean brings periodic and often sustained droughts as well as occasional extreme precipitation that can lead to flooding. Climate change will likely exacerbate the wet and dry extremes. A lack of preparedness or inadequate adaptation strategies could result in critical situations in parts of the region. On the basis of recent migration trends, this may lead to an increased flow of people from the drier and poorer southern Mediterranean to (most likely) northern Mediterranean and northern European countries (Magnan et al. 2009).

However, the projections of GCMs have serious limitations because of their poor spatial resolution, and thus the projected patterns of the rainfall drivers of hazards such as droughts are not consistent between GCMs especially for areas with high precipitation variability over time and space, as in the case of Mediterranean climatic regions. Drought is also highly spatially variable and unpredictable, due to its correlation with unpredictable precipitation events, as has been pointed out for Turkey (Komuscu 1999) and in Spain (Vicente-Serrano et al. 2004).

Using climate projections available at the time, Jones et al. (1996) projected that by the end of the twenty-first century, Europe will face increases in the intensity, duration and spatial area of drought in the Mediterranean basin. The regional consequences might be severe owing to the paucity of water resources in this area coupled with the high demand for water for agricultural, industrial and tourism activities and the impact of erosion and land degradation associated with agriculture (López Bermúdez and Sánchez 1997).

There are also significant hydroclimatic impacts that might originate outside the region, for example all of the population and agriculture in Egypt is concentrated in a narrow strip along the banks of the Nile and in the Nile Delta. Disruption of the Nile flow regime, whether due to climate change or human activity within the ten African countries contributing to the flow of the Nile, may present serious consequences for the Egyptian population. Droughts could have serious impacts on water and food security with very rapid onset and difficult issues of conflict and benefit sharing around ecosystem service provision.

Impacts on Heatwaves

Heatwaves in Europe have become more frequent in recent years (Della-Marta et al. 2007). Fischer and Schär (2010) analysed high-resolution regional climate simulations in Europe and projected that the most pronounced changes in heat wave frequency and duration will occur in the southern-most areas of Europe, and the projections of related health impacts are most severe for low-altitude river basins in southern Europe and for the Mediterranean coasts, affecting many densely populated urban centres. Heatwaves exacerbate air pollution and can often lead to higher morbidity and mortality predominantly in the elderly, infants and persons with pre-existing cardiovascular and respiratory disease. An increase in frequency and intensity of summer heat waves can have serious economic consequences, such as on agricultural yield and energy demands for cooling.

Impacts on Sea Level Rise

Scenarios for sea level rise are highly uncertain. Climate change-driven thermal expansion of the oceans is thought to account for only a few cm of the expected rise over the coming decades. Rising sea level will mainly affect coastal and deltaic regions, but in the Mediterranean most of the population live in precisely these areas. Much of the Nile Delta is already at or below sea level and is one of the worlds most densely populated "megadeltas". Therefore even a marginal sea-level rise combined with storm surges could create disastrous flooding, for example in Alexandria, Egypt's second city. Increased Erosion of the Nile Delta has already been attributed to upstream engineering projects such as the Aswan High Dam and

Fig. 3.15 Inundation resulting from catastrophic 2 m sea level rise/surge. (Source: Mulligan (2007))

as a result the land levels of the delta are falling at the same time as sea levels are rising (Bohannon 2010).

Even if we ignore the few cm of sea level rise expected through ocean warming, It is prudent to quantify the potential impact of very high and catastrophic sea level rises and inundation as a result of storm surges, polar melting or tsunamis such as the historically documented event in eastern Mediterranean of AD 365 (Shaw et al. 2008). The World Bank (2009) estimates that a rise in sea level of 1 m would affect just under 10 % of the Nile Delta population (around six million people) and cause a loss of 10 % of arable land. Using the Space Shuttle Radar Topography Mission (SRTM) data at 90 m spatial resolution (Farr and Kobrick 2000), Mulligan (2007) produced a crude global analysis of inundation under a series of catastrophic sea level rise scenarios. A 2 m rise would inundate the coastal zone areas below 2 m elevation shown in Fig. 3.15 as well as a number of other areas around the world. Some of these European areas (especially in the northern Mediterranean and northern Europe) are highly populated with significant infrastructure and commercial activity, meaning this inundation could significantly affect the high level of GDP in those areas.

Significant population densities are also present in some of these areas, such that inundation could have very serious consequences for forced migration within

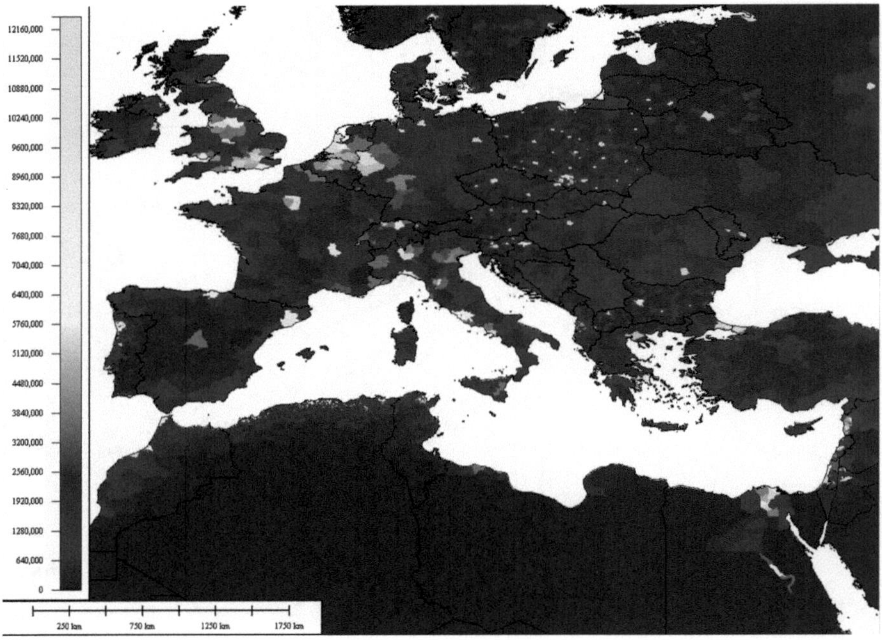

Fig. 3.16 Index of flood risk exposure (mean for administrative units)

northern Europe (UK, Netherlands), the northern Mediterranean (Venice) and the southern Mediterranean deltaic areas (Nile). The terrain datasets used here are too crude (90 m spatial resolution) for any more detailed analysis and do not fully incorporate significant investments in sea defences made in the Netherlands for example, but nevertheless the main spatial patterns of risk are clear.

Impacts on Flooding

Flood risk, particularly in low lying coastal cities, is a serious threat in Europe. Llasat et al. (2010) reviewed a total of 185 flood events for 18 countries across the Mediterranean for the period 1990–2006 and calculated a reported material damage that exceeded €29 billion. Risk of flooding may also change as a result of climate change. It is widely acknowledged that climate change may severely alter the risk of hydrological extremes and that changes in precipitation, especially the increased intensity and frequency of rainstorms, are likely to be reflected in increased runoff.

Figure 3.16 shows a crude index of flood risk exposure (excluding the effects of mitigation measures that may be in place). This is an index of upstream cumulated rainfall (a measure of river system magnitude and thus risk) and local population density (a measure of exposure). The data are calculated on a 1 km pixel basis but are presented here as means for administrative units. The effects of mitigation

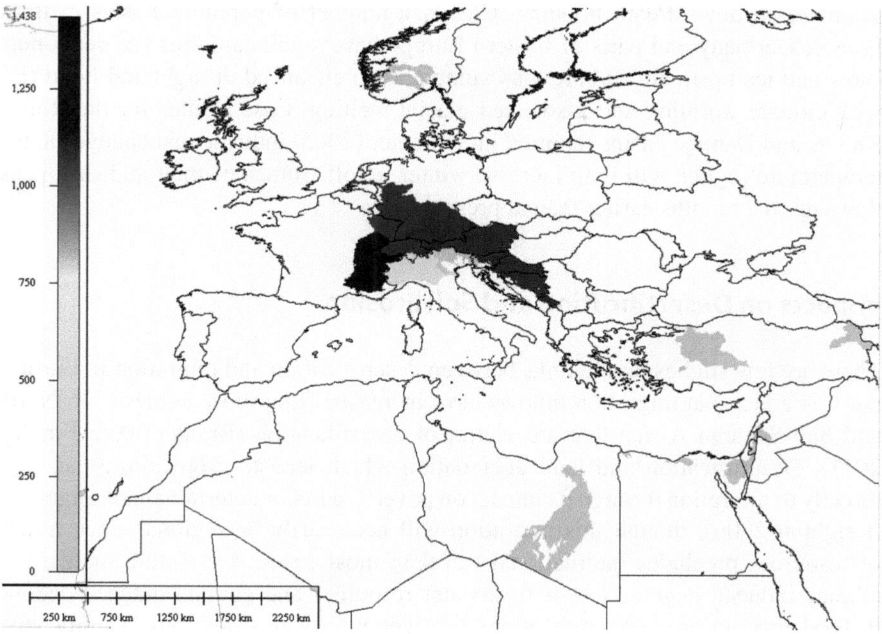

Fig. 3.17 Area of permanent snow and ice (according to GlobCover) in river basins (km^2)

measures are not factored in, but the pattern of risk is clear (this risk will be reduced where investments in flood defence are made and maintained, though at a cost). Risks are greater in northern Europe and parts of the north and east Mediterranean because of the high population densities and large, humid catchments. Population, agriculture, and economic activity in the north and east Mediterranean and parts of Northern Europe are disproportionately concentrated in areas at risk of flooding. Although the flood risk is lower, flooding in the southern Mediterranean countries is of particular concern because flooding has historically been most devastating in this region in terms of casualties because of its rapid onset nature (flash flooding) and lower investments in protection and preparedness.

Climate change may reduce risks overall in regions that receive lower rainfall but may increase them locally if extreme rainfall and snowmelt events become more frequent or of greater magnitude (Fig. 3.17). Moreover snowmelt dynamics may also affect both flood and (seasonal) drought risk. Figure 3.17 shows the total area covered by permanent snow and ice (according to the GlobCover remote sensing product). The permanent snow and ice area is summed in hydrologic basins as an indication of (a) the dependence of water resources on snowmelt and thus the potential for enhanced drought if snow were to disappear and no longer sustain flows in the dry season and (b) the potential for rapid onset snowmelt leading to enhanced downstream flooding. Clearly a number of populous basins in Italy, France, Germany and parts of Eastern Europe have significant areas of permanent snow and ice upstream and are thus vulnerable to enhanced drought and flood risk with climate warming and associated glacial melting. Case studies for the Rhine, Rhone, and Danube basins reported in Schroeter (2005) indicate that changes in the temperature regime will both increase winter runoff from snowmelt and shift peak flows up to 2 months earlier than at present.

Impacts on Desertification and Soil Erosion

There are few studies on the links between desertification and migration in Europe, but it is a fact that migration inflows have increased from those countries in North and Sub-Saharan Africa that are victim of desertification (Brauch 2006; Findlay 2011). Desertification and land degradation which lead to crop failure can lead directly to migration through its impact on poverty, a major determinant of migration (Leighton 2006), though this migration will necessarily be regional since a lack of resources precludes international travel in most cases. Any future increase in climate-induced degradation of freshwater resources and climate-induced decline in food production with implications for economic wellbeing could significantly increase environmentally-driven migration (Brauch 2010) within and between countries.

Desertification is a result of both physical processes (land cover change, soil erosion, climate change) and human processes (unsustainable agricultural practices, rural depopulation). The impact of climate change on desertification will depend on a number of factors; the expected increase in storm frequency may increase soil erosion, a general decrease in annual rainfall may decrease vegetation cover and soil moisture impacting upon soil production, productivity, runoff and groundwater recharge. Those catchments with the steepest land may be most at risk, particularly in the drier and more intensively farmed parts of Spain, Italy, Greece, Turkey and Sardinia. The EACH-FOR project on Environmental Change and Forced Migration highlighted a case study in the south-eastern regions of Spain (Murcia and Almeria) (Fermin 2009) where desertification processes are well documented. It concluded that the increase of intensive agricultural production following an investment in greenhouses and irrigation technology has encouraged migration to the area, but the horticultural economy remains dependent on EU subsidies. The area remains vulnerable to desertification and water shortages.

Given the projected combined impacts of climate change, dryland population growth and the pressure on drylands brought about by a projected global decline in good cultivable land, enhanced Mediterranean desertification may represent a risk to food security and livelihood security, all of which impact on decisions to migrate. In North Africa food security issues are particularly important where there is already a dependence on cereal imports and thus international market dynamics (Yang and Zehnder 2002). Nevertheless, desertification is a slow process that can be prepared for and responded to, rather than a catastrophic event, and thus the direct links between desertification and migration events within the Mediterranean Basin have yet to be demonstrated (Safriel 2009).

Impacts on Wildfire Risk

Fire risk depends on human, ecological, biophysical and climatic factors, such as whether it is a wet or dry summer, wind speed and direction, sources of ignition and the amount of dead wood available to propagate combustion (Pausas 2004). The recent and continued abandonment of land by European farmers, and a reduction in livestock pressure have resulted in high fire risk scrub species moving into the previously grazed areas. Four successive dry years would appear to constitute a critical threshold in the resistance of forests to fire. If summers become warmer and drier, forest ecosystems will become more vulnerable. Current burned areas are much more frequent and extensive in the eastern Mediterranean and, given that precipitation is projected to decline and temperature to increase in this area, wildfires may become more frequent and their consequences for landscape, air pollution and agriculture may worsen. However, some of these impacts may be manageable through vegetation management and changes to agricultural practices (reducing or better-controlling the use of fire in farming).

We have shown that in addition to its direct impacts, climate change may have significant knock on effects for environmental hazards and the provision of ecosystem services affecting: food and water security, health and morbidity and quality of life. Many of these impacts may have implications for either forced migration or economic migration driven by the resulting political, social and economic changes.

Conclusions

Europe is a geographically complex region with strong international and intra-national gradients in climate, landscape and socio-economic conditions. The enormous diversity of the region means that generalisations on climate, landscape, agriculture, ecosystem services, hazards, migrations and the impacts of environmental change upon them cannot be made. Rather, one must take a spatially explicit approach to understand the distribution of state and change within and between countries in order to better understand the interactions between environment and inter-regional or inter-national migrations. Most of the proximal drivers for migration are political, social, economic and cultural and thus it is very difficult to find clear relationships with environmental factors, though environmental factors surely have a role to play in setting the context for the socio-economic landscape to play out.

Population, GDP and infrastructure have developed over Millennia in Europe and both the level of population and the magnitude of GDP are strongly conditioned by overall climate suitability. The SRES scenarios project significant change in population and GDP throughout Europe with the greatest population growth in the southern and eastern Mediterranean and the greatest GDP increase in the north of Europe, potentially further separating the income and well-being gradient that currently exists from North to South. Much of the population increase is expected to take place in urban areas and will thus generate increasingly spatially concentrated pressures on ecosystem services and exposure to hazards. Continued land degradation, and agricultural intensification or abandonment, industrialisation and post-industrialisation may lead to continued rural–urban flows within and between countries. Since GDP increases are focused on the urban and industrial areas of the north Mediterranean and northern Europe, cities in these regions may be particularly attractive to migrants from inside and outside Europe and may also be much more open to migration given a potential surplus of employment over demand.

Climate changes will be strongly spatially variable within countries both latitudinally, continentally and altitudinally. According to an ensemble mean for 17 GCMs, climate is expected to warm throughout Europe, but especially in the south and rainfall is expected to decrease throughout the Mediterranean but increase in the extreme south and also in northern Europe. These changes will have a mixture of positive and negative impacts for agriculture, water resources and other ecosystem services as well as for hazards. There may be improvements in water resources and ecosystem services in parts of the Mediterranean and northern Europe. Agricultural productivity may increase in parts of the Mediterranean basin and northern Europe although strongly limited by water availability and the impact of drought and extreme rainfall events. Short term increases may lead to longer-term declines if temperatures continue to increase. Productivity in northern Europe will most-likely increase with the increasing temperature, growing season length, and rainfall.

The risk of environmental hazards can be expected to increase throughout Europe but exposure to this risk will likely be much greater in the more populous and urbanised northern Mediterranean and northern Europe, and also where mountains are present and where existing infrastructure and GDP is greatest. However, since significant investments in mitigation and emergency planning measures exist in these countries the resulting disasters (for all but the most extreme events) may be, as currently, less damaging to life and property than they are in the southern Mediterranean. This is a priority for future research especially for those hazards with a high probability of leading to migration (landslides, flooding, coastal inundation).

The complexity of impacts of climate change on agricultural productivity, ecosystem service provision, hazard and disaster are poorly known at the regional scale, through a number of case studies within the region point to potentially serious negative impacts especially in the northern Mediterranean and some potentially positive impacts in parts of the southern Mediterranean. Such case studies should be extrapolated with caution given the heterogeneous nature of Europe as indicated by the spatial analysis presented. Further geographically detailed analyses of this type are a priority for future research.

Overall, whilst the south of Europe and north Africa is clearly currently the economically poorer area, climate change may improve productivity, agriculture, water resources and ecosystem services in parts of that region whilst creating greater risks for natural hazards in the populous and infrastructurally sensitive north of Europe. If, as expected, population and urbanisation continues to grow in the south, exposure to risk in this region is likely to increase. The distribution of climate impacts in relation to the current distribution of population and the land that provides populations with key provisioning and regulating ecosystem services will be critical. Any positive changes in the south may be cancelled out by increased human pressure on the landscape as populations grow. The areas that may suffer the greatest potentially negative impacts include those agriculturally intensive and populous regions of the northern Mediterranean, whereas areas with potentially positive impacts of climate change are currently both agricultural and population deserts.

There are significant uncertainties as to how agriculture and human behaviour will respond to environmental change since the responses are dependent upon the population density, economics, politics, culture and adaptive capacity. These characteristics are not easily mapped at the European scale and they also change over time in unpredictable ways so further research is necessary. The wealthier northern countries are, on the face of it, more adaptable but since these areas carry the greatest populations and intensive agriculture which is near the limits of water sustainability, these areas are also more vulnerable to loss of ecosystem services and increase in the frequency and magnitude of hazards through climate change.

In conclusion, whilst environmental change of one form or another is inevitable, its impact on the *status quo ante* will vary according to the complexity and magnitude of the current human infrastructure. Significant environmental change over coming decades may lead to enhanced economic or even environmentally forced migration within Europe and from outside of Europe to areas where environmental and economic conditions are more amenable. There will be significant pressures for movement within the Mediterranean especially south to north, south to east and rural to urban as well as potentially between rural and urban areas. Given the complexity of Mediterranean landscapes and the spatial heterogeneity of the impacts of climate change, internal migration may be a smart adaptation mechanism to make use of new agricultural and other opportunities, especially in the extreme southern Mediterranean.

Whilst push factors for migration may increase in the coastal southern Mediterranean areas and parts of eastern Europe, so some pull factors in the north Mediterranean and northern Europe may also decline as a result of climate change and its impacts, particularly given the significant current flows of food from southern to northern Europe. Indeed, the populations most vulnerable to climate change impacts are likely to be those very highly urbanised and infrastructurally complex societies in parts of the Northern Mediterranean that are already close to agricultural and water resource limits and in hazard prone environments, rather than the relatively dispersed and potentially more resilient populations of the southern Mediterranean.

Acknowledgement Some of the research for this chapter was carried out as part of the UK Government Office for Science Foresight Programme on Migration and Global Environmental Change and is described in Burke et al. (2011).

References

Bade, K. (2003). *Legal and illegal immigration into Europe: experiences and challenges* (43 pp). Meijboomlaan, Wassenaar: Netherlands Institute for Advanced Study in the Humanities and Social Sciences.

Bleu, P. (2009). *State of the environment and development in the Mediterranean*. Athens: UNEP/MAP-Plan Bleu.

Bohannon, J. (2010). The Nile Delta's sinking future. *Science, 327*(5972), 1444–1447.

Brauch, H. G. (2006, October 25–27). *Desertification and migration: From Almeria I to Almeria II: Achievements and policy tasks*. International symposium on desertification and migration. Almeria.: Afes Press. http://www.afes-press.de/html/download_hgb.html

Brauch, H. G. (2010). *Climate change and Mediterranean security: International, national, environmental and human security impacts for the Euro-Mediterranean region during the 21st century*. Proposals and perspectives. European Institute of the Mediterranean, PAPERSIEMed. http://www.iemed.org/publicacions/papers9.pdf

Burke, S. M., Mulligan, M., Parks, K.,& van Soesbergen, A. (2011). *DR8a: Quantifying change in ecosystem services and exposure to hazards in the Mediterranean basin over the next 50 years that might be relevant to migration*. Report to Global Environmental Migration Programme, Foresight Programme, Government office for Science. Available at http://www.bis.gov.uk/assets/foresight/docs/migration/drivers/11-1177-dr8a-ecosystem-services-and-hazards-in-mediterranean-basin.pdf

Cemagref. (2009, May 26). Repeated fire and drought: A menace for Mediterranean forests. *ScienceDaily*. www.sciencedaily.com/releases/2009/05/090519080045.htm

Center for International Earth Science Information Network (CIESIN). (2002). *Country-level population and downscaled projections based on the B2 Scenario 1990–2100 [digital version]*. Palisades: CIESIN Columbia University. Available at http://www.ciesin.columbia.edu/datasets/downscaled

Center for International Earth Science Information Network (CIESIN), Columbia University; International Food Policy Research Institute (IFPRI), The World Bank, & Centro Internacional de Agricultura Tropical (CIAT). (2004). *Global Rural–urban Mapping Project (GRUMP), Alpha Version: Urban Extents*. Palisades: Socioeconomic Data and Applications Center (SEDAC), Columbia University (downloaded 2008).

Cornelius, W. (1994). Spain: The uneasy transition from labor exporter to labor importer. In *Controlling immigration: A global perspective* (pp. 331–370). Stanford: Stanford University Press.

Crawley, H. (2011). Past UK experience of the arrival of populations displaced by extreme events. In Foresight (Ed.), *Migration and global environmental change*. London: Government Office for Science.

de Haas, H. (2011). Mediterranean migration futures: Patterns, drivers and scenarios. *Global Environmental Change, 21*(1), S59–S69. http://www.bis.gov.uk/foresight/our-work/projects/published-projects/global-migration/reports-publications

Della-Marta, P. M., Haylock, M. R., Luterbacher, J., & Wanner, H. (2007). Doubled length of western European summer heat waves since 1880. *Journal of Geophysical Research, 112*.

Easterlin, R. (1961). Influences in European overseas emigration before World War I. *Economic Development and Cultural Change, 9*(3), 331–351.

Farr, T. G., & Kobrick, M. (2000). Shuttle Radar Topography Mission produces a wealth of data. *American Geophysical Union Eos, 81*, 583–585.

Fassmann, H., & Munz, R. (1992). Patterns and trends of international migration in western Europe. *Population and Development Review, 18*(3), 457–480.

Fermin, A. (2009). *Spain: Case study report for the EACH-FOR project. Case study Environmental Change and Forced Migration Scenarios (EACH-FOR).* http://www.each-for.eu/documents/CSR_Spain_090328.pdf

Findlay, A. M. (2011). Migrant destinations in an era of environmental change. *Global Environmental Change*. doi:10.1016/j.gloenvcha.2011.09.004.

Fischer, E. M., & Schär, C. (2010). Consistent geographical patterns of changes in high-impact European heatwaves. *Nature Geoscience, 3*, 398–403.

Fitzpatrick, D. (1980). Irish emigration in later nineteenth century. *Irish Historical Studies, 22*(86), 126–143.

Foresight: Migration and Global Environmental Change. (2011). *Final project report*. London: The Government Office for Science. http://www.bis.gov.uk/assets/foresight/docs/migration/11-1116-migration-and-global-environmental-change

Goria, A. (1999, December 9–10). Impacts of climate change on migration in the Mediterranean basin: Environmental dimension of South–north migration and directions for future policies. In *Proceedings of the conference on the impacts of climate change on the Mediterranean Area: Regional scenarios and vulnerability assessment* (pp. 10–25), Venice.

Guinnane, T. (1994). The great Irish famine and population: The long view. *The American Economic Review, 84*(2), 303–308.

Hansen, R. (2003). Migration to Europe since 1945: Its history and its lessons. *The Political Quarterly, 74*, S1.

Hansen, R. (2011). The European Union's role in migration up to 2030 and then, 2060. In *Migration and global environmental change*. London: Government Office for Science.

Hatton, T., & Williamson, J. (1994). What drove the mass migrations from Europe in the late nineteenth century? *Population and Development Review, 20*(3), 533–559.

Heisbourg, F. (1991). Population movements in a post-Cold War Europe. *Survival, 33*, 31–43.

Hijmans, R. J., Cameron, S. E., Parra, J. L., Jones, P. G., & Jarvis, A. (2005). Very high resolution interpolated climate surfaces for global land areas. *International Journal of Climatology, 25*, 1965–1978.

Huntoon, L. (1998). Immigration to Spain: Implications for a unified European Union immigration policy. *International Migration Review, 32*(2), 0423–0450.

Jokisch, B., & Pribilsky, J. (2002). The panic to leave: Economic crisis and the 'new emigration' from Ecuador. *International Migration, 40*(4), 75–102.

Jones, P. D., Hulme, M., Briffa, K. R., & Jones, C. G. (1996). Summer moisture availability over Europe in the Hadley centre general circulation model based on the Palmer drought severity index. *International Journal of Climatology, 16*, 155–172.

Komuscu, A. U. (1999). Using the SPI to analyze spatial and temporal patterns of drought in Turkey. *Drought Network News, 11*, 7–13.

Koser, K., & Lutz, H. (1998). Introduction. In K. Koser & H. Lutz (Eds.), *New migration in Europe: Social constructions and social realities*. Basingstoke: Macmillan.

Krings T. (2012). EU Labor Markets after Post-Enlargement Migration. *Journal of Ethnic and Migration Studies,* 1–2 first article.

Laczko, F., & Aghazarm, C. (2009). *Migration, environment and climate change: Assessing the evidence*. International Organization for Migration (IOM). http://publications.iom.int/bookstore/free/migration_and_environment.pdf

LandScan™. (2007). *Global population database 2007*. Oak Ridge: Oak Ridge National Laboratory. Available from: http://www.ornl.gov/landscan/

Lehner, B., Verdin, K., & Jarvis, A. (2008). New global hydrography derived from spaceborne elevation data. *Eos, Transactions of the American Geophysical Union, 89*(10), 93–94.

Leighton, M. (2006). Desertification and migration. In P. M. Johnson, K. Mayrand, & M. Paquin (Eds.), *Governing global desertification: Linking environmental degradation, poverty and participation* (p. 292). Aldershot: Ashgate.

Leitner, H. (1997). Reconfiguring the spatiality of power: The construction of a supranational migration framework for the European Union. *Political Geography, 16*(2), 123–143.

Llasat, M.C., Llasat-Botija, M., Prat, M.A., et al. (2010). High-impact floods and flash floods in Mediterranean countries: the FLASH preliminary database. *Advances in Geoscience, 23L*, 47–55. Available from: www.adv-geosci.net/23/47/2010/

López Bermúdez, F., & Sánchez, M. C. (1997). Las sequías y su impacto en el riesgo de desertificación de la cuenca del Segura. Apuntes para la gestión y sostenibilidad del agua. *Areas, 17*, 155–168.

Magnan, A., Garnaud, B., Bille, L., Gemenne, F., & Hallegatte, S. (2009). *The future of the Mediterranean: From impacts of climate change to adaptation issues*. Paris: IDDRI.

Mahroum, S. (2001). Europe and the immigration of highly skilled labour. *International Migration, 39*(5), 27–43.

Mulligan, M. (2007). *Global sea level change analysis based on SRTM Topography and Coastline and Water Bodies Dataset. (SWBD) (Version 1.0)*. Available from: http://www.ambiotek.com/sealevel

Mulligan, M. (2009). *SimTerra: A consistent global gridded database of environmental properties for spatial modelling*. Available from: http://www.ambiotek.com/simterra

Mulligan, M. (2010). *Global cumulative burn frequency mapping*. Available from: http://www.ambiotek.com/fires

Mulligan, M., Saenz-Cruz, L., van Soesbergen, A., Smith, V.T., & Zurita, L. (2009). *Global dams database and Geowiki* (Version 1). Available from: http://www.ambiotek.com/dams

Myers, N. (2002). Environmental refugees: A growing phenomenon of the 21st century. *Philosophical Transactions of the Royal Society London B, 357*, 609–613. doi: 10.1098/rstb.2001.0953. http://rstb.royalsocietypublishing.org/content/357/1420/609.full.pdf

Nakicenovic, N., & Swart, R. (Eds.). (2000). *IPCC special report on emissions scenarios*. Cambridge: Cambridge University Press.

Narusova-Schmitz, A., & Lilleorg, K. (Eds.). (2009). *Climate change, environmental degradation and migration: Addressing vulnerabilities and harnessing opportunities*. Switzerland: International Organization for Migration (IOM). http://publications.iom.int/bookstore/index.php?main_page=redirect&action=url&goto=publications.iom.int%2Fbookstore%2Ffree%2FHSN_Quadrilingual_Report.pdf

Nicholls, R. J. (2010). Impacts of and responses to sea level rise. In John. Church, Philip L. Woodworth, T. Aarup, S. Wilson (Eds.), *Understanding Sea-Level Rise and Variability* (p. 456). Chichester: Wiley.

Nicholls, R. J., Hoozemans, F. M. J., Marchand, M. (1999, October). Increasing flood risk and wetland losses due to global sea-level rise: Regional and global analyses. *Global Environmental Change, 9* (Supplement 1), S69–S87.

Pausas, J. G. (2004). Changes in fire and climate in the eastern Iberian Peninsula (Mediterranean basin). *Climatic Change, 63*, 337–350.

Roy, D. P., Boschetti, L., Justice, C. O., & Ju, J. (2008). The collection 5 MODIS burned area product – Global evaluation by comparison with the MODIS active fire product. *Remote Sensing of Environment, 112*, 3690–3707.

Safriel, U.N. (2009, December 10–11). Status of desertification in the Mediterranean region. In J. L., Rubio, U., Safriel, R., Daussa, W. Blum, F., Pedrazzini (Eds.), *Water scarcity, land degradation and desertification in the Mediterranean region* (Book Series: *NATO science for peace and security series C* – Environmental security, pp. 33–73). *Proceedings of the NATO Advanced Research Workshop on Water Scarcity, Land degradation and desertification in the Mediterranean region*, Valencia.

Schröter, D., Cramer, W., Leemans, R., et al. (2005). Ecosystem service supply and vulnerability to global change in Europe. *Science, 310*, 1333–1337.

Shaw, B., Ambraseys, N. N., England, P. C., et al. (2008). Eastern Mediterranean tectonics and tsunami hazard inferred from the AD 365 earthquake. *Nature Geoscience, 1*, 268–276.

Siebert, S., Döll, P., Feick, S., Frenken, K., & Hoogeveen, J. (2007). *Global map of irrigation areas* (Version 4.0.1). Frankfurt/Rome: University of Frankfurt (Main)/FAO.

Vicente-Serrano, S. M., González-Hidalgo, J. C., de Luis, M., & Raventos, J. (2004). Drought patterns in the Mediterranean area: The Valencia region (eastern Spain). *Climate Research, 26*, 5–15.

World Bank. (2007). *Development actions and the rising incidence of disasters*. Evaluation Brief 4. Available from: http://www.worldbank.org/ieg

World Bank. (2009). *Adapting to climate change in Europe and Central Asia*, Washington, DC. Available from: www.worldbank.org/eca/climate/ECA_CCA_Full_Report.pdf

Yang, H., & Zehnder, A. J. B. (2002). Water scarcity and food import: A case study for southern Mediterranean countries. *World Development, 30*, 1413–1430.

Zomer, R. J., Trabucco, A., Bossio, D. A., van Straaten, O., & Verchot, L. V. (2008). Climate change mitigation: A spatial analysis of global land suitability for clean development mechanism afforestation and reforestation. *Agricultural Ecosystems and Environment, 126*, 67–80.

Environmental Change and Human Migration in Sub-Saharan Africa

4

James Morrissey

Acronyms

IOM International Organization for Migration
IPCC Intergovernmental Panel on Climate Change
NDVI Normalized Difference Vegetation Index

Introduction

Literature on the environment–migration nexus frequently makes mention of the limited claims that can be made given the dearth of empirical work on the topic. As a counter to such claims, this chapter offers a review of over 30 studies, all of which have been conducted in sub-Saharan Africa and detail in various ways the mobility responses by groups and individuals in the context of a variety of environmental stresses. These studies have been chosen due to their empirical nature and because they (a) offer documented insights into the mechanics of the environment–migration nexus and (b) supersede the speculative claims of crude predictive accounts that have been embroiled in the politics of environmental protection and migration. (For a detailed account of these politics, see Morrissey 2009, 2012a.)

This chapter differs from other syntheses of empirical work in Africa because it covers all of sub-Saharan Africa and because the information it contains is up to date. In addition, this chapter is distinct in that it attempts to move on from contested claims about the relative role of the environment in migration and the meaningfulness of labels that may be applied to groups and individuals

J. Morrissey (✉)
Department of International Development, University of Oxford, Oxford, UK
e-mail: morrissey.james1@gmail.com

E. Piguet and F. Laczko (eds.), *People on the Move in a Changing Climate*,
Global Migration Issues 2, DOI 10.1007/978-94-007-6985-4_4,

moving in a context of environmental stress.[1] In this regard, the chapter differs from works such as those of Morrissey (2009) and Jonsson (2010), which bring a critical lens to understanding which claims have been made about the role of environmental change, based on different empirical accounts. Instead, this chapter offers a synthesis of different studies and what they might say about the relationship between environmental stress and human mobility, focusing on what these studies reveal, rather than the political claims they enable.

With the above in mind, it is necessary to acknowledge that the recent growth of interest in the environment–migration nexus is not simply the outcome of slowly expanding interest in the human condition. Rather, it is the product of growing concern over the potential impacts of future climate change and what these might mean for both demographics and human well-being. As such, this chapter also contains a discussion of climate change projections that highlight impacts that are shown to have important implications for decisions about human mobility. In addition, because migration has been shown to be one of many potential livelihood responses to environmental stress, the chapter also details established mobility trends and patterns in sub-Saharan Africa as it is thought that these offer a meaningful context for understanding mobility as a livelihood response to climate change.

This chapter thus opens with a brief account of the different methodological approaches that have been used in the study of the environment–migration nexus. This is done in order to delineate the chapter and the studies that will be reviewed. Next, there is a discussion of the predicted impacts of climate change in sub-Saharan Africa, followed by an exploration of the empirical data linking such change to human migration. The chapter closes with an account of the dominant contemporary features and trends of mobility on the subcontinent.

The Environment–Migration Nexus: Methodological Approaches

Studies of the relationship between environmental change and migration can be broken down into two major types: (a) empirical studies, which aim to document the linkages between environmental change and migration; and (b) predictive models. In both cases, such studies can be further broken down into two approaches. In terms of (a) empirical studies, these are broadly constituted by (i) palaeoclimatological and archaeological approaches, and (ii) contemporary approaches.

Palaeoclimatological and archaeological approaches look into the archaeological record and invoke palaeoclimatological methods to identify correlations between evidence of change in human settlement and evidence of change in the climate.

[1] Note that the description "moving in a context of environmental stress" is used in an attempt to supersede simplistic accounts of, and debates over, the degree to which environmental stress may or may not "cause" migration.

Contemporary approaches focus on biophysical events that are analogous to the impacts of climate change but that are recent enough to allow for their detailed exploration.

While palaeoclimatological and archaeological approaches are important in their own right (providing insights into the processes by which changes in material culture and technology occur, and providing nuance to accounts of where people have historically lived and moved), they are also important because they highlight the potential link between climate change and human migration. Their value in addressing concerns about the impacts of future climate change is becoming increasingly limited. This is because palaeoclimatological and archaeological methods currently lack the temporal resolution required to address contemporary policymaking concerns about the implications of climate change for both demographics and human livelihoods. In addition, such approaches are not currently able to provide sufficient detail on the role of non-environmental factors that empirical studies have shown to be of fundamental importance.

On the other hand, contemporary analogous approaches are thought to be particularly useful as they (a) offer a means to explore the important interactions between environmental and non-environmental factors impacting on migration in a context of environmental stress and (b) provide useful analogies to the impacts expected to accompany future climate change. The major limitation of such studies, however, is that there is only so much that one can learn about the future through the study of analogous, past events. Most notably, such approaches cannot account for non-linearity in the environment–migration nexus.

Regarding predictive models – compared with the empirical studies described above – the two constitutive approaches are less easy to distinguish. That said, they quite clearly comprise, on the one hand, generalized, simplistic models, and on the other, sophisticated attempts at understanding interactions between the multitude of factors that shape mobility responses in a context of environmental stress. In terms of simplistic accounts, the general approach has been to overlay climate change impacts onto predicted populations at some discrete time-step in the future. Migration is then inferred from some proportional rate of outmigration based on the percentage of predicted change in the biophysical environment. Such approaches, which have simply assumed that migration is a linear response to environmental change, have been heavily critiqued elsewhere for their simplicity, determinism, invocation of neoclassical migration theory, and objectives of raising concern over environmental change rather than accurately representing the dynamics of the environment–migration nexus (Castles 2002; Mortreux and Barnett 2009; Tacoli 2009; Morrissey 2012a).

On the other hand, sophisticated works aimed at modelling mobility decisions in a context of environmental stress have recently emerged. Such attempts distinguish themselves from simplistic models by their efforts at dealing with a host of non-environmental factors – the moderating impacts of which have been identified in existing contemporary analogous studies – and their focus on understanding the dynamic interactions between such factors, rather than on predicting numbers of "environmental refugees."

With the above in mind, it should be noted that this chapter limits itself to a discussion of the contemporary, analogous works and sophisticated attempts at modelling that have been applied to the environment–migration nexus in sub-Saharan Africa. It is thought that much of the concern cited in simplistic modelling approaches will be covered by this chapter's overview of future climate change impacts, while an assessment of the archaeological data is beyond the scope of this work, and of only limited value. A final note in this regard is that this chapter does not detail works exploring the impact of migration on the biophysical environment (such as that by Black and Sessay 1997). This is because such works have historically fallen outside of concern regarding climate change, pertaining more to the politics of migration and asylum. That said, this work is becoming important in the discussion of climate change impacts as concern about the impacts of outmigration (undertaken in a context of environmental stress) on sending regions grows. Here, the concern pertains to labour shortages potentially compromising attempts at agricultural intensification or efforts at environmental protection (see Afifi 2011).

With the scope of this chapter now clearly delineated, the discussion moves to the predicted impacts of climate change as a means for framing the review of empirical literature on the relationship between environmental change and migration.

Climate Change in Sub-Saharan Africa

Climate change predictions for sub-Saharan Africa reflect the size and diversity of the subcontinent. Predictions for changes in rainfall show high levels of variability in both time and space, while predictions for expected temperature change, although suggesting a more uniform rise, show significant levels of variation in terms of impact due to the localizing effects of topography and moisture availability (IPCC 2012). Efforts at modelling such changes, as well as their likely impacts, are further frustrated in the African case by the limited access to initial conditions data – relevant for both climate models and biogeographical response models – and the limited effort that has been directed at downscaling Global Climate Model projections for the subcontinent (Thornton et al. 2010).

With these limitations in mind, a starting point for this discussion is the manner in which crude predictions of future human displacements have focused on how changes in the average climate might render what are currently populated parts of the globe uninhabitable. Such concerns have revolved around the impacts that changes in temperature and rainfall might have on food production, and the impacts of sea-level rise and extreme weather events, which could destroy the physical infrastructure upon which people's livelihoods are currently reliant (Jacobson 1988; Myers and Kent 1995; Myers 2001; Conisbee and Simms 2003; Christian Aid 2007; Rajan 2008).

In addition to such concerns, it should be noted that empirical work on the relationship between environmental change and migration (much of which is detailed below) has highlighted that migration is not simply an outcome of intolerable vulnerability, but rather a complex product of a particular set of livelihood options that are framed by a complex mix of biophysical conditions and social, political and economic contexts. In this regard, the undermining of livelihood security, in general, has been shown to constitute an important part of the decision-making process regarding migration. Since human health is a central component of livelihood security – being the point at which livelihood security is compromised – this chapter explores the impact climate change is predicted to have on human health through its impact on changing access to food and water (as well as the quality of that water) and changing the distribution and transmission characteristics of certain diseases. In this regard, human health is linked to mobility strategies as areas in which diseases are prevalent change, and therefore change habitability characteristics, and as health impacts place an extra burden on the household – both in terms of cost of treatment and lost labour – which can reasonably be hypothesized to increase the imperative to migrate. Finally, the uneven distribution of health facilities (usually clustered in urban areas) means that health impacts might bring people to places where those facilities are located. The impacts of this could be significant on persons with ailments requiring long-term treatments. As such, the next section of this chapter deals with climate change predictions in terms of impacts on food production, implications for extreme weather and health concerns.

Predictions for temperature increases across Africa, not taking into account increases due to land cover changes,[2] suggest an increase of between 3 and 4 °C, while rainfall predictions vary significantly in both time and space (Boko et al. 2007; IPCC 2012). Predictions for West Africa include both a drying and wetting of the Sahel, with increased rainfall expected to accompany an increase in vegetative cover (Boko et al. 2007; IPCC 2012). In East Africa and the Horn of Africa, rainfall predictions are variable due to the large latitudinal extent of the region and the range of topography that characterizes the area. That said, increasing marginality is expected in coastal areas of the region, while changes in highland areas are less certain. Southern Africa is expected to see increases in both summer and winter temperatures, while areas south of 10° latitude are expected to see a decrease in rainfall, particularly in the western part of the region (Boko et al. 2007; Goodess 2011).

Notably, while climate change is thought likely to increase heavy precipitation events, there is only a weak signal for such an increase on the continent, with the exception of East Africa (Tebaldi and Sansó 2009; Shongwe et al. 2009; Goodess 2011; IPCC 2012), where a fairly strong and consistent indication of increased heavy precipitation is present and where climate models tend to agree (Shongwe et al. 2011). Both East Africa and West Africa are predicted to see an increased tendency towards drought (Goodess 2011; IPCC 2012), while more wet years are also predicted in both areas (Boko et al. 2007; Shongwe et al. 2011). Southern Africa

[2]Reduced vegetative cover, which is expected with the conversion of land, is thought likely to intensify this temperature increase.

is predicted to see an increase in heavy precipitation in the east (IPCC 2012), with a shift in return period for flooding changing from 100 to 30 years by the end of this century (Goodess 2011). Tropical cyclones also present a significant destructive force on the east coast of Southern Africa and although evidence for changes in their frequency and intensity remains inconclusive (Boko et al. 2007; Goodess 2011; IPCC 2012), it appears that there will be an increase in the strength of the most intense events (Goodess 2011).

Food production, which is already compromised by arid conditions on much of the continent, is expected to experience negative impacts due to both decreases in the length of the growing season and currently marginal areas being forced out of production (Boko et al. 2007). In particular, the production of maize, sorghum, millet, groundnuts and cassava, under a medium–high emissions scenario (A1B),[3] is predicted to show decreases across the board with the greatest losses in maize and the lowest in cassava (Schlenker and Lobell 2010). Of concern here is that maize, although badly affected by temperature rises, is still relatively resilient to temperature when compared with other important dietary crops, most notably legumes (Jones and Thornton 2003).

West Africa is predicted to see agricultural losses of between 5 and 7 % in the Sahara, and between 2 and 4 % in Western and Central Africa (Boko et al. 2007).[4] In East Africa and the Horn of Africa, some parts of the Ethiopian highlands might see an increase in the length of the growing season as a result of changes in temperature and rainfall (Boko et al. 2007). In this regard, it is thought that mixed rain-fed temperate/tropical highland regions will, in general, do better, while mixed rain-fed humid–sub-humid and mixed rain-fed arid–semi-arid lowlands will do worse (Thornton et al. 2010). In Southern Africa, crop production is expected to drop by between 0.4 and 1.3 %. It should be noted, however, that impacts will not be uniform across regions; notably, Mozambique will potentially experience an increase in the length of the growing season due to changes in temperature and rainfall (Boko et al. 2007).

Losses in crop producing areas might be offset by moving away from growing crops and focusing on small livestock instead, with such farms predicted to have an increase in production of between 26 and 58 %, based on a warming of between 2.5 and 5 °C, respectively. Large livestock farms, on the other hand, are expected to experience a decrease in profits of approximately 22 % with a warming of 2.5 °C and 35 % with a warming of 5 °C. However, these processes are complicated by the fact that increases in precipitation are expected to have negative impacts on all livestock farming (Boko et al. 2007). Further to this, crop and animal production should not be seen as simply exchangeable as reductions in crop production will

[3]Scenarios mentioned in this text refer to the *Special Report on Emissions Scenarios*, published by the Intergovernmental Panel on Climate Change (IPCC 2000).

[4]Despite not being a part of the focus of this chapter, Central Africa is included here because the delineation of different regions of the subcontinent is not consistent among different studies. In this case, the study was aggregating data from the Western and Central African region.

impact livestock herds in mixed farming areas, where stover forms an important component of animal feed (Jones and Thornton 2003).

Sea-level rise, which will likely affect livelihoods through both permanent inundation (Boko et al. 2007) and an increase in the impacts of storm surge and flooding (Nicholls and Lowe 2004), is thought likely to reach between 14 and 44 cm within this century (Kundzewicz et al. 2007). All of East, West and Southern Africa are thought vulnerable; however, the West African coast is thought to be the most vulnerable due to the large populations present there[5] and the lack of protective infrastructure (Nicholls and Mimura 1998).

In addition to infrastructural damage and direct impacts on production (which themselves impact on human health), extreme weather events could impact on health through their influence on water quality. With much of Africa reliant on unimproved water sources, flooding poses a risk of contamination, while drought could both concentrate pathogens and reduce access to water sources (Boko et al. 2007). At the same time, temperature increases have important impacts on water sources, changing oxygen regimes, redox potentials, lake stratification and mixing rates (Kundzewicz et al. 2007). Finally, changes in the extremes and ranges of both rainfall and temperatures will impact on exposure to and transmission of malaria. In this regard, under medium to high emission scenarios, the expansion of malaria is thought likely to increase by 5 to 7 % with an increase of between 16 % and 28 % in person–months of exposure (Tanser et al. 2003; Boko et al. 2007).

In West Africa, predicted drying is thought likely to reduce person–months of exposure to malaria in a host of countries (such as Burkina Faso, Ghana and Mali). That being said, regions where rainfall is the limiting factor for the transmission of malaria are thought to be particularly susceptible to malaria epidemics (Tanser et al. 2003). This could be of particular concern in West Africa, with its predictions for both a drying trend and an increase in wet and dry years. Because large parts of East Africa and the Horn of Africa are close to the climatic threshold needed for the transmission of malaria, there appear to be potentially large increases in exposure to malaria across a number of emission scenarios. For example, Ethiopia is projected to show increases of more than 100 % in person–months of exposure towards the end of the twenty-first century (Tanser et al. 2003). Similarly, parts of Southern Africa are expected to be particularly impacted by increased exposure to malaria, while other parts of Southern Africa will see a decrease in person–months of exposure. In this regard, Namibia and Mozambique are both expected to see a decrease, while Zimbabwe and South Africa are expected to see an increase in excess of 100 % of person–months of exposure by the end of the century (Tanser et al. 2003).

With a sense of the sorts and scales of changes expected to accompany climate change, as well as its impacts in sub-Saharan Africa, a discussion of the empirical work documenting the relationship between environmental stress and human migration follows.

[5]The population between Luanda and Rabat is expected to rise from 21 million to 51 million by the end of the century.

Migration and Environmental Change: Documenting the Relationship

Works documenting the relationship between environmental change and mobility in sub-Saharan Africa have tended to focus on the subcontinent's dryland areas. This is because drylands have highly variable climates and constitute marginal areas for sedentary agriculture. As such, they are areas where the impacts of environmental stress on mobility decisions can be documented historically. In addition, these areas are both marginal and thought to be particularly vulnerable to the impacts of future climate change. They have therefore been the focus of analogous studies seeking to understand the impacts of climate change on human mobility.

Analogous studies in dryland areas have taken two dominant forms: (a) one looks at the impact of the iconic droughts of the 1970s and 1980s, which affected both the West African Sahel and the Horn of Africa; and (b) the second looks at other conditions of climate variability and associated environmental stresses, notably rainfall stress, soil erosion and depletion, and related processes of forest clearance. Below, these studies are referred to separately, but general conclusions are revealed by both of them.

Following this, the chapter includes a description of works that have taken a longer term view on the impacts of environmental change, looking at the impacts of drying in the Sahel over the latter half of the previous century. This deals with both change and acute stresses, and does the opposite of focusing on time as a discrete variable, as early predictions of environmental refugees tended to do (Morrissey 2012a). Instead, they take an holistic approach by focusing on the iterative processes by which migration is enacted as well as its impacts on livelihoods in both the sending and receiving regions.

This chapter then addresses a body of work subtly different from that dealing with environmental change or stress by exploring literature that has examined how endowments of "environmental quality" impact on mobility decisions. Such work differs from that on environmental stress as it focuses less on change and more on environmental quality as a relatively stable variable. As such, this work is more like that of neoclassical migration theory: testing the hypothesis that people will move from areas with poor environmental endowments to those with better ones. In distinguishing these bodies of work, I should, however, be careful not to overstate my case. This is both because the biophysical environment can never be meaningfully thought of as static and because work on environmental endowments has, as a result of being empirical, found itself asking questions about how environmental stress has a differential relationship with mobility in areas with better and worse endowments of environmental capital. As such, there is a clear overlap between these studies and those focusing on stress and change. I have separated them here because I think that their findings have different implications.

A final emergent body of work that is outlined in this chapter focuses on agent-based modelling of the relationship between environmental change and human migration. Notable in distinguishing this work from the crude "common-sense models" already described is, firstly, the level of sophistication involved and, secondly, the intention of the work, which is more about increasing understanding of the interactions between variables in the model (notably identifying non-linearity) than it is about trying to accurately predict future human movements.

Major Droughts and Other Environmental Stresses

The first and most prominent finding from empirical work on the relationship between environmental stress and human migration is that such stress certainly does act to influence, if not drive, human mobility. This has been documented in a host of countries and in response to a variety of stresses, including droughts of the 1970s and 1980s in northern Nigeria (van Apeldoorn 1981), the western Sahel (Mortimore 1989), Ethiopia (Kidane 1989; Ezra 2001; Ezra and Kiros 2001), Mali (Findley 1994; Pedersen 1995) and northern Sudan (Haug 2002). The mechanisms by which this relationship operates are multiple and have been documented as famine, where migrants leave in search of food (Ezra 2001), or where they are forcibly resettled by government with drought and famine being used as justifications for the move (Ezra and Kiros 2001). More prominent in these works, however, are accounts of migration being undertaken as a means to supplement failing production, reduce the burden on the sending household (Mortimore 1989; Findley 1994; Pedersen 1995) and/or attain alms (van Apeldoorn 1981; Haug 2002).

More recent works, looking at more generic and potentially more subtle processes of environmental stress in sub-Saharan Africa, provide similar conclusions, with migration being documented as a response to a mixture of rainfall stress, reductions in soil quality, flooding, water stress, and deforestation across sub-Saharan Africa (Barrios et al. 2006), as well as specifically in western Botswana (Sporton et al. 1999), northern Mali (de Haan et al. 2002), Burkina Faso (Hampshire 2002; Wouterse and van den Berg 2004), Ghana (Carr 2005), Niger (Afifi 2011), western Sahara (Gila et al. 2011), Benin (Doevenspeck 2011), Ethiopia (Gray 2011; Morrissey 2012b), Mozambique (Stal 2011), and Kenya and Uganda (Gray and Mueller 2012). Similar to these works focusing on drought, such studies highlight the importance of migration as a means for dealing with the reduction in incomes generated by worsening environmental conditions.

While these works thus make the case for viewing migration as a response to environmental stress, many of them similarly point out that migration is but one of a host of livelihood responses undertaken by households during a drought (van Apeldoorn 1981; Mortimore 1989). It has been noted that other strategies might be enacted concurrently, or instead of migration, which can be an undesirable strategy because it is too costly to undertake and thus too risky to be incorporated into the existing portfolio of livelihood strategies (Hampshire 2002). Other livelihood strategies documented as responses to environmental stress include changes in fertility, nuptiality and family separation as responses to drought in Ethiopia (Kidane 1989); while engaging in off-farm labour (Ezra 2001; Ezra and Kiros 2001), selling cattle and calling upon community networks (Gray and Mueller 2012) have also been identified in Ethiopia. In addition to migration, it has been documented in Niger that people engage in savings cooperatives, work as hired labour and forage for wild foods in response to water stress, pollution and drought (Afifi 2011). Notably, Gray and Mueller (2012) have suggested that, like migration, these alternative livelihood strategies were potentially undermined by drought, which can affect whole community networks and distort the price of cattle, thereby undermining their effectiveness in securing livelihoods. As such, Gray and Mueller (2012) consider migration to constitute a highly contextual response to drought in Ethiopia.

There is disagreement in the literature about what non-migratory responses might manifest in a context of environmental stress. In this regard, Findley (1994) hypothesized that, because migration responses to drought were more about reducing the food burden on a household than they were about the economic benefit of alternative employment, a host of other strategies that reduce the food burden on the household might similarly be pursued. For example, Findley (1994) suggested that ceremonies such as marriage could be encouraged as they take members out of the household. This may be achieved by practices such as waiving dowry fees (Findley 1994). As a counter to this, however, Ezra (2001) found that marriage was being postponed and occurring at increasingly older ages during the drought years in Ethiopia,[6] with the number of marriages increasing in bumper years. Further complicating this picture, Pedersen (1995) found that migration was the dominant strategy. He documented this in the Malian Gourma by noting how census-type data showed no evidence of the major droughts of the 1970s and 1980s other than by an increase in mobility.

While it thus seems to be commonly appreciated that migration is but one of a host of livelihood responses to environmental stress, it needs to be noted that accounts of migration in a context of environmental stress tend to highlight the fact that it is not only environmental stress that is responsible for generating migration. In Ethiopia, land tenure, population pressure, family size and structure, crop varieties, soil conditions, household assets, availability of off-farm work, locations of educational and health facilities, exposure to social stresses (including conflict), and access to other natural resources (fuel and water) and agricultural services have all been identified as playing a role in shaping the conditions of livelihood (in)security, and thus in the propensity for people to move (Kidane 1989; Meze-Hausken 2000; Ezra 2001; Morrissey 2012b). In West Africa, Doevenspeck (2011) similarly identified population pressure and the availability of transport infrastructure as playing an important role in people's decision to move from the north of Benin to the south of the country under conditions of worsening soil quality. In the western Sahara, changes in agriculture, urbanization, efforts at sedentarization, the enforcement of political borders, a lack of facilities and insecurity contribute to people's decision to migrate to the refugee camps in the

[6] Note that Ezra (2001) suggested that part of this dynamic could have been the increasingly small land holdings in Ethiopia, which were reducing the number of children that people were wanting to have.

Polisario-controlled region of the western Sahara (Gila et al. 2011). Expanding on this, Gila et al. (2011) identified the prohibition on the sale of humanitarian goods in the refugee camps, support from non-governmental organizations for migration to Spain and the urge to inform the world about the plight of the people of the western Sahara as contributing, along with rainfall, to generating the migratory imperative to leave the camps. Finally, Afifi (2011) cited colonial policies and the droughts of 1913, which undermined cattle stocks, for creating a poverty trap that has institutionalized migration as a means for ensuring contemporary livelihoods in Niger.

While it is clear that migration is considered a viable response to environmental stress, but one that is neither exclusive nor solely attributable to that stress, it is also clear from empirical accounts that migration is generally an established livelihood strategy that (a) has been shaped by historical processes such as colonial taxes, the slave trade and forced labour schemes (de Haan et al. 2002; Afifi 2011; Doevenspeck 2011) and (b) constitutes a long-standing response to environmental stress in West Africa (van Apeldoorn 1981; Mortimore 1989; de Haan et al. 2002; Hampshire 2002), East Africa (Gray and Mueller 2012) and the Horn of Africa (Ezra 2001). In this regard, Mortimore (1989) pointed out that circular migration was in fact a privileged response to livelihood stress due to its flexibility, which means that it can account for fluctuations in both the rural economy and the urban economy.

A host of studies also document how established mobility strategies can be altered during conditions of significant environmental stress. For example, van Apeldoorn (1981) noted that the intensity of the droughts of the 1970s in Nigeria meant that the established pattern of single male migration had to be changed so that more people left, whole families moved together, itineraries became irregular, people travelled further and to new areas, they spent longer periods away, they moved more often to urban areas and increasingly they did not return. Further, he documented that such migration became more about survival than about economic betterment (1981). Findley (1994) similarly found that the droughts of the 1980s in Mali meant that migration was internal and away from established international destinations, which require greater capital. Similarly, in northern Sudan, the droughts of the 1980s were so severe that people's established migration strategies were not sufficient and they had to seek support in the relief camps (Haug 2002), while the same droughts in Burkina Faso resulted in a novel form of migration among the Fulani, which was remarkable for the fact that it was directed towards cities outside of the country and completely removed people from the sphere of agropastoral livelihoods (Hampshire 2002). As an extreme example of mobility constituting a novel response to environmental stress, Stal (2011) found that most migrants in the relief centres set up in Mozambique in response to the 2008 flooding of the Zambezi River described this movement as the first time they had had to leave their places of established residence.

Identifying that migration strategies can change during times of significant environmental stress is really only a smaller piece of a larger phenomenon whereby migration forms a dynamically responsive process to changing socio-economic conditions in the sending and receiving areas. The clearest evidence of this process of identifying migration documented in a context of environmental stress comes from Niger, where migrant destinations have shifted from Nigeria and Ghana to Côte d'Ivoire and then Libya, following the changing political and economic fortunes in these countries (Afifi 2011).

If Niger captures the longer term changing dynamics of migrants, empirical work on the impacts of environmental stress on migration also illustrates that migrants' decisions about their destinations are contingent upon shorter time scales as well. For example, Mortimore (1989) found that migration decisions in the Sahel were directed away from traditional destinations during drought and were instead oriented towards areas where work and alms were perceived to be available. In northern Nigeria, this process was evidenced in the Islamic town of Kano, which turned into an important receiving area due to the Islamic tradition of giving alms. Similarly, in the Malian Gourma, migration was directed towards the town of Gossie during the droughts of the 1970s and 1980s because the town had recently gained importance as a market town and because Norwegian Church Aid was operating a food distribution centre there (Pedersen 1995). In a more generalized sense, de Haan et al. (2002) noted that migration from two Malian villages experiencing environmental stress was altered from established destinations to other locations based on perceptions of the conditions in different receiving areas.

Migration thus appears as one of any number of responses to a host of environmental and other livelihood stresses. It is an established yet dynamic response that appears to be context specific. In this regard, an important outcome of empirical work on the environment–migration nexus has been an appreciation of those individuals and groups who do not migrate and of the factors that distinguish such groups. Dealing with the dynamics of environmental stress, the most marked account of those who do not migrate comes from Haug (2002), who explained that, while approximately 20,000 people left their areas of regular residence in response to the droughts in the 1980s in northern Sudan, approximately 6,000 people remained behind. While explanations of who migrates and who does not have tended to cluster around gender, class and ethnicity (discussed below), a host of barriers to migration have been identified in the empirical literature looking at the impacts of environmental stress.

Significant barriers to migration have been identified as a lack of education in Botswana (Sporton et al. 1999), Ethiopia (Morrissey 2012b), Kenya and Uganda (Gray 2011); attachment to land in Ethiopia (Morrissey 2012b) and Niger (Afifi 2011); levels of market integration in Kenya and Uganda (Gray 2011); and the availability of food aid and microcredit, as well as the high costs of urban living and the saturated urban job market, in Ethiopia (Morrissey 2012b). In addition to such social structures, Barrios et al. (2006) identified legislative barriers to migration when they noted how rural–urban migration as a response to rainfall stress across the continent was historically constrained by colonial restrictions on movement. In Botswana, physical barriers to movement have been identified in the form of fences on ranch land, which limited movement in response to drought and thereby undermined short-distance moves back to the ranch land from which people had been expelled (Sporton et al. 1999).

Regarding barriers to migration, it is also worth pointing out that climate change might also act to undermine existing barriers to migration. For example, Morrissey (2012b) found that, in the highlands of Ethiopia, concerns about exposure to malaria in the lowlands inhibited mobility from high to lowland areas, where frost stress was not a problem and both rural and urban work were more available. As climate change is expected to increase the range of malaria into the highlands, such barriers to migration could be removed as resistance to malaria is developed in the currently malaria-free highland areas of Ethiopia. Thus, climate change might not only act to encourage mobility by compromising livelihoods, it might also alter barriers to migration.

Keeping the above barriers to migration in mind, literature on the environment–migration nexus has frequently focused on gender, class and ethnicity as the major barriers. Regarding gender, the literature is conflicting, with empirical accounts documenting instances in which both men and women are involved in migration (Findley 1994; Haug 2002; de Haan et al. 2002; Morrissey 2012b) and in which migration remains a male-dominated undertaking (Sporton et al. 1999; Hampshire 2002; Afifi 2011; Gray and Mueller 2012). Notably, in cases where men and women both migrate it has been increasingly recognized that they do so in a particular fashion, engaging in distinct work and having particular intentions. Making this point, Haug (2002) found that, while men responding to the droughts of the 1980s in northern Sudan would work anywhere, including in relief camps and in the Gulf States, women tended to focus on domestic labour in Sudanese towns.

In their study focused on highlighting the agency of migrants responding to environmental stress in Mali, de Haan et al. (2002) offered a more nuanced account of the gendered components of migration. They argued that, since much of migration in a context of environmental stress was undertaken with the intention of maintaining the institution of the household, men and women, with different roles in the household, migrated for different reasons. For men, a dominant feature of this process is the fact that migration allowed one to attain material and social status without having to leave the household completely. For women, a dominant component of the decision to migrate is the desire to attain the wealth necessary for marriage. As such, the places to which men and women go and the type of work they do differ. A related account comes from Ghana, where Carr (2005) described how mobility in a context of worsening rainfall and dwindling forest resources was fundamentally about maintaining and contesting gendered access to, and control over, the resources that are altered by changes in the biophysical environment. As such, Carr (2005) argued that a full understanding of mobility decisions taken in a context of environmental stress required an appreciation that migration was as much about power within the household as it was about survival, and as such it should be expected to have a gendered component.

Regarding class, it has been noted that nobles and slaves had different propensities to migrate in response to the droughts of the 1970s and 1980s in the Malian Gourma (Pedersen 1995). Similarly, in Burkina Faso (Hampshire 2002), Ethiopia (Ezra 2001; Gray and Mueller 2012), Kenya and Uganda (Gray 2011), household wealth, specifically stocks of cattle (Hampshire 2002; Gray and Mueller 2012) and the availability of labour (Ezra 2001), were found to be important in differentiating who migrated from who did not. In this regard, Hampshire (2002) described migration in a context of environmental stress as being the purview of the relatively wealthy rather than an act undertaken by those who are intolerably vulnerable. Again, however, the literature is potentially contradictory, with Meze-Hausken (2000) finding that, in Ethiopia once the major droughts had set in, the propensity to migrate was equal across households, irrespective of wealth.

In terms of ethnicity, the Tuareg, Shonghay and Fulani groups in the Malian Gourma were each found to have a different propensity to migrate in response to the droughts of the 1970s and 1980s (Pedersen 1995). Such differences are potentially explained by other findings in Mali that suggest that the propensity to migrate in response to environmental stress is structured around different livelihoods, which overlap with ethnicity (de Haan et al. 2002). A further component of ethnicity in migration dynamics, however, is observed in Niger, where Afifi (2011) documented how Tuareg groups distinguished migration to Libya, which connoted social respect, from migration to the south of the country, which was perceived as being about survival.

In addition to these social factors that appear to shape mobility in a context of environmental stress, other factors, operating on a more individual scale, have also been found to discern people's propensity to move. Age and access to networks in Burkina Faso (Hampshire 2002), as well as marital status and household size (Gray 2011) in Kenya and Uganda, have all been documented as important factors in determining who migrates. In Ethiopia, Gray and Mueller (2012) found there were lower rates of migration in households with children and among children of the household head. There were greater rates in larger households. Women's mobility increased when there were problems in attaining agricultural inputs and was higher among women living in female-headed households (Gray and Mueller 2012). Finally, Morrissey (2012b) argued that such individual-specific features were central to distinguishing who migrates and who does not in Ethiopia. This is because, although rural areas are in a state of decline, which is exacerbated by environmental stress, urban areas are not a panacea to these problems. As such, for whom migration is a viable livelihood strategy is highly context specific, depending on the individual's relative position within the power and material structures of society (Morrissey 2012b).

While the above discussion somewhat explains why people do not migrate, an important contribution from the empirical literature on environmental stress and migration is the identification of the important role played by migrant networks in facilitating migration in a context of environmental stress. In this regard, kin groups have been found to be very important for ensuring the well-being of migrants in receiving areas. This has been found to be the case even for international moves (van Apeldoorn 1981) and in relief centres (Haug 2002).

In addition to simply affecting migrant welfare, migrant networks also seem to be centrally important in determining where migrants go. In the Kalahari region of Botswana, Sporton et al. (1999) found that migrants responding to a combination of drought and structural changes in land ownership did not move in a haphazard or speculative fashion. Instead, choices of destination were informed by migrant networks comprising of kin groups that had invited migrants to come to certain areas. The clearest account of the importance of migrant networks, however, comes from Doevenspeck (2011), who reported that 71 % of the migrants sampled in central Benin said that they were related to someone in the settlement in which they had arrived.

While much of the literature described above justifies the international concern that has arisen around the environment–migration nexus, it should be pointed out that a number of studies have also found that migration as a response to environmental stress was frequently temporary and its impacts could be transient. In this regard, in northern Nigeria, van Apeldoorn (1981) noted that, unlike the refugees from Mauritania and Mali, those migrants responding to the droughts of the 1970s left no discernible mark on the country. Similarly, Mortimore (1989) noted that, although the droughts of the 1970s were linked with 2 years of harvest failure and several years of food shortages in the western Sahel, there was relatively little permanent redistribution of people as a result. Haug (2002) came to similar conclusions, finding that most migrants in northern Sudan returned to their place of established residence when the drought came to an end. Notable in Haug's (2002) account, however, is her argument that external assistance and secure land tenure were important in facilitating return. Among those who did not return, the important determinants were a lack of assets to make the move, a lack of secure tenure in the area of origin and successful livelihood establishment – and in particular having children in school – in the receiving area (Haug 2002). Finally, Stal (2011) found that people were leaving the resettlement centres set up in Mozambique in the wake of the 2008 floods of the Zambezi River and returning to their lands as the resettlement centres were prone to drought.

A final feature of the environment–migration nexus that is apparent in the empirical literature and that is worth discussing here pertains to the extent to which migration in a context of environmental stress constitutes an adaptive or maladaptive response to environmental stress. In this regard, a convincing case for the degree to which migration should be considered as adaptation comes from Pedersen's (1995) demographic study of the droughts of the 1970s and 1980s, which found that other than possibly reducing fertility, the only discernible sign of the droughts in the Malian Gourma was increased mobility. Such an account suggests that migration constitutes an effective means for reducing drought-related mortality.

Making the case for migration as adaptation more theoretically explicit, Scheffran et al. (2012) pointed to the important role played by remittances in dealing with climate stress, as well as the positive externalities surrounding the knowledge and technology transfers that migrants generate. To this end, Scheffran et al. (2012) have distinguished three relationships between adaptation and migration: (a) adaptation as preventing migration, (b) migration as adaptation and (c) migration for adaptation. While the first of these is self-explanatory, the third is distinguished from the second in terms of the degree to which migration forms either the means or the ends of adaptation. To make this clear, Scheffran et al. (2012) have cited migration in response to the droughts and longer term drying of the Sahel as exemplary of migration as adaptation. With regard to migration for adaptation, they highlighted the role of international migrant remittances in development initiatives in the sending region. These strategies constitute migration for adaptation as they are thought to be effective in securing livelihoods under conditions of subsequent environmental stresses (Scheffran et al. 2012).

While Scheffran et al. (2012) have made the theoretical case for migration for adaptation, empirical accounts offer less clarity on the role played by remittances. For example, Findley (1994) pointed out that, although the droughts of the 1980s had led to an increase in internal migration and a decrease in international migration, those households with international migrants at the time of the drought became more reliant on remittances from them. Undermining a simple account of this process, however, Findley (1994) argued that migration in response to the droughts was oriented more at reducing the food burden on the household than at attaining remittances. In Botswana, Sporton et al. (1999) provided a similar account, noting that, although remittances were important to communities in the Kalahari, migration during the drought did not contribute a large component of those remittances.

Potentially further undermining the notion of migration as adaptation are studies highlighting that drought undermines the ability to migrate altogether, due to the manner in which it undermines the necessary capital stocks (Findley 1994; Gray 2011; Gray and Mueller 2012). Lastly, a more concerning account from Niger found that outmigration as a response to a worsening environment for established agricultural activities was undermining efforts at environmental rehabilitation in the sending areas (such as rainwater harvesting) due to the lack of available labour (Afifi 2011).

Having detailed the dominant components of the environment–migration nexus as illuminated by studies of analogous stresses, the discussion now turns to accounts that have been based on longer term observations of environmental change and their impacts on communities affected by such change. Notably, these sorts of accounts offer a relatively inexpensive means for attaining similar results to those that can be achieved through longitudinal studies.

Environmental Change and Migration

Studies exploring the impacts of longer term changes in the biophysical environment on mobility reveal the complex ways in which climate change might alter mobility and demography in sub-Saharan Africa. Three studies, all concerned with the drying of the Sahel over the latter half of the last century, stand out in this regard. The first explored the plight of nomadic Kel Adrar, a community of ethnic Tuareg, in northern Mali (Swift 1973). The second examined the fortunes of the Fulbe, also a pastoralist group in Mali (de Bruijn and van Dijk 2003). While the third described the southerly expansion of pastoralist activity into the humid Sudanic region of the Sahel (Bassett and Turner 2007).

Detailing the changing patterns of movement among the Kel Adrar, Swift's (1973) findings are in accordance with those of more contemporary works, noting that migration constituted an established livelihood strategy central for ensuring livelihood security in a context characterized by biophysical uncertainty. Similar to the work already discussed, Swift (1973) distinguished regular nomadic movements in response to rainfall variability from two other kinds of movement enacted by the Kel Adrar: (a) under conditions of acute environmental stress; and (b) in response to a longer term worsening of the rainfall and resultant conditions of water stress. The former of these is characterized by longer distance migrations, while the latter involves a more permanent and southerly shift of the Kel Adrar. Swift's (1973) account is further in line with the findings of the works already described in this chapter highlighting the importance of migrant networks, which take the form of a loose set of political alliances between major confederations of nomadic groups. Finally, Swift's (1973) study identified that mobility was but one of a number of livelihood strategies employed to deal with environmental stress. Other strategies include keeping larger herds of animals than subsistence requires to allow for barter or slaughter, renting animals for milk, sharing animals and hunting wild animals (Swift 1973).

Based on 15 years of engagement in the Sahel, de Bruijn and van Dijk (2003), looking at the case of Fulbe migrants in the Dogon-inhabited Bandiagara escarpment in southern Mali, identified a complex process by which the droughts of the 1980s shaped mobility responses. The authors described how the droughts of the 1970s and 1980s forced the Fulbe, in the north of the country, to sell their cattle to wealthier sedentary Dogon farmers, in the south. This generated a demand for livestock herders among the Dogon in the south, which attracted young Fulbe men who were looking for both opportunity and work in order to make up for the livelihood losses caused by the drought. The loss of labour in the Fulbe-sending communities, as well as the poor wages in the south, meant the Fulbe-sending regions suffered further. As such, the sending regions experienced another wave of outmigration as those who had remained behind headed for the cities in search of work. In the Dogon regions of the south, problems are made worse as recent worsening terms of trade and pressure on land undermine the sedentary livelihoods on which the migrants are reliant. De Bruin and van Dijk (2003) undermined the notion of migration as adaptation or migration for adaptation, describing groups of Fulbe who are now scattered across the country as inhabiting positions at the social and economic margins.[7]

The account by de Bruijn and van Dijk (2003) clearly highlighted that migration was not a new or exceptional phenomenon but rather one that was part of several flexible livelihood responses enacted in times of environmental stress. They reiterated that environmental change and stress were among a number of factors shaping an increasing trend in mobility, including competition for land, urbanization, uneven economic development and economic depression (de Bruijn and van Dijk 2003). Lastly, the authors highlighted the dynamic nature of migration,

[7]The description of the situation in the Dogon regions is from the time of writing of the study by de Bruijn and van Dijk.

noting that, although migration was an established livelihood strategy among the Fulbe, during the last century the motivations for that movement had changed from being principally political to being economic and ecological (de Bruijn and van Dijk 2003).

Finally, Basset and Turner (2007) offered a longer term account of the environmental dynamics of mobility in the Sahel, describing the expansion of pastoralist activity into the humid Sudanic region that had previously been avoided due to the prevalence of the tsetse fly. In their account, Basset and Turner (2007) critiqued the characterization of this process as a sudden shift of pastoralism to the south after the droughts of 1973 and 1984 compromised livelihoods in the north, while desiccation reduced exposure to the tsetse fly in the Sudanic region of the Sahel. Instead, they provided an account whereby this movement was shaped by environmental change, but in which it was characterized more by a "migratory drift" than a sudden shift. In their study, this process of migratory drift on the part of pastoralists included a number of preparatory strategies, such as undertaking reconnaissance on potential rangelands and cross-breeding of animals with hardier southern herds (Bassett and Turner 2007). Similarly to de Bruijn and van Dijk (2003), Basset and Turner (2007) also described the process by which animals were transferred to the south through sale to sedentary farmers. Notably, Basset and Turner (2007) argued that the migration they described would not have been possible if it had been undertaken as a sudden shift and, as such, accounts of migration in response to environmental change should consider that short-term migrations might be part of a larger "drift in population" (Bassett and Turner 2007).

Thus, it seems that studies of longer term changes in the biophysical environment confirm that migration is an established strategy for dealing with environmental stress. They also confirm that it is a flexible one that can be altered in response to more severe changes in the environment. Longer term studies further reveal the complexity of the processes by which mobility is enacted, highlighting the manner in which market forces and iterative mobility strategies need to be taken into consideration in making sense of the environment–migration nexus. Finally, such studies suggest that, as much as migration might constitute an effective means of adaptation to climate stress, migration might also end with migrants finding themselves in conditions of significant marginalization.

This chapter now moves on to describe a third type of study: that which focuses on environmental endowments. As mentioned above, such studies are meaningfully distinct from those looking at environmental stress in that they treat the environment as static and investigate the hypothesis that migration might be undertaken in the direction of better-endowed environments.

Environmental Endowments and Migration

Works looking at the relationship between environmental endowments and migration seek to test whether an environmental component of migration is manifest through migrants' choices to move away from marginal environments in the direction of better-endowed areas. These works draw loosely on the neoclassical understanding of migration by suggesting that environmental quality drives mobility, where differential environmental quality has been substituted for wage differentials. As mentioned above, while this is not a reason to disregard this research,[8] if one tests the hypothesis that migration has an environmental component, one should exercise similar caution when interpreting these findings. This is similar to the way contemporary migration theory appreciates the importance of wage differentials, without arguing that they lie at the heart of migration decisions, or that they might explain all movement. Thus, favourable environmental conditions might be considered to be important to mobility without being determinant of it.

Using statistical methods to explore the role of environmental endowments in Burkina Faso, Henry et al. (2003) explored the relative influence of land degradation, land availability and climatic variability on mobility decisions. These factors were compared with a host of sociodemographic factors, including the percentage of the population who are male, literacy and economic activity rates, and the presence of a resettlement policy. The results revealed that sociodemographic features offered slightly more explanatory power, but that both they and environmental factors were important for explaining mobility decisions. Interestingly, the study found that the explanatory power of environmental variables was greater in provinces with worse measures of environmental quality (Henry et al. 2003). As such, the authors suggested that, while the simple common-sense models have overstated the strength of the relationship between the biophysical environment and migration, predictive and explanatory works on migration should start to incorporate ecological variables.

Expanding on this approach, Henry et al. (2004a) conducted a study in Burkina Faso in an attempt to determine whether migrants would leave a place with an unfavourable environment more quickly than they would an area with a favourable environment, and whether they would have a greater propensity to migrate to areas with favourable environmental conditions than they would to marginal areas. Notably, in light of previous findings, this work found that fewer people migrated out of areas with unfavourable climatic conditions when compared with those with favourable ones. This was explained in terms similar to those described in studies on environmental stress: households working in unfavourable environments were less able to attain the capital necessary for investment in migration. Supportive of this finding was the fact that higher percentages of the migrants leaving favourable areas travelled to foreign destinations (mainly Côte d'Ivoire). Also notable in this work is that land degradation was found to be a more compelling migratory force than simply unfavourable climatic conditions. Finally, highlighting a potentially iterative component of migration, the work found that individuals were more likely to move

[8]The value of these studies is that they may reveal an ecological component of mobility, even though the environment is never truly static. In addition, such studies may be revealing of an environmental component in migration as they highlight how environmental stress has different implications for mobility when it occurs in well-endowed, as opposed to marginal, environments.

again after moving from their original place of residence. This was thought to be the result of such migrations not being retarded by attachments to land (Henry et al. 2004a).

Blurring the distinction between endowments and stress somewhat, a study by Henry et al. (2004b), also in Burkina Faso, explored the impacts of rainfall stress (used as a proxy for household food production), land availability and road access on the first decision to leave an area.[9] The study found that the impacts of poor rainfall were small and insignificant on mobility decisions. Instead, individual factors, such as level of education, ethnic group, livelihood type (cattle rearing or not) or gender, have a greater impact (Henry et al. 2004b). Such factors were found to be important in determining not only who moves but also where people go and with what intentions. For example, men were found to move for economic reasons while women moved for reasons of marriage. Levels of education also had different impacts among men and women, and also impacted on the migrant's likely choice of destination (rural or urban) (Henry et al. 2004b).

Finally, working in Ghana, van der Geest et al. (2010) used Normalized Difference Vegetation Index (NDVI) data, census data and rainfall data to explore the environmental dimensions of migration. Their work described three major migration routes in Ghana: from the densely populated southerly regions to Accra; and from the densely populated northern regions to either the sparsely populated central region or the western cocoa frontier. Their findings suggested that the two latter movements had significant environmental dimensions, which were also shaped by population density (van der Geest et al. 2010). In Ghana, outmigration dominates in the dry and sparsely vegetated northern regions, while in-migration dominates the more humid central and western regions, which have higher NDVI values. Notably, in-migration has an inverse relationship with NDVI, while outmigration is positively linked to vegetative cover. As such, the authors concluded that migration to the central region was undertaken with the aim of attaining "greener pastures," but noted that this relationship was also shaped by historic colonial exploitation of the north. Movement is thus thought to be driven by environmental "push" factors but not "pull" factors – as there are areas with higher NDVI values that don't attract migrants. The dominant pull factor in this movement is thought to be the low population densities in the centre of the country and the nature of land tenure there. Migration to the cocoa frontier shows migration's dynamic character, responding to the establishment of the cocoa industry in the 1980s. Again, it is the existence of low population densities that allows for settlement in this region and that accounts for reductions in NDVI.

Therefore, works on environmental endowments and mobility appear to confirm, and even clarify, the relative roles played by environmental and non-environmental factors in mobility decisions. This certainly is of relevance to understanding the environment–migration nexus; however, in terms of understanding what this means

[9]By "first decision" the authors mean the first decision made by an individual living in the place of their birth to leave that area.

for migration in response to environmental stress or change, and how mobility might manifest under conditions of future climate change, the relevance of these results remains unclear.

This chapter turns now to a discussion on the fourth and final body of work exploring the relationship between environmental stress and human mobility: agent-based modelling. As mentioned above, a distinguishing feature of this approach, compared to the crude early models, is that its focus is on understanding the interaction between features in the model, and not on providing predictions of the future numbers of people who will move as a result of climate change.

Agent-Based Modelling

Moving to attempts at building an agent-based model of migration in a context of environmental change, in Burkina Faso, Kniveton et al. (2011) built a model in which migration is a product of both the attributes of the location in which the potential migrants find themselves and the social network in which they exist. The accuracy of the model was tested through hindcasting,[10] which produced a correlation coefficient of 0.93. The model was then run using four different scenarios with different narratives for demographic, economic, social, political and environmental change, running to 2060. From this account, the model suggests that the largest total and international migration from Burkina Faso will result under a future climate that is dry and combined with low demographic growth as well as inclusive and connected social and political governance (Kniveton et al. 2011). The lowest total migration flows take place in a future that is characterized by high demographic growth, low economic growth, and exclusive and diverse social and political governance. Notably, this scenario holds irrespective of climate change. Finally, the lowest rates of international migration are produced by future scenarios characterized by a move to a wetter climate with high demographic growth, as well as exclusive and diverse social and political governance. This last scenario occurs irrespective of economic growth. Significant in this model is the fact that different contexts result in migration decisions being more or less impacted by environmental change (Kniveton et al. 2011).

In a similar study, Kniveton et al. (2012) constructed an agent-based model, also based on empirical studies from Burkina Faso, highlighting the non-linear relationships between environmental stress, human migration and population growth. Again the model took into consideration a host of non-environmental features, including microscale capitals and social interactions, as well as economic and demographic changes. Notably, this work did not focus on distress moves, but instead on circular migration and established empirics for determining individual propensities to move. The model was hindcast over 25 years to determine its accuracy. The findings of the model were that, under a climate scenario tending towards a wetter climate,

[10]For the period 1970 to 1999.

migration falls relative to what could be expected under the contemporary climate. A drier climate results in model outputs that show slightly higher rates of migration relative to those that would be produced under current climatic conditions (Kniveton et al. 2012). However, these changes in migration with drying and wetting are increased when population growth is incorporated into the model. As such, the model shows that demography can combine with the impact of climate on individual behaviour to produce non-linear and emergent changes in total migration that are not apparent when only climate change is considered (Kniveton et al. 2012).

While the findings of agent-based models are certainly interesting, and while it seems that agent-based modelling is an important frontier for understanding the impacts of future environmental (and other) changes for mobility, the models currently suffer from treating individual preferences for migration as a black box. In this regard, individual preferences are determined empirically (for example, men are more likely to migrate than women, young people are more likely to migrate than older people) and then placed in the model. In addition, the models currently do not incorporate institutions of power – describing propensities to migrate simply in terms of such probabilities without exploring the structures informing such probabilities (such as gender or the household). Given the contradictory findings around these empirics (which were described above) and the degree to which empirical work shows the decision to migrate as being highly individualized and rooted in dynamic power structures, it seems important to be wary of extrapolating these findings too liberally and to refrain from interpreting them as predictions.

Having outlined and discussed the empirical findings on migration in response to environmental stress in sub-Saharan Africa, the discussion now turns to the context in which this migration is taking place through a description of some contemporary migration trends in sub-Saharan Africa.

Migration Trends in Sub-Saharan Africa

In attempting to provide some context for the dominant migration processes in which mobility strategies might be enacted in a context of environmental stress, it should be noted that neither intercontinental migration nor conflict-induced displacement is discussed here. This is because the bulk of evidence described above suggests that the dominant forms of mobility will comprise cyclical, cross-border, rural–rural and rural–urban movements. In this regard, it is possible that migration might operate in a stepwise fashion so that migrants moving from rural areas to towns replace workers who have moved abroad, before moving abroad themselves. There is, however, no empirical work documenting such a relationship, and as such it is not thought important to include as a discussion here.

Looking at the literature documenting the dominant migration trends in sub-Saharan Africa, one cannot but be struck by the degree to which it mirrors the accounts provided by the literature on environmental change and migration. In this regard, most migration literature insists that an understanding of contemporary mobility in sub-Saharan Africa requires some contextualization in the historical processes that have shaped mobility. This is particularly the case in West Africa, where colonialism is identified as interrupting patterns of mobility through the establishment of political and economic structures, most prominently taxes and national borders. In addition, the colonial State is identified with shaping migration through its forced labour policies aimed at completing colonial infrastructure projects (Adepoju 2005; IOM 2005; Jonsson 2009).

While the literature confirms the important role played by historical structures, it also highlights the dynamic nature of migration, which is shown to shift in response to changing political and economic fortunes. Most notably, these have included the increasing prominence of South Africa and Botswana as destinations, which is a result of the ending of apartheid in the former and rapid and stable economic growth in the latter, and a shift in the choice of destination in West Africa from Nigeria and Ghana towards Côte d'Ivoire and then Libya and the Gulf States (Adepoju 2000).

Further to this trend, and resonating with accounts that migration under conditions of environmental stress is increasingly varied in terms of extent, direction and duration, contemporary research on migration in sub-Saharan Africa reveals that "there are more people moving into and within the African continent; that the trajectories of migrants in Africa are less straight forward than they were in the past; and that an increased variety of actors and places of departure and destination are involved in African migration" (Jonsson 2009:2).

While contemporary literature highlights the responsive nature of migration in sub-Saharan Africa, it also highlights its somewhat paradoxically transient character. For example, although the impact of the global financial crisis in 2008 was expected to have significant effects on migrant remittances and therefore on migration, it appears that the impact was relatively short-lived, resulting in only "a modest reduction in remittance flows ... [which] ... increased again in 2010 ... [to reach] ... the level recorded in 2008" (IOM 2011:62). Similarly, while the conflict in Côte d'Ivoire was thought to have a major impact on outmigration (IOM 2011), Shimeles (2010) noted that it remained the major destination for Malian migrants despite its recent troubles.[11]

Finally, in a manner similar to that of the literature on environmental change, contemporary migration literature highlights the increasing feminization of labour migration in sub-Saharan Africa (IOM 2000, 2005; Adepoju 2000, 2005; Jonsson 2009), although it notes that women were still underrepresented in labour migrant populations (IOM 2005). Explanations for this trend have been offered in terms of a worsening economic trend in sending countries (Adepoju 2000) and the impact of structural adjustment programmes. These programmes have affected the formal sector to the relative detriment of men's employment opportunities, while leaving opportunities in the informal sector, where women tend to cluster, relatively unaffected (IOM 2005).

[11]Note that this account is referring to the "recent troubles" in 2003, not 2011. Given the date of Shimeles' (2010) paper, this situation may no longer be the case.

One focus of contemporary migration literature that is missing in writings on environmental change and migration is the important role played by regional political blocks and their negotiations regarding freedom of movement. In this regard, the Economic Community of West African States agreement on free movement and the East African Community protocol on the common market are considered important political contexts for contemporary migration (IOM 2011). Similar sentiment is directed towards the relaxing of visas within the Common Market for Eastern and Southern Africa and the increasing openness of the Southern African Development Community to the free movement of people and goods in the region (IOM 2000, 2011).

With the above in mind, it is finally worth mentioning the dominant migration trends on the continent. Movement in sub-Saharan Africa continues to be dominated by movement that is across borders and circular. It also involves an important component of unregistered movement (IOM 2008, 2011), with rural–urban migration accounting for a significant component (IOM 2005). Major receiving countries in West Africa are Burkina Faso (IOM 2011) and Côte d'Ivoire (Shimeles 2010). In East Africa, they are Kenya (IOM 2011) and the Sudan (Shimeles 2010), while in Southern Africa, South Africa (IOM 2000; Adepoju 2000, 2005; Shimeles 2010) and Botswana (IOM 2008) dominate.

Notwithstanding the above, two contemporary points of caution are worth highlighting. The first comes from de Haas (2011), who has warned against focusing on push factors in determining migration decisions. He has suggested, instead, that a more compelling determinant of migration lay in the political and economic conditions of potential receiving States. In this regard, he has warned against simply assuming that increased environmental stress would lead to increased outmigration (de Haas 2011). The second cautionary note comes from Parnell and Walawege (2011), who have argued that a focus on rural–urban migration as the determining feature of the future African city was misplaced. This is because rates of natural increase account, and will account, for a greater component of African urbanization than will rural–urban migration. As such, they warn against focusing on environmental stress and rural–urban migration at the expense of other urbanizing forces that could be crucial in making African cities places of acute vulnerability to future climate stress (Parnell and Walawege 2011).

Conclusion

The studies described above suggest that climate change would have significant impacts on livelihood security on the subcontinent, and that, despite claims to the contrary, there was in fact a substantial and growing body of evidence of the links between human migration and environmental change/stress. Conclusions from the works described in this chapter make it clear that mobility forms an important response to environmental change and stress, but that, at the same time, migration is an established strategy that is but one of a host of livelihood options. Furthermore, migration is a dynamic phenomenon that is shaped by history, responsive to the changing social and economic context in which it occurs, and frequently transient in terms of its impacts. Regarding the dynamic nature of migration, it must be appreciated that changes in livelihoods resulting from environmental stress can generate novel forms of mobility and might even result in mobility strategies being undertaken in communities and among individuals where it had not occurred previously. Such issues should be of particular note given the potential for non-linearity in the relationship between environmental stress, sociopolitical and economic forces, and migration.

From the above studies, it is also clear that migration is more than the outcome of poverty or intolerable vulnerability. Instead, mobility appears to manifest as a highly strategic response to which significant barriers operate. As such, mobility appears to be a highly context-specific response, which is shaped by a host of institutions operating at a variety of scales. In this regard, a clear message from the empirical work on the subject is that there are major barriers to migration and that these take both generic forms (notably gender, class and ethnicity, as well as others) and idiosyncratic forms, which operate at the scale of the household/individual.

As such, it seems that understanding the demographic impacts of climate change will require a more nuanced conception than simply expecting people to leave their places of established residence to search for food under conditions of environmental stress. Instead, empirical work suggests that it would require an iterative account of the complex suite of environment–society dynamics through which institutions determine access to, and power over, material resources. This in turn will require a multiscalar account of the socio-economic and political context in which such features are embedded and by which they are stabilized and challenged. Finally, while some accounts of the impoverished and marginal state of migrant livelihoods do not bode well for hopes that migration will simply provide a means for adapting to climate change, other accounts describing migrant innovation and premeditation are a reminder of the dynamism with which mobility strategies can be enacted and of their potential importance as an adaptive strategy. In this regard, it should be noted that policy responses that focus solely on mobility as the end point of vulnerability could well prove to be misplaced as they ignore groups and individuals who remain behind or for whom mobility is not a viable livelihood option.

With the above in mind, it seems that a research agenda for understanding the relationship between environmental change and human migration in Africa cannot be based on single ideal-type studies. Instead, a thorough understanding of the phenomenon requires undertaking a variety of studies and approaches. Qualitative analogous approaches are vital for exploring and explaining the mechanisms and pathways that link environmental stress to decisions about human mobility, at multiple scales. Such approaches are the only way to reveal novel interactions between environmental stress and human mobility and their multiscalar nature. They can be effectively undertaken by looking at both singular moments of environmental stress or through longitudinal studies. The current dearth of the latter increases their relevance to contemporary knowledge, especially given the current interest in the degree to which migration forms an effective means for adapting to climate change.

Notably, this question can only be answered by looking at the impacts in both sending and receiving areas, and the impacts on the migrants themselves, in the long term.

All that being said, qualitative studies are limited in what they can say about the environment–migration nexus. Notably, they cannot reveal much about the relative role played by such stress in mobility decisions, or describe, with any rigour, how one area compares with another. This is because: such works are usually carried out using relatively small samples of respondents, a systematic approach cannot be applied to engaging with respondents and they are not based on random samples. Similarly, analogous approaches, being based on historical events, are limited in terms of what they can relate about future events. This is especially relevant given the likelihood of complex environment–society systems displaying non-linear behaviour.

As such, while analogous, qualitative approaches can relate what matters in mobility decisions, quantitative approaches are required in order to explain how much those features matter. In this regard, such approaches can be effectively informed on what to measure, and what might constitute meaningful proxies by a close reading of the exploratory qualitative work.

Finally, an important part of this portfolio of research lies in building agent-based models. Such models are likely the best way for exploring how environmental and non-environmental factors will interact under future conditions of environmental change. Given that existing work on agent-based models has identified non-linear relationships in these systems and the fact that complex systems tend to contain non-linear relationships, the value of agent-based models is further increased. That being said, black-box elements of such models – pertaining to the pathways by which environmental stress manifests in mobility and different social groups' propensities to migrate – mean that they need to be informed by robust bodies of knowledge constructed from well-designed analogous approaches of both a qualitative and quantitative nature. Further, determining means by which agent-based models can begin to account for social institutions as well as the relations of power they create (and which themselves create the institutions) would improve their insights significantly.

Having described this portfolio of research approaches, it is now worth mentioning that analogous approaches, of both a qualitative and quantitative nature, and agent-based models need to be generated in an iterative process with one another. While crude predictive models have been the starting point for a discussion on the relationship between environmental stress and human migration, such models have been refined through the provision of findings from qualitative and quantitative analogous studies. In this regard, qualitative findings have informed quantitative studies, and interesting findings in quantitative studies have been interrogated in subsequent qualitative approaches. These findings have then made their way into agent-based models, and the important relationships such models reveal should, in turn, be scrutinized in specially designed qualitative and quantitative analogous approaches, with the nuanced findings being fed back into such models.

References

Adepoju, A. (2000). Issues and recent trends in international migration in Sub-Saharan Africa. *International Social Science Journal, 52*(165), 383–394.

Adepoju, A. (2005). *Migration in West Africa.* Paper prepared for the Policy Analysis and Research Programme of the Global Commission on International Migration.

Afifi, T. (2011). Economic or environmental migration? The push factors in Niger. *International Migration, 49*(s1), e95–e124.

Barrios, S., et al. (2006). Climatic change and rural–urban migration: The case of Sub-Saharan Africa. *Journal of Urban Economics, 60*(3), 357–371.

Bassett, T., & Turner, M. (2007). Sudden shift or migratory drift? Fulbe herd movements to the Sudano–Guinean region of West Africa. *Human Ecology, 35*(1), 33–49.

Black, R., & Sessay, M. (1997). Forced migration, environmental change and woodfuel issues in the Senegal River Valley. *Environmental Conservation, 24*(3), 251–260.

Boko, M., et al. (2007). Africa. In M. Parry et al. (Eds.), *Climate change 2007: Impacts, adaptation and vulnerability – contribution of Working Group II to the fourth assessment report of the Intergovernmental Panel on Climate Change* (pp. 433–467). Cambridge: Cambridge University Press.

Carr, E. (2005). Placing the environment in migration: Environment, economy, and power in Ghana's Central Region. *Environment and Planning A, 37*(5), 925–946.

Castles, S. (2002). *Environmental change and forced migration: Making sense of the debate.* (New issues in Refugee Research, Working Paper No. 70). United Nations High Commissioner for Refugees, Geneva.

Christian Aid. (2007). *Human tide: The real migration crisis – a Christian Aid report.* London: Christian Aid.

Conisbee, M., & Simms, A. (2003). *Environmental refugees: The case for recognition.* London: New Economics Foundation.

de Bruijn, M., & van Dijk, H. (2003). Changing population mobility in West Africa: Fulbe pastoralists in central and south Mali. *African Affairs, 102*(407), 285–307.

de Haan, A., et al. (2002). Migration, livelihoods and institutions: Contrasting patterns of migration in Mali. *Journal of Development Studies, 38*(5), 37–58.

de Haas, H. (2011). Mediterranean migration futures: Patterns, drivers and scenarios. *Global Environmental Change, 21*(s1), s59–s69.

Doevenspeck, M. (2011). The thin line between choice and flight: Environment and migration in rural Benin. *International Migration, 49*(s1), e50–e68.

Ezra, M. (2001). Demographic responses to environmental stress in the drought- and famine-prone areas of northern Ethiopia. *International Journal of Population Geography, 7*(4), 259–279.

Ezra, M., & Kiros, G. (2001). Rural out-migration in the drought prone areas of Ethiopia: A multilevel analysis. *International Migration Review, 35*(3), 749–771.

Findley, S. (1994). Does drought increase migration? A study of migration from rural Mali during the 1983–1985 drought. *International Migration Review, 28*(3), 539–553.

Gila, O., et al. (2011). Western Sahara: Migration, exile and environment. *International Migration, 49*(s1), e146–e163.

Goodess, C. (2011). *How is the frequency, location and severity of extreme events likely to change up to 2060?* Paper prepared for the Foresight project: Migration and Global Environmental Change. United Kingdom Government Office for Science, London.

Gray, C. (2011). Soil quality and human migration in Kenya and Uganda. *Global Environmental Change, 21*(2), 421–430.

Gray, C., & Mueller, V. (2012). Drought and population mobility in rural Ethiopia. *World Development, 40*(1), 134–145.

Hampshire, K. (2002). Fulani on the move: Seasonal economic migration in the Sahel as a social process. *Journal of Development Studies, 38*(5), 15–36.

Haug, R. (2002). Forced migration, processes of return and livelihood construction among pastoralists in northern Sudan. *Disasters, 26*(1), 70–84.
Henry, S., et al. (2003). Modelling inter-provincial migration in Burkina Faso, West Africa: The role of socio-demographic and environmental factors. *Applied Geography, 23*(2–3), 115–136.
Henry, S., et al. (2004a). Descriptive analysis of the individual migratory pathways according to environmental typologies. *Population and Environment, 25*(5), 397–422.
Henry, S., et al. (2004b). The impact of rainfall on the first out-migration: A multi-level event-history analysis in Burkina Faso. *Population and Environment, 25*(5), 423–460.
Intergovernmental Panel on Climate Change (IPCC). (2000). In N. Nakicenovic & R. Swart (Eds.), *Special report on emissions scenarios*. Cambridge: Cambridge University Press.
Intergovernmental Panel on Climate Change (IPCC). (2012). In V. Fields et al. (Eds.), *Managing the risks of extreme events and disasters to advance climate change adaptation: A special report of Working Groups I and II of the Intergovernmental Panel on Climate Change*. Cambridge: Cambridge University Press.
International Organization for Migration (IOM). (2000). *World migration report*. New York: United Nations and IOM.
International Organization for Migration (IOM). (2005). *World migration report 2005: Costs and benefits of international migration*. Geneva: IOM.
International Organization for Migration (IOM). (2008). *World migration report 2008: Managing labour mobility in the evolving global economy*. Geneva: IOM.
International Organization for Migration (IOM). (2011). *World migration report 2011: Communicating effectively about migration*. Geneva: IOM.
Jacobson, J. (1988). *Environmental refugees: A yardstick of habitability* (Worldwatch Paper No. 86). Washington, DC: Worldwatch Institute.
Jones, P., & Thornton, P. (2003). The potential impacts of climate change on maize production in Africa and Latin America in 2055. *Global Environmental Change, 13*(1), 51–59.
Jonsson, G. (2009). *Comparative report: African migration trends. African perspectives on human mobility programme*. Oxford: International Migration Institute.
Jonsson, G. (2010). *The environmental factor in migration dynamics: A review of African case studies* (Working Paper No. 21). Oxford: International Migration Institute.
Kidane, A. (1989). Demographic consequences of the 1984–1985 Ethiopian famine. *Demography, 26*(3), 515–522.
Kniveton, D., et al. (2011). Agent-based model simulations of future changes in migration flows for Burkina Faso. *Global Environmental Change, 21*(s1), s34–s40.
Kniveton, D., et al. (2012). Emerging migration flows in a changing climate in dryland Africa. *Nature Climate Change, 2*(6), 444–447.
Kundzewicz, Z., et al. (2007). Freshwater resources and their management. In *Climate change 2007: Impacts, adaptation and vulnerability – Contribution of Working Group II to the fourth assessment report of the Intergovernmental panel on Climate Change*. Cambridge: Cambridge University Press.
Meze-Hausken, E. (2000). Migration caused by climate change: How vulnerable are people in dryland areas? *Mitigation and Adaptation Strategies for Global Change, 5*(4), 379–406.
Morrissey, J. (2009). *Environmental change and forced migration*. Background paper for the workshop on environmental change and displacement (pp. 1–49), Oxford.
Morrissey, J. (2012a). Rethinking the "debate on environmental refugees": From "maximilists and minimalists" to "proponents and critics. *Journal of Political Ecology, 19*, 36–49.
Morrissey, J. (2012b). Contextualising links between migration and environmental change in northern Ethiopia. In K. Hastrup & K. Fog Olwig (Eds.), *Climate change and human mobility: Global challenges to the social sciences*. Cambridge: Cambridge University Press.
Mortimore, M. (1989). *Adapting to drought: Farmers, famines and desertification in West Africa*. Cambridge: Cambridge University Press.
Mortreux, C., & Barnett, J. (2009). Climate change, migration and adaptation in Funafuti, Tuvalu. *Global Environmental Change, 19*(1), 105–112.

Myers, N. (2001). Environmental refugees: A growing phenomenon of the 21st century. *Philosophical Transactions of the Royal Society B: Biological Sciences, 357*(1420), 609–613.
Myers, N., & Kent, J. (1995). *Environmental exodus: An emergent crisis in the global arena.* Washington, DC: Climate Institute.
Nicholls, R., & Lowe, J. (2004). Benefits of mitigation of climate change for coastal areas. *Global Environmental Change, 14*(3), 229–244.
Nicholls, R., & Mimura, N. (1998). Regional issues raised by sea-level rise and their policy implications. *Climate Research, 11*, 5–18.
Parnell, S., & Walawege, R. (2011). Sub-Saharan African urbanisation and global environmental change. *Global Environmental Change, 21*(s1), s12–s20.
Pedersen, J. (1995). Drought, migration and population growth in the Sahel: The case of the Malian Gourma: 1900–1991. *Population Studies, 49*(1), 111–126.
Rajan, S. (2008). *Blue alert: Climate migrants in South Asia – Estimates and solutions*. Vancouver: Greenpeace.
Scheffran, J., et al. (2012). Migration as a contribution to resilience and innovation in climate adaptation: Social networks and co-development in Northwest Africa. *Applied Geography, 33*, 119–127.
Schlenker, W., & Lobell, D. (2010). Robust negative impacts of climate change on African agriculture. *Environmental Research Letters, 5*(1), 1–8.
Shimeles, A. (2010). *Migration patterns, trends and policy issues in Africa* (African Development Bank Group, Working Paper No. 119), Tunis.
Shongwe, M., et al. (2009). Projected changes in mean and extreme precipitation in Africa under global warming. Part I: Southern Africa. *Journal of Climate, 22*(13), 3819–3837.
Shongwe, M., et al. (2011). Projected changes in mean and extreme precipitation in Africa under global warming. Part II: East Africa. *Journal of Climate, 24*(14), 3718–3733.
Sporton, D., et al. (1999). Outcomes of social and environmental change in the Kalahari of Botswana: The role of migration. *Journal of Southern African Studies, 25*(3), 441–459.
Stal, M. (2011). Flooding and relocation: The Zambezi River Valley in Mozambique. *International Migration, 49*(s1), e125–e145.
Swift, J. (1973). Disaster and a Sahelian nomad economy. In D. Dalby & R. Church (Eds.), *Drought in Africa: Report of the 1973 symposium*. London: University of London.
Tacoli, C. (2009). Crisis or adaptation? Migration and climate change in a context of high mobility. *Environment and Urbanization, 21*(2), 513–525.
Tanser, F., et al. (2003). Potential effect of climate change on malaria transmission in Africa. *The Lancet, 362*(9398), 1792–1798.
Tebaldi, C., & Sansó, B. (2009). Joint projections of temperature and precipitation change from multiple climate models: A hierarchical Bayesian approach. *Journal of the Royal Statistical Society: Series A (Statistics in Society), 172*(1), 83–106.
Thornton, P., et al. (2010). Adapting to climate change: Agricultural system and household impacts in East Africa. *Agricultural Systems, 103*(2), 73–82.
van Apeldoorn, G. (1981). *Perspectives on drought and famine in Nigeria.* London: George Allen & Unwin.
van der Geest, K., et al. (2010). Migration and environment in Ghana: A cross-district analysis of human mobility and vegetation dynamics. *Environment and Urbanization, 22*(1), 107–123.
Wouterse, F., & van den Berg, M. (2004). Migration for survival or accumulation: Evidence from Burkina Faso. In *85th EAAE seminar: Agricultural development and rural poverty under globalization: Asymmetric processes and differentiated outcomes*. Wageningen.

Climate Change, Extreme Weather Events, and Migration: Review of the Literature for Five Arab Countries

5

Quentin Wodon, Nicholas Burger, Audra Grant, George Joseph, Andrea Liverani, and Olesya Tkacheva

Introduction

By the next century, global mean annual temperatures may increase by 3–5 °C, which could lead to dramatic negative effects, especially for developing countries (World Bank 2012). In the Arab world climate change may manifest itself through reduced rainfall, greater seasonal temperature variability, and a rise in the Mediterranean sea level, all of which constitute threats to agricultural production and economic security. Agriculture employs about half of the active population in many countries, especially among the poor. Declining precipitation is likely to affect availability and usage of water, causing agricultural productivity to decrease (e.g., UNDP 2009; World Bank 2011a, b). Climate change is also associated with a higher likelihood of extreme temperatures, floods, and droughts, and thereby with risks of substantial displacement after extreme weather events (e.g., IPCC 2012; Elasha 2010; McSweeney et al. 2009).

Estimates of the number of people likely to be affected by climate change and the extent to which they are likely to be affected are notoriously difficult to provide. But in the region as a whole, some 80–100 million people may be expected to experience water stress by 2025 (Warren et al. 2006).[1] By 2050 water availability per capita is expected to decline by 50 % (IFAD 2010). Furthermore,

The authors are with Rand and the World Bank. This work was prepared with funding from AFD (the Agence Française de Développement) and the Trust Fund for Environmentally and Socially Sustainable Development at the World Bank. The opinions expressed in the chapter are however only those of the authors, and need not represent those of AFD, Rand, or the World Bank.

[1]The authors define the population exposed to water stress as the population living in watersheds where water availability falls below 1,000 m^3 per capita per year.

Q. Wodon (✉) • N. Burger • A. Grant • G. Joseph • A. Liverani • O. Tkacheva
World Bank, Washington, DC, USA
e-mail: Qwodon@worldbank.org

E. Piguet and F. Laczko (eds.), *People on the Move in a Changing Climate*, Global Migration Issues 2, DOI 10.1007/978-94-007-6985-4_5,

while the literature is mixed on this issue, aside from the economic impacts climate change may be associated with local conflict as groups compete for access to scarce natural resources (Reuveny 2007; Nordås and Gleditsch 2007). Climate change also presents unique gender challenges. Most migrants in the region, whether internal or external, are male, so that women are left to assume the burden of increased workload at the place of origin. And for women who do migrate, job opportunities tend to be less attractive than for men.

This paper, which was prepared for a study by the World Bank, AFD, and Rand (Wodon et al. 2013) reviews the evidence on the link between climate change and migration in the Arab world. Migration is considered as one of several coping and adaptation mechanisms used by households (Foresight 2011). It responds to both push (from the area of origin) and pull (from the area of destination) factors. Changes in the climate (e.g. warming temperatures, heat waves, declining rain fall, and rising water levels) are part of push factors because they lead to a deterioration of the environment. Extreme climactic conditions, such as weather shocks like severe floods or droughts, can lead to temporary migration and displacement. Climate and subsequent environmental degradation decreases agricultural crop production, livestock or water availability, which adversely affects economic activity. By contrast pull factors are those conditions that attract migrants or potential migrants, mainly to urban areas. They also shape and guide human choice. Thus the existence of economic opportunities such as employment, well-established community networks that help reduce uncertainty and risk, and available housing may all be variables that pull migrants to a particular area. Schools, health care, electricity, clean water, functioning sewage systems and other infrastructural features inform decisions as well.

In migration models the environment is often considered as a push factor because of threats to household livelihood and the increase in uncertainty over payoffs in a rational actor model (see Kniveton et al. 2009 for a review). But more recent models suggest that the relationship between environment and migration is more nuanced than income-maximization models suggest because the environment interacts with other drivers of migration: the decision to migrate is filtered through such household characteristics as socio-economic status and political context (Black et al. 2011). Since there are many intervening variables, the effect of environmental change is not uniform across households, and household member-specific responses—including the type of migration—depend on socio-economic context (Carr 2005). For example, migration can be permanent or temporary, it can be undertaken by a single individual or by an entire household, over a long or short distance. Similarly, environmental impacts can take an acute form, such as flooding or sand storms, or occur gradually via deforestation or changes in temperature and precipitation. These various dimensions render the analysis of migration complex (Kniveton et al. 2009).

The complex temporal and spatial dimensions of migration are a challenge for data collection aiming to better understand the causes and consequences of migration. Adaptation assessment requires pre- and post- treatment data, but collecting panel survey data on migration is difficult because migration implies that respondents change their place of residence, making respondent tracking difficult.

Only a handful of authors have used panel data in their analysis (Pereira and Caravajal 2008; Gray and Mueller 2012; Bohra-Mishra and Massey 2011). One alternative involves surveying households about family members who moved to another location (Dillon et al. 2011). The caveat with this approach, however, is that the results on migration decisions by some members of the household may not generalize to the migration by the entire household. Households that decide to send only one person to a different location may be different from households that migrate together. Furthermore, migration by one or more household members is distinct from migration by the entire household as the latter harder to reverse and requires higher risk tolerance by household members (see Piguet 2010, on alternative methodological approaches). While some studies use macro-level data on international migration (Beine and Parsons 2012; Reuveny and Moore 2009), this is not a panacea since international migration statistics document only cross-border migration and fail to capture movements of people within country. Many of the limits of data sets on international migration flows also apply to data for internal migration.

Despite these challenges, the empirical literature seeking to estimate the effect of changing environment has been growing rapidly (see Laczko and Aghazarm 2009, for a review). This literature suggests that the relationship between migration and environmental change is nuanced, so that careful attention should be paid to the type of climate or weather event and the type of migration. For example, using census data for Yemen, Joseph and Wodon (2013a, b) find that while climate factors do play a role in migration flows between districts, that role is much smaller than the effect of socio-economic variables. There is also a growing consensus that migration requires a minimum level of resources, and liquidity constraints impede internal migration by the poorest households leading to an "immobility paradox" (Meze-Hausken 2000; Foresight 2011; Gray 2009; Halliday 2006; Hammer 2004) which has gender implications[2].

Most micro-level studies measure climate change either by the incidences of extreme weather events or by variation in temperature or rainfall. The evidence is weaker on household responses to incremental environmental degradation such as pollution, soil degradation, or deforestation. Since these factors are incremental, it is harder to isolate their effects on migration from other drivers. Studies by Rappaport and Sachs (2003) and Rappaport (2007) indirectly address the issue of migration as a response to the quality of living by looking at the weather-related moves in the United States. They suggest that in countries with higher per capita income weather becomes a location amenity that affects choice of residence.

[2]Gray and Mueller (2012) show that droughts in Ethiopia increased the probability of long-distance out-migration by male members of the household but reduced the probability of marriage-related migration among females because households could not finance wedding related expenses. Similarly, Dillon et al. (2011) find that *ex post* and *ex ante* weather related risk affected probability of out-migration by male household members but not by females. There is also empirical evidence that suggests that urban centers do not always serve as magnets for migration and individuals from affected rural areas migrate to other villages rather than to cities (Henry and Schoumaker 2004).

Table 5.1 Selected findings in the literature on climate patterns and migration

	Domestic	International
Extreme weather (droughts, flooding, etc.)	Induce migration, but not among the poorest (Hammer 2004; Halliday 2006; Meze-Hausken 2000) Have different effects on female and male outmigration (Gray and Mueller 2012) Responses depend on the social context (McLeman and Smit 2006)	Discourage international out-migration (Henry and Schoumaker 2004; Findley 1994), but encourage rural-to-urban migration (Beine and Parsons 2012) Encourage outmigration after extreme weather events, but only from non-democratic countries (Reuveny and Moore 2009)
Changing climate (change in precipitation or dryer climate)	Induces both rural to urban and rural to rural migration (Barrios et al. 2010; Kniveton et al. 2009; Dillon et al. 2010)	Increases the probability of international migrations, but the effects are specific to the policies and other community-level factors (Munshi 2003; Kniveton et al. 2009; Feng et al. 2010; Andersen et al. 2011)
Environmental change	Changes in the quality of life induce migration (Rappaport and Sachs 2003; Rappaport 2007)	–

Source: Authors

Further research, however, is needed on how households in developing countries respond to the quality of environment (Findlay 2011; Findlay and Geddes 2011). Some studies that focus on international migration have also identified environment as a driver. Changes in temperature and rainfall have been found to induce out-migration from rural communities in Mexico to the United States (Feng et al. 2010; Munshi 2003; Andersen et al. 2011). Consistent with the liquidity constraint hypothesis, international migration declines immediately after extreme weather events but internal migration increases because affected households cannot afford long-distance relocation (Henry and Schoumaker 2004; Findley 1994; Beine and Parsons 2012) (Table 5.1).

While there is a burgeoning literature on climate change and migration, the evidence for the Arab world remains limited. As noted by Icduygu and Sert (2011), migratory movements in the region have environmental causes and consequences, among others in terms of water scarcity, aridity, drought, infertility, declining food self-sufficiency, deforestation, and rising insufficiency in infrastructure and overcrowding in cities. But it has been difficult to disentangle causal effects due to data scarcity as well as lack of interdisciplinary research and lack of theory. A review on the potential impact of climate change on a range of outcomes from floods to agricultural production is provided for Europe and North Africa by Burke et al. (2011), but it is based essentially on simulation-based scenarios, as opposed to data on how households react to extreme weather events and changing climate patterns.

In the report on the Mediterranean workshop conducted by Foresight (2011) for its climate change and migration study, Zihnioglu (2011) identifies factors that may affect migration, including economic, political, social, demographic, environmental, and technological factors. Yet while the workshop report discusses some of the conceptual frameworks which can be used to discuss the issue of environmental migration, as well as some of the policy challenges that it creates, it does not provide much new data on the links between climate change and migration from the point of view of households.

This study aims to provide a broader review of the evidence on climate change and migration in the Arab world, including insights from a recent study carried by AFD and the World Bank (Wodon et al. 2013). The objective is not to provide policy recommendations on how to adapt to climate change in the region (for a recent study doing just that, see Verner 2012; on the evidence and policies worldwide, see also World Bank 2010a, b). In conducting the review, we focus on five countries: Algeria, Egypt, Morocco, Syria, and Yemen. Much of the review for each of the five countries is adapted from Burger et al. (2013), and the idea behind this focus is that limiting the scope of the review to a subset of countries makes it easier to go a bit more in-depth in the discussion for each of the countries selected. In turn, this helps in showing how the realities on the ground tend to differ between the various countries, so that some broad or uniform messages for the region as a whole may not necessarily apply to all individual countries. In what follows, section "Choice of focus countries" explains the choice of the five focus countries. The next five sections discuss the evidence for each of the five countries. A brief conclusion follows.

Choice of Focus Countries

Three main considerations led to the choice of the five countries for this review and for the broader study for which this review was prepared (Wodon et al. 2013). The aim was to select countries which (1) had a large population to ensure that findings would be illustrative (but not representative) of the relationships between climate change and migration for a large share of the Arab world population; (2) were substantially affected by climate change and extreme weather events, especially in terms of droughts; (3) represented various levels of development, but with an emphasis on middle income and low income countries where the impact of climate change on households may be largest and the ability of households to cope weakest.

Consider first population size. According to United Nations population data, Egypt, Algeria, and Morocco are the three most populated countries in the Arab world. Yemen and Syria rank seventh and eighth in population size. Iraq was not chosen because of on-going tensions within the country which limited the feasibility of new data collection. Sudan was not selected because administratively, it belongs to the sub-Saharan Africa region of the World Bank, and not to the Arab world, itself part of the Middle East and North Africa Region. Saudi Arabia was not considered because it is much wealthier and less agrarian than the other countries, and thereby has much more resources at its disposal to cope with and adapt to climate change. Iran was also not selected because it is usually not considered as an Arab country[3].

Consider next vulnerability to climate change. An analysis of vulnerability to droughts was recently conducted by the Arab Center for the Studies of Arid Zones and Dry Lands and the United Nations Secretariat of the International Strategy for Disaster Reduction Regional Office for Arab States (ACSAD and ISDR 2011). On the basis of data on drought frequency and consecutive droughts, the authors estimate that 69.1 % of the area of the Arab region is not vulnerable to droughts, 14.45 % has a low level of vulnerability, 10.98 % is moderately vulnerable and 5.47 % is highly vulnerable. Some of the countries that are most vulnerable to droughts are Syria (especially the north eastern area), Sudan (southern area), Tunisia (northern area), Algeria (northern area), Morocco (northern area), Somalia (north eastern area), Iraq (north eastern area), Saudi Arabia (north eastern area), and Yemen. The focus countries were chosen in part in order to represent conditions in different parts of the region, with Algeria and Morocco being part of the Maghreb, Egypt belonging to the Central Region, Syria located in the Mashreq and Yemen located in the Arabian Peninsula. Four of the five countries are among the most vulnerable to droughts in the region, and Egypt was chosen as the largest country in the Central Region and more generally in the Arab world, as discussed earlier.

Consider third levels of socio-economic development. The principle guiding the choice of countries was to focus on middle and low income countries, because these countries tend to have fewer means at their disposal than high income countries to be able to cope with and adapt to climate change (higher income countries are also of course affected by climate change, and often in dramatic ways, but their resources make it easier to protect the population from some of the damaging effects of extreme weather shocks and climate change). Also, middle and low income countries are more likely than high income countries to have net emigration flows internationally. Our five focus countries include four middle income countries – Algeria (with a per capita GDP of US$ 7,643 in 2011[4]), Egypt (per capita GDP US$ 5,547 in 2011), Syria (per capita GDP of US$ 4,741 in 2010) and Morocco (per capita GDP of US$ 4,373 in 2011), but the sample also include a much poorer country – Yemen, with a per capita GDP of only US$ 2,060 in 2011 (all figures are from the World Bank's World Development Indicators).

Algeria

A basic message from this review is that while climate change and extreme weather events are contributing to migration, in most cases they are not the main drivers of current or past migration patterns. Consider first Algeria. France has long been a

[3]It is customary to consider four criteria when defining the Arab world: language, ancestry, religion, and culture. On all four counts, Iran does not conform to the characteristics associated with the Arab world.

[4]In 2005 US$, adjusting for purchasing power parity.

primary destination for Algerian emigrants for reasons not at all related to climate patterns. Migration to France continued to be true throughout the 1954–1962 War of Independence, and after the war, over one million Algerians who had fought on the French side moved to France. In 1973, the Algerian government formally suspended migration to France as it viewed migration as a form of post-colonial dependence and anticipated sufficient labor demand from the petroleum sector (de Haas 2007), and in addition France also decided to stop the recruitment of Algerian workers at about the same time (Meynier and Meynier 2011). While mass labor migration was curtailed as a result of these policy changes, family reunification migration continued.

Over the last decade, more Algerians, in particular highly skilled individuals, have moved to a more diverse set of destination countries, including Spain (Di Bartolomeo et al. 2010). But in addition, Algeria itself has become a destination for immigrants from sub-Saharan Africa (Sekkai 2008) and special immigrant groups, such as Chinese workers working for Chinese companies (Temlali 2010). There were around 100,000 foreigners living legally in the country in 2008 and probably three times as many illegal immigrants and refugees, most of them Sahrawis who fled from West Sahara in 1975 and 1976. Probably around 40 % of foreign migrants see Algeria as their final destination, another 40 % see it as a transit station on their way to Europe, and the remainder are refugees or immigrants who would like to return to their home country but currently don't have the means to do so (Temlali 2010).

As for internal migration, it now takes place primarily between rural areas and mid-sized towns, rather than towards large urban areas (Gubert and Nordman 2009). In 2010, 66 % of the population was living in urban areas, up from 60 % in 2000. Urban areas have been growing at around 2.5 % over that period, while the number of people living in rural areas has been declining (World Bank 2011a, b). In addition to voluntary moves, during the Algerian Civil War from 1992 to 2002, at least one million Algerians were forced to move within the country. Many of these refugees have not yet been able to return home (EACH-FOR 2008).

As in other countries, unemployment is probably the main driver of migration flows, including for highly skilled individuals (Di Bartolomeo et al. 2010; de Haas 2007). Although unemployment declined from 29.3 % in 1999 to 10.2 % in 2009, rates among under-30 year olds are typically twice as high as the national and even higher among young highly educated individuals (Ministry of Territorial Management and Environment [MATE] 2010). It is worth noting, however, that unemployment is on average higher among Algerian-born individuals residing abroad than in Algeria itself (European Commission 2010).

Shifting the discussion to climate change, there is no doubt that environmental conditions and population movements are related. The Algerian territory encompasses three main geographic regions: the Tell in the north, the high planes in the center, and the Sahara in the South of the country. The littoral region of the north is the most densely populated, with 63 % of the population on 4 % of the territory. By contrast, one tenth of the population lives in the Sahara, which represents 85 % of the territory. Slow-onset environmental degradation is an issue throughout the country and since 1975 precipitation has gradually decreased, with currently only 600 m^3 of fresh water per habitant and year, a level below the U.N. defined scarcity threshold of 1,000 m^3 (MATE 2010). Based on FAO data and World Population Prospects data, Fargues (2008) estimates that this amount will not drastically decrease until 2030, but it may decline after that. Apart from water scarcity, agriculture is affected by soil erosion and desertification, with much of the territory at risk or already affected (MATE 2010). Cline (2007) estimates potential shortfalls in agricultural yields by 2080 ranging between 28 and 36 % versus the baseline of 2003, while Rousset and Arrus (2006) predict that by 2020, agricultural yields could decrease from 5 to 14 % as a result of environmental problems exacerbated by climate change.

Extreme climate events are also frequent, with parts of the northern region of the country highly vulnerable to droughts (such as the several-year long drought of the early 1990s), cold waves (January 2005), flash floods (possibly associated with mud slides) due to heavy rainfall (in November 2001, October 2002 and October 2008), and heat waves (summer 2003). The consequences of these events range from large harvest losses and destruction of homes to deaths. For instance, in November 2001 the municipality of Bal El Oud in the region of Algier registered a rainfall of 290 mm, 40 % of the mean annual precipitation, in less than 17 h. In the wake of this event, in this municipality alone, 712 people died, 115 had disappeared, 311 were injured, and 1,454 families lost their home. Since 1975, both droughts and flooding have become more common, and there are concerns that this trend will continue in the future (MATE 2010).

Yet while there is some evidence that climate and migration patterns are related, to our knowledge, no studies are available on how past environmental shocks have affected migratory movements in Algeria, or how they may be related in the future. A conference hosted by the National Intelligence Council actually concluded that Algeria might be less vulnerable to climate change because a large percentage of its GDP (30 %) is derived from oil and natural gas and because agriculture and tourism play a limited role for the country's economy (NIC 2009), both in terms of GDP and labor force share (only 14 % of the labor force is employed in the agricultural sector according to the CIA, 2011). This may suggest that climate-induced out-migration will not drastically increase due to climate stress. However, Algeria already has to import 45 % of its food, a percentage that may increase in the future due to increasing desertification, soil erosion, and water shortages, so that climate change may exacerbate the food security situation, leading to refugee flows (Brown and Crawford 2009). As to sea level rises, Algeria is predicted to have the highest rise in sea levels in North Africa, but the population living in affected areas may be smaller than in other countries (Sowers and Weinthal 2010).

Egypt

Egypt also has a long migration tradition. It is estimated that 2.7 million Egyptians reside abroad (IOM 2010), 70 % of whom are in other Arab countries, and especially Saudi Arabia (with 0.92 million Egyptians), followed by Libya, Jordan, and Kuwait.

Among non-Arab countries, the United States is the most common emigrant destination, with 0.3 million Egyptian immigrants. Highly-educated individuals, such as doctors and engineers, are most likely to emigrate permanently (EACH-FOR 2009). In total, around 5 % of the tertiary educated were living abroad in 2000 (World Bank 2010a, b). There are no reliable estimates of the extent of illegal emigration, but the EACH-FOR report states that *"there are tens – if not hundreds – of Egyptians who have been caught illegally attempting to cross the Mediterranean,"* and the 2007 EU-Egypt Action Plan contains some general clauses on cooperation on migration control (European Union 2007). As to migration to Egypt itself, from 2000 to 2010, the number of foreign-born individuals living in Egypt increased from 170,000 to 245,000 (World Bank 2011a, b). However, these figures are likely to be underestimates. Immigrants include labor migrants from Arab and African countries, Europe and the United States, and refugees from Palestinian, Somali, Ethiopian and Eritrean nationalities.

The most important internal migration flows are from the south to the north, from both south and north to the Canal Zone, and from rural areas to Cairo and Alexandria. Over the past decade the urban population growth was at 1.9%, which is similar to the national average, so that urbanization remains just below 43 %. Reasons cited for historic migration include rising population density in areas where agriculture could not be expanded further geographically, decreasing economic opportunities, and unemployment. In addition, involuntary migration occurred in response to the 1967 Arab-Israeli war, which displaced 750,000 people, and to the construction of the Aswan Dam, which displaced 100,000 people, primarily of the Nubian ethnicity. While temporary and circular migration traditionally very important continues to occur, it used to be driven by seasonality, with primarily agricultural workers moving to cities when there was no work for them in rural areas. Today, the seasons appear to have much less of an influence on these circular moves (Zohry 2005).

Egypt has a primarily desert climate. During the winter there are some rains and mild temperatures in the coastal areas of lower Egypt, while upper Egypt is dry with warm weather. In the summer, the entire country experiences hot and dry weather (EEAA 2010). From 1961 to 2000, mean temperatures were steadily rising, and in Upper Egypt and the Western Desert, the number of days on which temperatures equaled or exceeded 45 °C (113 F) increased. The severity and frequency of extreme weather events—including sand storms, dense haze, and flooding—have also been increasing over the 1972–2002 period.

Some 80 % of the water is used in the agriculture sector, which accounted for 13.8 % of GDP in 2006/2007 and employed 55 % of the Egyptian labor force (EEAA 2010). Because 95 % of Egyptian agriculture is irrigated and all but 5 % of Egypt's water supply comes from the Nile River, the impacts of climate change on temperature and precipitation levels and on Nile flows are of vital importance. Under three global circulation models, it is estimated that Nile flows could either decrease by between 10 and 75 % or increase by one third relative to the 1996 level of 84 billion m^3 (EEAA 2010). In another paper, ten out of ten scenarios predicted losses from 5 to 50 % in 2020; and nine of ten scenarios predicted long-term decreases between 10 and 90 % by 2095 (Strezpek et al. 2001). While Egypt is relatively well equipped to deal with 1-year droughts through the water stored by the High Aswan Dam, long-term decreases in Nile flows would be a major problem (Agrawala et al. 2004). In addition, it is estimated that solely through the expected rise in temperature the production of major crops will decrease – for instance for rice by 15 % by 2050 and by 36 % in 2100 (Abou-Hadid 2006). Land degradation through pollution and salinization may further negatively affect agricultural output (EACH-FOR 2009).

With a 3,500 km coastline and much of the economy and population being concentrated in these areas, sea-level rises will also have an impact. For instance, El-Raey et al. (1999, in EACH-FOR 2008) estimates that a 0.18 m rise would lead to a loss of employment of 32,000 and a rise in half a meter would lead to a loss of 195,000 jobs in Alexandria; and in a recent presentation by Shalaby (2010), it is argued that sea level rises may reduce the Egyptian agricultural area in the Nile delta by 12–15 %. An environmental factor that is already displacing people today—and which is likely to become more severe in the future—is desertification. Advancing sand dunes in the Western desert are swallowing entire villages, forcing their inhabitants to move (Warner et al. 2008). Thus, Warner et al. found that more than 70 % of internal migrants that they interviewed in the Nile Delta and Valley, newly reclaimed desert and slums in Old Cairo named land degradation and water shortages as root causes in their migration decision. Another environment factor that may cause future migratory movements is a rise of the sea level. Dasgupta et al. (2007) predict that a 1-meter sea-level rise would displace around 10 % of the Egyptian population.

Demographic pressures may further influence future migration flows. It is projected that that Egypt's population, which in 2006 was approximately 77 million, will increase to 92–100 million by 2020 and 104–119 million by 2030 (EEAA 2010). This population growth will put strains on the environment, for instance reducing the available water per capita from 1,000 m^3 in the early 1990s to 468 m^3 in 2030, assuming that Egypt will continue to be able to get its current fixed share of Nile water as was laid out in a water sharing agreement in 1959 (EEAA 2010).

Morocco

International migration by Moroccans has traditionally been and remains primarily directed towards Western Europe. It is estimated that two million Moroccans, 85 % of all emigrants, are currently living in Europe. During the 1960s and 1970s, emigrants were mainly guest workers; from the 1970s to the 1990s, family reunification was the main form of migration. Since then, the share of illegal emigration in migration flows has been on the rise, partly driven by increasing activities in the agricultural and construction sectors in Southern Europe (EACH-FOR 2008). The majority of emigrants are unskilled; but in 2000, nearly 20% of tertiary-educated individuals were living abroad (Gubert and Nordman 2009; World Bank 2011a, b). For decades the Moroccan government encouraged emigration for economic and political reasons. More recently, the government has started cooperating with the European Union on the prevention of illegal migration of Moroccans and sub-Saharan Africans, for whom Morocco is an important transit country on their way to Europe. For instance, Morocco participates in joint navy patrols, and takes in both Moroccans and other nationals who are apprehended by border forces. In return, the country receives development assistance. It is estimated that there are 10–20,000 illegal immigrants in Morocco (IOM, undated).

Internal migration flows are larger than international flows. These movements are predominately rural to urban, which explains why cities have been growing at more than four times the rate as rural areas over the 2000–2010 period. During that time, the urbanization rate increased from 53 to 56 %. However, as rural areas and small and medium-sized towns have become more developed partly due to remittances from international migrants, they have become more attractive as destinations in their own right (GCIM 2005).

Morocco's inhabitants are already experiencing deteriorating environmental conditions and shocks that may put their livelihoods at risk, and it is expected that climate change will exacerbate this trend. Shifts include a decrease in precipitation, increasing risks of droughts, more dry areas towards the North of the country, and less ground water. Despite a low per-capita water usage of 500 m^3, Morocco is experiencing water stress. These water shortages are predominantly caused by climate-related factors in the south and the pressure of population growth in the northern parts of the country (EACH-FOR 2008). More than 80 % of water is used in agriculture, even though only 13 % of cropped land is irrigated. The Moroccan government estimates that at least until 2020, subject to sufficient investments in infrastructure, water needs will be met (Kingdom of Morocco 2001). However, the EACH-FOR study suggests that by 2025, per capita available water resources will have decreased by 30 %. Under different climate scenarios developed by a consortium of international (World Bank, Food and Agriculture Organization) and national organizations (VMN, INRA), it is expected that agricultural output will not be drastically affected by climate change until 2030. Subsequently, however, there may be a strong decrease in agricultural output particularly in the northern and center-west parts of the country, where rainwater-based agriculture is currently predominant. The study also points out that the country faces uncertain declines in revenues from irrigation-based agriculture.

In a country where 40 % of the population is employed in agriculture, and where nearly 70 % of the poor live in rural areas, environmental shocks and resultant decreases in agricultural output are likely to negatively affect the livelihood of many individuals, and this creates the potential for migration movements. In addition, Morocco has 3,500 km of coastline, so that an increasing number of floods and as well as a rising sea level may also induce future migration movements. A recent report prepared for the U.S. National Intelligence Council (CENTRA Technology and Scitor Corporation 2009) suggests that "*climatic stress will add to the already substantial movement of population from rural areas into cities*" in Northern Africa, and speculates that the cyclical movements in agricultural output that have been accompanied by cyclical movements of population between rural and urban areas in Morocco in the past might be replicated in the future. Another study points out that following a severe drought in 2007, two thirds of the illegal migrants arrested in Spain were from the farming and mining region of Khouribga (EACH-FOR 2008). This suggests a link between environmental shocks and migration exists, and that at least some of the induced migration will be international. A case study by Hamza et al. (2009) based on a survey of 30 households in two oasis villages in the Middle Draa Valley of Southern Morocco found that environmental degradation was a major factor for either past or intended migration, with lack of access to services and opportunities also being major contributing factors.

Additional information on perceptions about the climate, weather shocks, and their impact on households is available for Morocco through special modules on climate change and shocks incorporated in a national survey implemented in 2009–10 (Nguyen and Wodon 2013a). In the survey, 28.1 % of households were involved in agriculture, and among those 92.1 % declared having been affected by deteriorating climate conditions in the last 5 years. The most likely shock was a reduction in agricultural yields due to inadequate rainfall, mentioned by 62.2 % of agricultural households. In a separate part of the survey, more than one in five households declared having been affected by a recent weather shock such as a drought or flood. Most of the households who were affected declared that they had not been able to recover from the shock, and this was especially the case among poorer households. According to regression analysis, households in the top quintile of wealth were 20 percentage points more likely to recover from weather shocks than households in the bottom quintile. Using this survey, Nguyen and Wodon (2013b) suggest through regression analysis that weather shocks increase the likelihood of temporary migration by slightly more than 1 percentage point. While the impact of weather shocks on permanent migration is not statistically significant, the impact of changing structural conditions such as reduced agricultural yields due to lack of water is, and it does contribute to higher permanent migration away from the affected areas.

Syria

International out-migration is high in Syria, although estimates of its extent vary. According to Kawakibi (2009), 18 % of Syrians have migrated abroad, and a study by the Syrian Commission for Family Affairs estimated that in 2007, 3.4 million people (15 % of the total population) lived abroad. But another report by the United Nations stated that in 2005, less than 750,000 Syrians were living abroad (Marzouk 2010). Still, it appears that throughout the 2000s, emigration has increased. While the majority of emigrants during the 1960s and 1970s moved to the Persian Gulf, circular migration with Lebanon, Jordan, and the Gulf states became more common in recent decades. There is also some evidence of brain drain: while only 0.9 % of Syrians live in OECD countries, the emigration rate of individuals with tertiary education to these countries in 2000 was 3.7 % (Dumont 2006).

Syria is also a receiving country for migrants and especially refugees, including Iraqis since 2003 and Lebanese in 2006 (Kawakibi 2009). As a result, while in 2000 5.6 % of the population was made of immigrants, this had increased to 10.2 % by 2010 (World Bank 2011a, b). In addition, Khawaja (2002) found that 14 % of the Syrian population migrated internally from one administrative unit to another at some point during their life, and that around 5 % migrated during the previous 5 years. The most common moves are from rural to urban areas, which in 2010 were home to 55 % of the population, up from 52 % in 2010 (World Bank 2010a, b). Khawaja sees this as a lower than expected internal migration rate, and suggests that the main reasons for this are the high percentage of people who own their apartment or houses, and the relative equal distribution of access to social services and even wages throughout the country. Surveys indicate that moves for family reasons are more common than those for work-related motives, but internal migration may have recently increased.

Syria is an arid to semi-arid country. A third (32 %) of the land is cultivated, of which 75 % is rainfed and 45 % is pastures. Since the 1980s, the country has been able to meet its own food needs and even export food (SMEA 2010). However, today the country is witnessing a deteriorating environment and a more adverse climate. Between 1980 and 2006, 13 % of agricultural land was downgraded because of human activity. Rainfall decreased by 10 mm per year (with an annual average of 300 mm per year) from 1956 to 2006 (El-Atrache 2010), with large regional variation. For example, the northern and northeastern parts of the country have experienced decreasing precipitation in winter, while the northern and central regions have experienced more rain in autumn (Ministry of State for Environmental Affairs [MSEA] 2010). Summer temperatures have risen significantly, particularly in the coastal zones (Qabbani 2010; MSEA 2010). From 2006 to 2009, Syria experienced a severe drought (Sowers and Weinthal 2010), and droughts may become more common in the future (MSEA 2010). Rises in the sea level will be another potential consequence of climate change, with an estimated 3.8 % of the coastal population potentially being affected. Among those affected, 2–6,000 farming families may lose their livelihoods and homes (MSEA 2010).

Syria uses 18 billion m^3 of water per year, in excess of the estimated 15 billion m^3 of renewable resources. The agricultural sector is responsible for 90 % of water usage. With an available 1,000 m^3 of water per capita and year, Syria is on the threshold of water poverty (MSEA 2010). Population growth, decreases in precipitation, pollution, and increased demand due to development are likely to decrease consumption, possibly to 500 m^3 per person per year in 2025. Syria's first National Communication to the United Nations Framework Convention on Climate Change reports that between 2010 and 2100, annual precipitation could decrease between 34 and 75 mm, and temperatures could rise by 2.8–7 % (Qabbani 2010). Based on the CROPWAT irrigation management model of the FAO, it was estimated that under these scenarios, irrigated wheat agriculture in the Hassakeh governorate could experience a reduction in yield of 16 % by the end of the century, while yields from non-irrigated land could drop by more than 20 % if no additional irrigation be provided (MSEA 2010).

There are few systematic studies on how past environmental changes have impacted migration in Syria. Khawaja (2002) did explicitly ask whether environmental problems were related to the decision to move, but the low incidence of individuals naming work or income as reasons for migration may suggest that at least in the late 1990s and early 2000s, climate factors were not a prime motive for migration. However, in a country where 45 % of the rural population work in agriculture, and 75 % of the cultivated land is not irrigated, as temperatures rise and precipitation falls, this may change in the future, especially in periods of droughts (Qabbani 2010). A study by the Canadian International Institute for Sustainable Development suggests that 160 villages in the north-east of the country were entirely abandoned during the 2007/2008 period of the current drought that has lasted since 2006 (El-Atrache 2010; Sowers and Weinthal 2010). A separate assessment mission of the Syrian government and the UN estimated that the drought has affected 1.3 million inhabitants of Eastern Syria. More than 800,000 persons lost their entire livelihood, with small-scale farmers and herders being particularly affected. Many initially tried to react to the drought by selling livestock below prices, reducing food intake, and selling assets (Alqusairi undated), but eventually, an estimated 40,000–60,000 households migrated in response to the drought (DREF 2009). Among these, 200,000–300,000 people were from the Al-Hasakeh governorate, representing 13–20 % of the total previous population and 41–62 % of the affected population (DREF 2009). Such massive moves put pressure on urban infrastructure, reduce economic growth, and lead to worsening educational and health outcomes among migrants. Unusually, migration was predominantly by entire families, rather than just men.

Yemen

Yemenis have been migrating out of the country to oil-producing Arab Gulf States, North America, and Europe for several decades. The current stock of Yemeni emigrants is estimated at 1,134,700, or 4.7 % of the population. Migrants constitute 6 % of the tertiary-educated population. The main destination countries for migrants remain Saudi Arabia, the UAE, and the United States (Ratha et al. 2010). Yemen is also a country of passage for migrants from Sub-Saharan Africa (Fargues and Bensaad 2007) and a destination for an increasing number of refugees (UNHCR 2011). Some 60 % of immigrants are Somali, 30 % Ethiopian, and 10 % Eritrean and from other nationalities (Al-Ariqi 2010). Currently, immigrations constitute 2.1 % of the population.

Internal migration is primarily directed towards urban areas, which constitute an increasing share of total population and are experiencing faster population growth than rural areas. In 2010, urban areas constituted 31 % of the total population, up from 26 % in 2000 (World Bank 2011a, b). Although urban population growth rates have fallen slightly between 2000 and 2010 from 4.9 to 4.7 % per year, urban population growth is more than twice that of the rural population, at 2.0 % per year. In addition to voluntary migration, 350,000 individuals have been displaced since 2004 by a violent conflict between insurgents and the government in Northern Yemen (IOM 2010).

Although weather data for Yemen may be less accurate than for other countries, making it more difficult to construct future climate change scenarios, annual mean temperature appears to have increased by 1.8 °C since 1960. Temperatures are expected to increase by 1.2–3.3 °C by 2060 and 1.6–5.4 by 2090. Monthly rainfall has fallen by 1.2 mm since 1960, a decrease of 9 %. In the future, sea-level rises may also threaten the coastal cities, in particular Aden (Al-Tholaya 2010).

Predictions about future rainfall trends vary by climatological model, with some models predicting increased rainfall while others predict decreased rainfall (McSweeney et al. 2008). Decreased rainfall has the potential to harm rainfed agriculture, which constitutes an important fraction of Yemen's cereal production (FAOUN 2009). Even without decreases in rainfall, water scarcity will be a challenge due to limited water resources, population growth, and the water use for the production of khat. In 1955, per capita water availability was 1,098 m^3, but by 1990 it had fallen to 460 m^3 per capita. Moreover, projections suggest that under business-as-usual scenarios per capita water availability will fall further to 150 ft^3 (Republic of Yemen 2008).

A study by McKinsey suggests that water shortages could lead to the loss of 750,000 jobs and a decrease in income of 25 % (Rudolf 2010). In the capital Sana'a in particular, water is currently being extracted at four times the replenishment rate. According to reports based on a World Bank-funded assessment of the Sana'a basin, the capital city's water resources are being depleted by a fast-growing population (IRIN 2010), and the expanding urban population is due in part to migration from rural to urban areas (Boucek 2009). Projections suggest Sana'a may deplete is economically-viable water supplies—primarily groundwater— in a decade (IRIN 2010; see also Rudolf 2010; Sowers and Weinthal 2010). Extreme weather events are also becoming more common. For instance, Wynter (2009) reports that flash floods are becoming increasingly more common, endangering the life of villagers that live close to wadis.

There is limited systematic research into how past environmental change has impacted migration in Yemen, but Yemen's population is quite vulnerable to climate shocks. While agriculture generates only 15 % of GDP, it employs more than 55 % of the active population. Decreases in agricultural yields are expected in the future, which would impact livelihoods. Some 84 % of the rural poor are dependent on rainfed subsistence agriculture, a form of agriculture that is highly vulnerable to the impact of climate change (World Bank 2010a, b). Environmental changes affecting the agricultural sector could increase rural to urban migration.

By combining census and weather station data from Yemen, Joseph and Wodon (2013a) find that climate variables do affect migration from some districts to other districts, but in a somewhat limited way, with socio-economic and cost factors playing a much more prominent role. This analysis, which is based on past data, suggest that migration flows are unlikely to increase sharply in the near term, but if conditions were to change drastically, the effect of weather variables on migration could become much larger (Joseph et al. 2013a).

Another set of interesting findings for Yemen relates to remittances. Joseph et al. (2013b) use a national household survey for Yemen combined with weather data to measure remittance flows, both domestic and international, and assess the likelihood of households receiving remittances as well as the amounts received. The question is whether households living in less favorable areas in terms of climate (as measured through higher temperatures, lower rainfalls, more variability or seasonality in both, and larger differences in a given year between extreme temperatures) are more likely to benefit from remittances. The results suggest that this is not the case in Yemen. In another paper Joseph and Wodon (2013b) use matching techniques and the same household survey for Yemen combined with weather data to measure the impact of remittances, both domestic and international, on poverty and human development outcomes (school enrolment, immunization, and malnutrition). The estimations are carried both nationally and in areas with favorable and unfavorable climate.

Four main results are obtained. First, remittances tend to have positive impacts on poverty measures, school enrollment, and measures of malnutrition. Second, the impact of international remittances tends to be larger than that of domestic remittances, probably because among beneficiaries, the amount of remittances received tends to be higher for international than for domestic remittances. Third, the impact of remittances – and especially international remittances – on measures of poverty and malnutrition tends to be larger in areas affected by high temperatures, and also to some extent in areas with lower levels of rainfall, which in both cases tend to be more vulnerable. Fourth, and by contrast, in areas with higher levels of rainfall or lower levels of temperatures, where issues of poverty and malnutrition may be less severe, remittances – and again especially international remittances – tend to have a larger impact on school enrollment. Thus, in areas with unfavorable climate, remittances help first for meeting basic needs in order to escape poverty and malnutrition, while in areas with more favorable climate, remittances may be used more for investments, including in the education of children.

Evidence from New Household Surveys

It was mentioned in the introduction that new household surveys were collected in the five countries, in order to strengthen the evidence base on climate change and migration in the region. Do households living in areas susceptible of being affected by climate change believe that changes in climate patterns are taking place? On the basis of an analysis of these surveys, Adoho and Wodon (2013a) suggest that this is indeed the case. In the combined sample for the five countries, more than three fourths of households declare that rain has become more erratic, and almost three quarters say that temperatures are higher. Between half and two thirds declare that there is less rain today than 5 years ago, that the land is dryer or less fertile, that the rainy season starts later, is shorter, or ends earlier, and that droughts are more frequent. The changes in climate in turn appear to lead to more diseases in animals and livestock, more insects and pests in crops, less water in boreholes, rivers, lakes or streams, more air pollution, more frequent crop failures and livestock loss, and more soil erosion. Some of the extreme weather events associated with climate change such as rain storms and floods are not perceived as more frequent, and in some cases, households do suggest that temperatures are becoming cooler, and that there is more rain, but this is a minority of households. But overall, while there are differences between households and areas or countries, there is a clear perception that the climate is worsening.

To what extent are households migrating away from climate affected areas, and is climate itself a key push factor in such migration? Regression results suggest that poor climate and extreme weather events lead to a higher probability of migration, but the role of climate is smaller than that of socio-economic characteristics and job prospects in cities (Adoho and Wodon 2013c). The authors construct two indices or factors that summarize household perceptions regarding changes in weather patterns and the environment. The first factor captures the extent to which households perceive that the climate is becoming dryer and warmer, and it is associated with droughts. The second factor captures the extent to which households suffer from excess water, and it is associated with floods. Higher values for both factors (i.e., worse climate conditions) result in higher rates of permanent migration, with the coefficients being statistically significant and the effects of each of the two factors of a similar order of magnitude. The effects for temporary migration are similar, although statistically significant only for the whole period, as opposed to the last 5 years. The estimates suggest that a significant deterioration of climate conditions would lead to an increase of about 1.5 percentage point for both types of migration. This increase would represent between one tenth and one fifth of the overall level of migration observed, and its magnitude is of an order of magnitude similar to that obtained when considering the reasons stated by households for the migration of some of their members.

There are however differences between areas and countries (Grant et al. 2013). The analysis of the surveys as well as additional qualitative work through focus groups and in-depth interviews suggest that on average, migrants in Morocco, and to some extent also in Egypt, tend to mention more frequently the opportunities that migration may offer at the places of destination in terms of jobs and freedom, while migrants in Syria, and to some extent in Algeria and Yemen, are quicker to relate their decision to migrate to the difficulty to survive in rural areas due to the deteriorating climate, droughts, and low agricultural yields.

Conclusion

The aim of this paper was modest: it was to provide a review of the evidence on the relationship between weather patterns, perceptions of climate change, and migration in the Arab world. We did not aim to provide specific policy recommendations, but it is worth pointing out that many of the conclusions reached in the recent Arab world study by Verner (2012) apply. That study suggested that a solid foundation for decision making related to climate change adaptation involves four iterative steps: (1) Assessing climate risks, impacts, and opportunities for action; (2) Prioritizing policy and project options; (3) Implementing responses in sectors and regions; and (4) Monitoring and evaluating implementation, then reassessing the climate risks, impacts, and opportunities. Our work falls squarely within the first of these four steps – we have reviewed the evidence on the extent to which households in vulnerable rural areas are affected by climate change and weather shocks, and whether this leads to migration away from affected areas. While the impact of weather shocks on migration may be smaller than often believed, the cost of climate change and weather shocks is already felt today by many rural households, and part of the migration observed today is already related to changing climate conditions.

Whether migration should be encouraged or not may not be the right question at this stage. Climate-induced migration is already taking place, and it must therefore be factored in the policy debate, especially at the national level given that climate-related migration flows appear to be mostly internal. We have focused most of this review on sending areas, but the qualitative work conducted in urban areas by Grant et al. (2013) suggests that the integration of migrants into major destination cities is not always working as well as it should. Concerns about employment and housing abound among migrants, with migration simply adding to existing pressures that can be dealt with only through broad-based economic policies not necessarily focused on migration per se. In other words, while dealing with climate-induced internal migration may require some interventions specifically aimed at migrants, the challenge is a much broader one. Effective climate change adaptation and migration policies will need to be integrated into a broader set of policy and economic reforms for growth and employment that require strong political leadership.

References

Abou-Hadid, A. F. (2006). *Assessment of impacts, adaptation and vulnerability to climate change in North Africa: Food production and water resources* (Final Report, Assessments of Impacts and Adaptations to Climate Change (AIACC), Project No. AF 90). "The International Start Secretariat" (see http://www.aiaccproject.org/Final%20Reports/Final%20Reports/FinalRept_AIACC_AF90.pdf in page 3).

ACSAD and ISDR. (2011). *Drought vulnerability in the Arab region: Case study – Drought in Syria, ten years of scarce water (2000–2010).* Damascus/Cairo: Arab Center for the Studies of Arid Zones and Dry Lands and United Nations International Secretariat for Disaster Reduction.

Adoho, F., & Wodon, Q. (2013a). Perceptions of climate change and impact on households in the MENA region. In Q. Wodon, A. Liverani, G. Joseph, & N. Bougnoux (Eds.), *Climate change and migration: Evidence from the Middle East and North Africa region.* Washington, DC: The World Bank.

Adoho, F., & Wodon, Q. (2013b). How do households cope with and adapt to climate change? Evidence from the MENA region. In Q. Wodon, A. Liverani, G. Joseph, & N. Bougnoux (Eds.), *Climate change and migration: Evidence from the Middle East and North Africa region.* Washington, DC: The World Bank.

Adoho, F., & Wodon, Q. (2013c). Do changes in weather patterns and the environment lead to migration? Survey-based evidence from the MENA region. In Q. Wodon, A. Liverani, G. Joseph, & N. Bougnoux (Eds.), *Climate change and migration: Evidence from the Middle East and North Africa region.* Washington, DC: The World Bank.

Agoumi, A. (2003). *Vulnerability of North African countries to climatic changes – Adaptation and implementation strategies for climate change*. Winnipeg: International Institute for Sustainable Development.

Agrawala, S., Moehner, A., El Raey, M., Conway, D., van Aalst, M., Hagenstad, M., & Smith, J. (2004). *Development and climate change in Egypt: Focus on coastal resources and the Nile*. Paris: OECD.

Al-Ariqi, Y. (2010, October 21). Fawzi Al-Zioud, office in charge of the IOM in Yemen, to the Yemen Times: 'IOM, through its cooperation with the EU, strives at empowering the government of Yemen and Civil Society to address many of the unique migration challenges Yemen faces today,' *Yemen Times*.

Alqusairi, D. (undated). *Risk Profile and Emergency Management of Syria, Mimeo*.

Al-Tholaya, A. (2010, May 3). Climate change in Yemen: Drastic changes in the coming 50 years. *Yemen Times*.

Andersen, L., Lund, L., & Verner, D. (2011). Migration and climate change. In D. Verner (Ed.), *Reducing poverty protecting livelihoods, and building assets in a changing climate: Social implications of climate change of Latin America and Caribbean*. Washington, DC: The World Bank.

Asian Development Bank. (2012). *Addressing climate change and migration in Asia and the Pacific*. Mandaluyong City: Asian Development Bank.

Beine, M. A. R., & Parsons, C. R. (2012). *Climatic factors as determinants of international migration* (CESifo Working Paper Series No. 3747).

Black, R., Kniveton, D., & Schmidt-Verkerk, K. (2011). Migration and climate change: Towards an integrated assessment of sensitivity. *Environment and Planning, 43*(2), 431–450.

Bohra-Mishra, P., & Massey, D. S. (2011). Environmental degradation and out-migration: New evidence from Nepal. In E. Piguet, A. Pécoud, & P. de Guchteneire (Eds.), *Migration and climate change*. Cambridge: Cambridge University Press.

Boucek, C. (2009). *Yemen: Avoiding a downward spiral* (Carnegie papers: Middle East program, Vol. 102). Washington, DC: Carnegie Endowment.

Brown, O., & Crawford, A. (2009). *Changements Climatiques et sécurité en Afrique – Une étude réalisée pour le Forum des ministres des Affaires éntrangères d'Afrique du Nord en 2009*. Winnipeg: International Institute for Sustainable Development.

Burger, N., Grant, A., Kups, S., Rana, Y., & Wodon, Q. (2013). Focus countries for the study on climate change and migration in the MENA region. In Q. Wodon, A. Liverani, G. Joseph, & N. Bougnoux (Eds.), *Climate change and migration: Evidence from the Middle East and North Africa region*. Washington, DC: The World Bank.

Burke, S. M., Mulligan, M., Parks, K., & Soesbergen, A. V. (2011). Quantifying change in ecosystem services and exposure to hazards in the Mediterranean basin over the next 50 years that might be relevant to migration. In Foresight (Ed.), *Migration and global environmental change*. London: Government Office for Science UK.

Carr, E. R. (2005). Placing the environment in migration environment, economy, and power in Ghana's central region. *Environment and Planning, 37*(5), 925–946.

CENTRA Technology, Inc., & Scitor Corporation. (2009). *North Africa: The impact of climate change to 2030*. Geopolitical implications, Conference report prepared for the National Intelligence Council.

Cline, W. R. (2007). *Global warming and agriculture: Impact estimates by country*. Washington, DC: Center for Global Development/Peterson Institute for International Economics.

Darwish, K. M., & Kawy, W. A. A. (2008). Quantitative assessment of soil degradation in some areas North Nile Delta, Egypt. *National Journal of Geology, 2*(2), 17–22.

Dasgupta, S., Laplante, B., Meisner, C., Wheeler, D., Yan, J. (2007). *The impact of sea level rise on developing countries: A comparative analysis, World Bank Policy Research* (Working Paper 4136). Washington, DC: The World Bank.

de Haas, H. (2007). *North African migration systems: Evolution, transformations and development linkages*. Oxford: University of Oxford International Migration Institute.

de Regt, M. (2007). *Migration to and through Yemen: The case of migrant domestic workers*. Cairo: American University Cairo.
Di Bartolomeo, A., Jaulin, T., & Perrin, D. (2010). *CARIM migration profile-Algeria: The demographic-economic framework of migration*. Florence: European University Institute.
Dillon, A., Mueller, V., & Salau, S. (2011). Migratory responses to agricultural risk in Northern Nigeria. *American Journal of Agricultural Economics, 93*(4), 1048–1061.
DREF Operation (2009). *Syria: Drought, International Federation of Red Cross and Red Crescent Societies*.
Dumont, J.-C. (2006). *Immigrants from Arab countries to the OECD: From the past to the future, United Nations Expert Group meeting on international migration and development in the Arab region*. Beirut: United Nations Secretariat Population Division.
EACH-FOR. (2008). *General overview study – Middle East and Northern Africa*. Bonn: United Nations University.
EACH-FOR. (2009). *Egypt case study report*. Bonn: United Nations University.
Economic and Social Commission for Western Asia and League of Arab States (ESCWA and LAS). (2011). *Arab region state of implementation on climate change* (Draft).
Egyptian Environmental Affairs Agency (EEAA). (2010). *Egypt second national communication under the United Nations Framework Convention on Climate Change*.
Eido, M. (2006). *Stakeholder analysis climate change in Syria*. Damascus: Ministry for Local Administration and Environment.
El Raey, M., Dewidar, K., & El Hattab, M. (2009). Adaptation to the impacts of sea level rise in Egypt. *Climate Research, 12*(27), 117–128.
Elasha, B. O. (2010). *Mapping of climate change threats and human development impacts in the Arab region, mimeo*. New York: United Nations Development Program.
El-Atrache, T. (2010, June 2). *Syria: 160 Villages Deserted 'Due to Climate Change'*. Arab Environment Watch.
European Commission. (2010). *Labour markets performance and migration flows in Arab Mediterranean countries: Determinants and effects* (Vol 2: National Background Papers Maghreb, Occasional Paper 60). Brussels: European Commission Directorate-General for Economic and Financial Affairs.
European Union. (2007). *EU-Egypt action plan*, Brussels.
Fargues, P. (2008). *Emerging demographic patterns across the Mediterranean and their implications for migration through 2030*. Washington, DC: Migration Policy Institute.
Fargues, P., & Bensaad, A. (2007). *Senders turned into receivers: Transit migration in the Middle East and North Africa*. Florence: European University Institute.
Feng, S., Kruegera, A. B., & Oppenheimer, M. (2010). Linkages among climate change, crop yields, and Mexico-U.S. cross-border migration. *Proceedings of the National Academy of Sciences, 108*(33), 14257–14262.
Findlay, A. M. (2011). Migrant destinations in an era of environmental change. *Global Environmental Change, 21*(1S), S50–S58.
Findley, S. E. (1994). Does drought increase migration? A study of migration from rural Mali during the 1983–1985 drought. *International Migration Review, 28*(3), 539–553.
Findlay, A., & Geddes, A. (2011). Critical views on the relationship between climate change and migration: Some insights from the experience in Bangladesh. In E. Piguet, A. Pecoud, & P. de Guchteneire (Eds.), *Migration and climate change*. New York: Cambridge University Press.
Food and Agriculture Organization [FAO]. (2009, August 11). *Syria drought response plan 2009*.
Food and Agriculture Organization of the United Nations (FAOUN). (2009). *FAO/WFP crop and food security assessment mission to Yemen*. Rome: Food and Agriculture Organization.
Foresight. (2011). *Migration and global environmental change*. London: The Government Office for Science.
Global Commission on International Migration (GCIM). (2005). *Morocco's migration transition: Trends, determinants and future scenarios* (Global Migration Perspectives, No. 28).

Grant, A., Burger, N., & Wodon, Q. (2013). Climate-induced migration in the MENA region: Results from qualitative fieldwork. In Q. Wodon, A. Liverani, G. Joseph, & N. Bougnoux (Eds.), *Climate change and migration: Evidence from the Middle East and North Africa region*. Washington, DC: The World Bank.

Gray, C. L. (2009). Environment, land, and rural out-migration in the Southern Ecuadorian Andes. *World Development, 37*(2), 457–468.

Gray, C. L., & Mueller, V. (2012). Drought and population mobility in rural Ethiopia. *World Development, 40*(1), 134–145.

Gubert, F., & Nordman, C. J. (2009). *Migration trends in North Africa: Focus on Morocco, Tunisia and Algeria*. Paris: OECD.

Haas, H. D. (2011). Mediterranean migration futures: Patterns, drivers and scenarios. *Global Environmental Change, 21S*(2011), S59–S69.

Halliday, T. (2006). Migration, risk, and liquidity constraints in El Salvador. *Economic Development and Cultural Change, 54*(4), 893–925.

Hammer, T. (2004). Desertification and migration: A political ecology of environmental migration in West Africa. In J. D. Unruh, M. S. Krol, & N. Kliot (Eds.), *Advances in global research: Environmental change and its implications for population migration*. Dordrecht: Kluwer Academic.

Hamza, A., Mohamed, Brahim El Faskaoui, & Alfons, F. (2009). *Migration and environmental change in Morocco: The case of rural oasis villages in the Middle Draa Valley* (EACH-FOR case study report).

Henry, S., & Schoumaker, B. (2004). The impact of rainfall on the first out-migration: A multilevel event-history analysis in Burkina Faso. *Population and Environment, 25*(5), 423–460.

Icduygu, A., & Sert, D. (2011). International migration in the Mashriq countries: Past trends, future potentials and determinants – Possible implications for environmental migration? In Foresight (Ed.), *Migration and global environmental change*. London: Government Office for Science UK.

International Organization for Migration (IOM). (2010). *Migration and development in Egypt*. Cairo: International Organization for Migration.

IPCC. (2012). Managing the risks of extreme events and disasters to advance climate change adaptation. In C. B. Field, V. Barros, T. F. Stocker, D. Qin, D. J. Dokken, K. L. Ebi, M. D. Mastrandrea, G.-K. Plattner, K. J. Mach, S. K. Allen, M. Tignor, & P. M. Midgley (Eds.), *A special report of Working Groups I and II of the Intergovernmental Panel on Climate Change*. Cambridge/New York: Cambridge University Press.

IRIN. (2010, March 23). *Yemen: Capital City Faces 2017 Water Crunch*.

Jager, J., Frühmann, J., Grünberger, S., Vag, A. (2009). *Environmental change and forced migration scenarios*, D.3.4. Synthesis Report.

Joseph, G., & Wodon, Q. (2013a). Is internal migration in Yemen driven by climate or socio-economic factors? In Q. Wodon, A. Liverani, G. Joseph, & N. Bougnoux (Eds.), *Climate change and migration: Evidence from the Middle East and North Africa region*. Washington, DC: The World Bank.

Joseph, G., & Wodon, Q. (2013b). Is the impact of domestic and international remittances on poverty and human development different in various areas depending on their climate? In Q. Wodon, A. Liverani, G. Joseph, & N. Bougnoux (Eds.), *Climate change and migration: Evidence from the Middle East and North Africa region*. Washington, DC: The World Bank.

Joseph, G., Wodon, Q., & Blankespoor, B. (2013a). Do remittances reach households living in unfavorable climate areas? Evidence from Yemen. In Q. Wodon, A. Liverani, G. Joseph, & N. Bougnoux (Eds.), *Climate change and migration: Evidence from the Middle East and North Africa region*. Washington, DC: The World Bank.

Joseph, G., Wodon, Q., Liverani, A., & Blankespoor, B. (2013b). Is climate change likely to lead to higher net internal migration in Yemen? In Q. Wodon, A. Liverani, G. Joseph, & N. Bougnoux (Eds.), *Climate change and migration: Evidence from the Middle East and North Africa region*. Washington, DC: The World Bank.

Kawakibi, S. (2009). *La Gestion de la Migration en Syrie: Une perspective globale*. Florence: European University Institute.

Khawaja, M. (2002). *Internal migration in Syria: Findings from a National Survey, Report 375*. Oslo: FAFO.

Kingdom of Morocco. (2001). *Communication Nationale Initiale à la Convention Cadre des Nations Unies sur les changements climatiques* [First national communication for the United Nations Framework Convention on Climate Change]. Rabat: Ministry for Regional Planning, Urbanism, Housing and the Environment.

Kingdom of Morocco. (2009). *Plan national de lutte contre le réchauffement climatique* [National Action Plan against climate change], Rabat.

Kniveton, D., Smith, C., Black, R., & Schmidt-Verkerk, K. (2009). Challenges and approaches to measuring the migration and environment nexus. In F. Laczko & C. Aghazarm (Eds.), *Migration, environment and climate change: Assessing the evidence*. Geneva: International Organization for Migration.

Laczko, F., & Aghazarm, C. (2009). *Migration, environment and climate change: Assessing the evidence*. Geneva: International Organization for Migration.

Marzouk, N. (2010). *L'émigration syrienne hautement qualifiée: les enjeux sociopolitiques*. Florence: European University Institute.

McLeman, R., & Smit, B. (2006). Migration as an adaptation to climate change. *Climate Change, 76*, 31–53.

McSweeney, C., New, M., & Lizcano, G. (2009). *UNDP climate change profiles: Documentation, Mimeo*. New York: United Nations Development Program.

Meynier, G., & Meynier, P. (2011). L'immigration algérienne en France: histoire et actualité. *Confluences Méditerranée, 77*, 219–234.

Meze-Hausken, E. (2000). Migration caused by climate change: How vulnerable are people in dryland areas? A case study in Northern Ethiopia. *Migration and Adaptation Strategies for Global Change, 5*, 379–406.

Ministère de l'Aménagement du Territoìre, et de l'Environnement (Ministry for Territorial Management and Environment; MATE). (2001). *Elaboration de la stratégie et du plan d'action national des changements climatiques*. Communication nationale initiale.

Ministère de l'Aménagement du Territoìre, et de l'Environnement (Ministry for Territorial Management and Environment; MATE) (2010). *Seconde communication nationale de l'Algérie sur les changements climatiques à la CCNUCC*.

Ministry of State for Environmental Affairs (2010). *Initial National Communication of the Syrian Arab Republic – Submitted to the United Nations Framework Convention on Climate Change*.

Munshi, K. (2003). Networks in the modern economy: Mexican migrants in the U.S. Labor Market. *Quarterly Journal of Economics, 118*(2), 549–599.

National Intelligence Council. (2009). *North Africa: The Impact of Climate Change to 2030 – Geopolitical Implications*. Conference Report prepared by CENTRA Technology, Inc., and Scitor Corporation.

Nguyen, M. C., & Wodon, Q. (2013a). Weather shocks, impact on households, and ability to recover in Morocco. In Q. Wodon, A. Liverani, G. Joseph, & N. Bougnoux (Eds.), *Climate change and migration: Evidence from the Middle East and North Africa region*. Washington, DC: The World Bank.

Nguyen, M. C., & Wodon, Q. (2013b). Weather shocks and migration in Morocco. In Q. Wodon, A. Liverani, G. Joseph, & N. Bougnoux (Eds.), *Climate change and migration: Evidence from the Middle East and North Africa region*. Washington, DC: The World Bank.

Nordås, R., & Gleditsch, N. P. (2007). Climate change and conflict. *Political Geography, 26*(6), 627–638.

Paul, B. K. (2005). Evidence against disaster-induced migration: The 2004 tornado in north-central Bangladesh. *Disasters, 29*(4), 370–385.

Pereira, I., & Caravajal, L. (2008). *Evidence on the link between migration, climate disasters and human development, environment, forced migration and social vulnerability international conference*, Bonn.

Piguet, E. (2010). Linking climate change, environmental degradation, and migration: A methodological overview. *Climate Change, 1*(4), 517–524.
Qabbani, A. (2010, November). State of Denial. *Syria Today.*
Ragnhild, N., & Gleditsch, N. P. (2007). Climate change and conflict. *Political Geography, 26*(6), 627–638.
Rappaport, J. (2007). Moving to nice weather. *Regional Science and Urban Economics, 37*(3), 375–398.
Rappaport, J., & Sachs, J. (2003). The United States as a coastal nation. *Journal of Economic Growth, 8*(1), 5–46.
Ratha, D., Mohapatra, S., & Silwal, A. (2010). *Migration and remittances factbook 2010.* Washington, DC: The World Bank.
Republic of Yemen. (2008). *National adaptation programme of action*, Sanaa.
Reuveny, R. (2007). Climate change-induced migration and violent conflict. *Political Geography, 26*(6), 656–673.
Reuveny, R., & Moore, W. H. (2009). Does environmental degradation influence migration? Emigration to developed countries in the late 1980s and 1990s. *Social Science Quarterly, 90*(3), 461–479.
Rousset, N., & Arrus, R. (2006). L'agriculture du Maghreb au défi du changement climatique : quelles stratégies d'adaptation face à la raréfaction des ressources hydriques?, Communication à WATMED 3, 3ème Conférence internationale sur les Ressources en Eau dans le Bassin Méditerranéen, Tripoli (Liban).
Rudolf, J. C. (2010, October 25). In Yemen, water grows scarcer, *New York Times Green* – A blog about Energy and the Environment.
Sekkai, R. (2008, January 2). Migrants target booming Algeria, BBC Focus on *Africa Magazine*.
Shalaby, T. (2010). *The impact of climate change on migration: The case of Egypt*. Presentation at the First Workshop on Climate-Induced Migration and Displacement in MENA.
Sowers, J., & Weinthal, E. (2010). *Climate change adaptation in the Middle East and North Africa: Challenges and opportunities* (Working Paper No. 2). Dubai: The Dubai Initiative.
Strezpek, K. M., Yates, D. N., Yohe, G., Tol, R. J. S., & Mader, N. (2001). Constructing "not implausible" climate and economic scenarios for Egypt. *Integrated Assessment, 2*, 139–157.
Syrian Ministry of State for Environmental Affairs (SMEA). (2010). *Initial national communication of the Syrian Arab Republic – Submitted to the United Nations Framework Convention on Climate Change*, Damascus.
Temlali, Y. (2010). *Algeria, land of E(Im)migration*. Mimeo.
United Nations Development Programme and Arab Fund for Economic and Social Development. (2009). *Arab human development report 2009: Challenges to human security in the Arab countries*. New York: UNDP.
Verner, D. (Ed.). (2012). *Adaptation to a changing climate in the Arab countries* (Directions in development). Washington, DC: The World Bank.
Warner, K. (2010). Global environmental change and migration: Governance challenges. *Global Environmental Change, 20*, 402–413.
Warner, K., Afifi, T., Dun, O., Stal, M., Schmidl, S. (2008). *Human security, climate change and environmentally induced migration*. United Nations University, Institute for Environment and Human Security.
Warner, K., Ehrhart, C., de Sherbinin, A., Adamo, S., & Chai-Onn, T. (2009). *In search of shelter: Mapping the effects of climate change on human migration and displacement*. Geneva: Mimeo.
Warren, R., Nicholls, N., Levy, R., Price, P. (2006). *Understanding the regional impacts of climate change: Research report prepared for the stern review on the economics of climate change* (Research Working Paper 90). Tyndall Centre for Climate Change.
Wodon, Q., Liverani, A., Joseph, G., & Bougnoux, N. (Eds.). (2013). *Climate change and migration: Evidence from the Middle East and North Africa region*. Washington, DC: The World Bank.
World Bank. (2003, April 23). *World Bank loan to fight poverty in rural Algeria with job creation* (Press Release No. 2003/314/MNA). Washington, DC: World Bank.

World Bank. (2007). *World Bank development report 2008. Agriculture for development.* Washington, DC: The World Bank.

World Bank. (2010a). *World development report 2010: Development and climate change.* Washington, DC: The World Bank.

World Bank. (2010b, May 5). *Project appraisal document on a proposed grant from the Global Environment Facility Trust Fund in the amount of US$4.0 million to the Republic of Yemen for an agro-biodiversity and climate adaptation project.* Washington, DC: World Bank.

World Bank. (2011a). *World development indictors.* Washington, DC: World Bank.

World Bank. (2011b). *Integrating climate change in the implementation of the Plan Maroc Vert.* Washington, DC: World Bank.

World Bank. (2012). *Turn down the heat: Why a 4° C warmer world must be avoided.* Washington, DC: World Bank.

Wynter, A. (2009, June 18). Yemen Red Crescent on climate-change Front Line, *Red Cross Red Crescent – News.*

Zihnioglu, O. (2011). Mediterranean workshop report, in Foresight, Editor, *Migration and global environmental change.* London: Government Office for Science UK.

Zohry, A. (2005). *Interrelationship between internal and international migration in Egypt: A pilot study, Development Research Centre on migration, Globalization and Poverty Research Report.* Brighton: University of Sussex.

Migration and Environmental Change in North America (USA and Canada)

6

Susana B. Adamo and Alexander M. de Sherbinin

Environmental Issues and Migration in North America: No Reasons for Concerns?

Consequences of climate and environmental change for human mobility are usually analyzed in the context of adaptation measures for developing countries (Black et al. 2011b). However, a glimpse at the consequences of past and recent climate and other environmental events shows that the developed world is not immune to these impacts. Cases in point are the depopulation of New Orleans after Hurricane Katrina (2005), and out-migration from the US Great Plains after the droughts and dust storms of the 1930s.

That said, there are relatively few environmental issues in North America that could be said to be grave enough to contribute to massive out-migration and emigration. Both Canada and the US struggle with some environmental issues, according to the 2012 Environmental Performance Index, an internationally comparative index of 132 countries (Emerson et al. 2012). Both countries score poorly on greenhouse gas emissions and the ecosystem effects of air pollution. In addition, the US scores poorly on the ecosystem effects of water management, a measure of the degree to which river flow is altered from its natural state owing to water withdrawals and reservoirs. Yet these environmental challenges do not rise to the level of representing potential drivers of migration.

Different types of *localized* environmental issues may contribute to population mobility. Examples include the collapse of the cod fisheries in Canada's Outer Banks (we come back to this example later in the chapter), "brownfields" (toxic waste dumping areas), strip mines, localized air pollution (especially in expanding southern cities such as Atlanta and Houston), land degradation/soil depletion in some regions, increasing infestations of pine beetles and other pests in the forests

S.B. Adamo (✉) • A.M. de Sherbinin
CIESIN, Columbia University, 61 Route 9W, Palisades, NY 10964, USA
e-mail: sadamo@ciesin.columbia.edu

E. Piguet and F. Laczko (eds.), *People on the Move in a Changing Climate*, Global Migration Issues 2, DOI 10.1007/978-94-007-6985-4_6,

of the Rocky Mountains (which may have a climate connection), heavy nutrient loads in the Mississippi River Valley and corresponding "dead zones" in the Gulf of Mexico, and biodiversity loss from land cover change and urbanization. Each of these issues may contribute to migration, but it is hard to separate the effects of industrial and economic decline from the environmental legacy of industrialization when it comes to understanding patterns of out-migration.

Natural disasters do have an impact on human mobility to a limited extent. Major flooding along the tributaries of the Mississippi River in recent years, the major drought that impacted the Southern US in 2011, beachfront erosion on East Coast, major hurricanes in the southeastern US, the 1998 ice storm in eastern Canada, and the increasing frequency and duration of wildfires (Westerling et al. 2006) may indeed have resulted in individual or household decisions to move locally or regionally, but the evidence for this is still limited.[1]

An Overview of the Main Current Internal and International Migration Flows

North America's population composition, dynamics and distribution play a relevant role in shaping vulnerability to disasters (Donner and Rodríguez 2008). Migration in North America (United States and Canada) is characterized by high levels of international immigration and substantial (but declining) internal population mobility, particularly from rural and small urban places to large metropolitan areas. These migration systems[2] already in place could offer important clues about the possible magnitude, directions and characteristics of future climate-induced population mobility (both migration and displacement) (Adamo and de Sherbinin 2011; Black et al. 2011a; Findlay 2011). This is relevant for environmental migration because (1) movements into fragile areas increase populations at risks; and (2) because movers could be more at risk or be more vulnerable that other groups in the population (Donner and Rodríguez 2008). Because of this, the following section outlines the main characteristics of United States and Canada's international and domestic migration patterns.

[1]In fact, one of the ironies of wildfires is that it is precisely in-migration and development in sensitive ecosystems that puts householders at risk (Westerling et al. 2006).

[2]Migration systems include two or more places connected by flows and counterflows of people (Faist 2004, page 50). Moving beyond early push–pull models, this dynamic approach allows for the reciprocal effects, multiple causation, and rapid changes that characterize different and interconnected forms of population mobility at various temporal and spatial scales (Boyd 1989; Fawcett 1989; Kritz and Zlotnik 1992).

Table 6.1 Summary of international migration stocks in the North America Region

Country	Indicator	1990	2000	2010
Canada	Estimated number of international migrants at mid-year	4,497,521	5,555,019	7,202,340
	International migrants as a percentage of the population	16.2	18.1	21.3
	Female migrants as a percentage of all international migrants	50.9	51.8	52.2
United States	Estimated number of international migrants at mid-year	23,251,026	34,814,053	42,813,281
	International migrants as a percentage of the population	9.1	12.1	13.5
	Female migrants as a percentage of all international migrants	51.1	50.3	49.8

Source: United Nations, Department of Economic and Social Affairs, Population Division (2011). Trends in International Migrant Stock: Migrants by Age and Sex (United Nations database, POP/DB/MIG/Stock/Rev.2011)

International Migration in US and Canada

North America is one of the main destinations for international migration flows (Table 6.1).[3] With 42.8 million immigrants in 2010, United States is the top immigration country, while Canada ranks 5th with 7.2 million immigrants (Ratha et al. 2011). Net migration rates are consistently positives for these two countries, and the number of international migrants increased from 40.4 million in 2000 to 50 million in 2010 (IOM 2010:150). The role of NA as region of immigrants implies that the impacts of climate change in other regions could eventually mean increased migration flows. In words of the IPCC Forth Assessment Report,

> the United States (U.S.) and Canada will experience climate changes through direct effects of local changes (e.g., temperature, precipitation and extreme weather events), as well as through indirect effects, transmitted among regions by interconnected economies and migrations of humans and other species (Field et al. 2007:619).

International migrants represented 20 % of the Canadian population and 13 % of the US population in 2010. The immigration rate[4] in the United States was 3.4 per thousand for the 1990s (United States Census Bureau 2011). In comparison, immigration rates were about 7.7 per 1,000 in 2006 in Canada, continuing a relatively stable trend over the last 20 years (Statistics Canada 2008). This immigration rate is more than double that of the United States, suggesting that despite their lower numbers immigrants are a larger presence in Canada than in the United States.

[3] A note on sources: the Unites States' 2010 Census did not include questions on the foreign-born population. After 2000 data on the foreign born are available only through the American Community Survey and the Current Population Survey.

[4] Annual rate per 1,000 US population.

Table 6.2 Net migration by geographic area, United States, 2000–2009

	Net migration			
Geographic area	Total	International[1]	Domestic	% International migration
United States	8,944,170	8,944,170	–	100.0
Northeast	−704,140	1,835,442	−2,539,582	20.5
Midwest	−593,753	1,158,438	−1,752,191	13.0
South	6,992,907	3,118,775	3,874,132	34.9
West	3,249,156	2,831,515	417,641	31.6

Source: Table 4. Cumulative Estimates of the Components of Resident Population Change for the United States, Regions, States, and Puerto Rico: April 1, 2000 to July 1, 2009 (NST-EST2009-04)
[1]Net international migration includes the international migration of both native and foreign-born populations

Latin Americans make up 49 % of the foreign population in the United States, (Ratha et al. 2011). Migration from Latin America to the United States is the classic example of South-North migration, notably because of its scale (Durand and Massey 2010). Mexico-United States was the main migration corridor in 2010 with 11.6 million migrants, followed by El Salvador (1.1 million) and Cuba (1 million) (Ratha et al. 2011). Other important corridors include China, Philippines, and India (1.7 million each), Vietnam (1.2), and South Korea (1.1). The United States ranked first among remittances-sending countries, with 48.3 billion dollars (Ratha et al. 2011), and it ranked 10th in number of refugees[5] in 2011, with 264,800 people (UNHCR 2012).

In terms of the regional distribution of international migrants in the United States (displayed in Table 6.2), the largest positive net migration balance in 2000–2009 corresponds to the South, with 35 % of the total, followed by the West (32 %), Northeast (20 %) and Midwest (13 %). This indicates that geographic distribution of international migrants within the United States is far from uniform, with some regions and states having more "pull" power than others.

Most (83 %) of the foreign population entered United States before 2005 (Walters and Trevelyan 2011), and some reports indicate that migration to USA is slowing down, with possibly an unprecedented (from a recent historical perspective) zero net migration between Mexico and the United States (Passel et al. 2012). Among the suggested reasons are the current economic crisis, stricter border enforcement, and improved conditions in sending countries (IOM 2010), highlighting once again the heavy influence of economic trends on migration patterns.

Turning now to Canada, the main countries of origin for the 2005–2007 period were China, India, Philippines, Pakistan and United States (IOM 2010). In contrast with the United States, immigrants from Latin America represent only 3 % of the immigrant stock in Canada. In terms of internal distribution of international

[5]Persons recognized as refugees under the 1951 UN Convention/1967 Protocol, the 1969 OAU Convention, in accordance with the UNHCR Statute, persons granted a complementary form of protection and those granted temporary protection (UNHCR 2012:46).

migrants, the majority of immigrants (about 80 %) to Canada go to just three provinces: Ontario, Quebec and British Columbia (Statistics Canada 2008). In 2010 Canada registered 164,900 refugees[6] (IOM 2010), and this population represents about 13 % of the immigrant population in any given year.

Uncertainty about migration stocks in North America arise from irregular or undocumented migration. Estimated unauthorized migration to the United States accumulated about 11.2 million in 2010 (Passel and Cohn 2011), or 30 % of the foreign born population. In Canada, where some authors consider that unauthorized migration as a non-issue (Goldring et al. 2009), numbers are much smaller and estimates show a broad variation, ranking from 20,000 to half a million (Magalhaes et al. 2010).

Emigration is far less significant for these two countries. It is estimated that less than 1 % of the total native population live outside the region, 60 % of them from the United States (IOM 2010). In 2000, emigration from the United States was estimated in 4.3 million people, and the main destinations were Mexico, Canada, Puerto Rico, United Kingdom and Germany. For the same year, there were 1.2 million emigrants from Canada, and the main destinations were the United States (75 %), United Kingdom, France, Australia and Germany (World Bank 2012).

Internal Migration

Another component of a country migration system is internal migration, or movements within national boundaries, very often involving far larger numbers than international migration, and usually the main force behind population redistribution. However, geographical mobility in United States has been declining over time: the mover rate was 12.5 % in 2009, down from 16.5 % in 1996–1997 and 20 % in 1947–1948. In addition, most (67 %) of the year 2009 moves were within the same county (short distance moves), and just 12.6 % of the movers crossed state boundaries. The top destinations for inter-state moves were California, Florida, New Jersey and New York.

The net balance of domestic migration in the US shows heterogeneities across regions and states, displayed in Table 6.2 and Fig. 6.1, with large intra-region variations. The Northeast and Midwest regions present a negative balance for the 2000–2009 period, with the largest negative balance corresponding to New York. The South and West regions, instead, display a positive balance for domestic migration, and the largest positive balance corresponds to Florida. It is appropriate to note that the negative balance for Louisiana is directly related to the displacement impact of hurricanes Katrina and Rita.

Inter-province migration in Canada has also fluctuated over time. After a decade of decreasing movements, it started to increase again in 2006. The largest flows were Alberta-to-British Columbia, Ontario-to-British Columbia, and Quebec-to-Ontario.

[6] Data on remittances is not available for Canada.

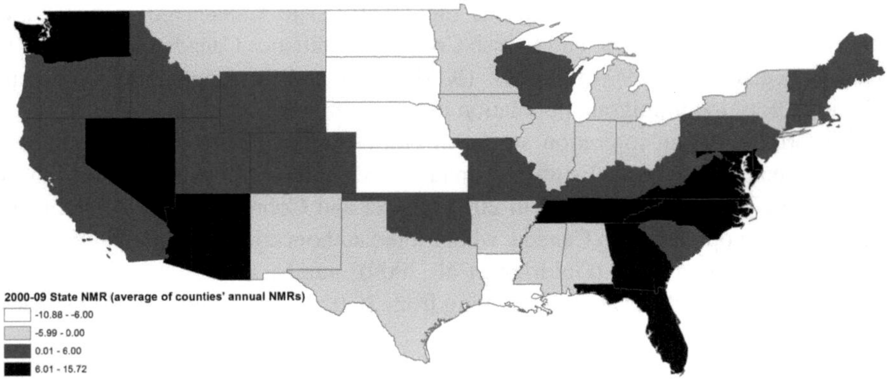

Fig. 6.1 Net migration rates by geographic area (state), United States, 2000–2009. Note: *NMR* net migration rate (Source: Own elaboration based on data downloaded from GEOCOMMONS http://geocommons.com/overlays/35162. Original data from US Census Bureau http://www.census.gov/popest/counties/counties.html)

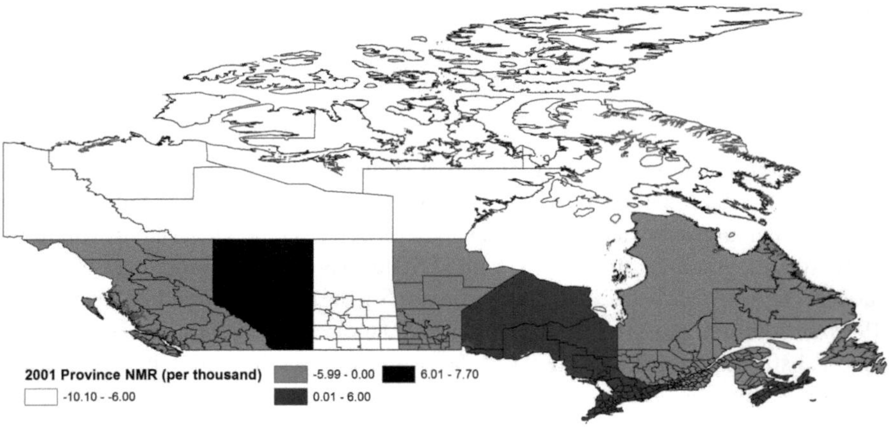

Fig. 6.2 Net migration rate by geographic area (province), Canada, 2001 (Source: Own elaboration based on Sharpe et al. (2007: appendix table 1))

In Fig. 6.2, the display of net migration by province in 2001 shows the distribution of positive and negative net migration rates. Most of the provinces and territories had negative net migration at that point, with the exception of Alberta, Ontario and Nova Scotia. At the census division scale, within-province heterogeneity is common, for example some areas in Quebec and British Columbia showed net gains and some in New Scotia presented net loses despite the overall outcome at the province scale (Statistics Canada 2008).

Historical Overview of Migrations Related to Environmental Events in the Past

Past environmental migration and displacement have been documented in North America in relation to several types of environmental events. This provides a base for postulating the possible effects of climate change impacts on population mobility, assuming that responses to climate variability in the past may highlight the determinants of the decision-making process (Orlove 2005).

Gutmann and Field have suggested to group environmental drivers of population mobility in the US in four-categories: environmental calamities, environmental hardship and bonanza, environmental amenities, and environmental barriers and their management (Gutmann and Field 2010). Here we provide an overview of some of these categories.

Hurricanes hold a prominent place among environmental calamities, but historically documenting their effect on population mobility beyond immediate displacement has been difficult (Gutmann and Field 2010). As an example, the death toll of the 1900 Galveston hurricane is better known that the number of displaced people. Closer in time, the analysis of the effects of hurricane Andrew (Smith and McCarty 1996) and the 2004 hurricane season in Florida (Smith and McCarty 2009) indicate that initial evacuation and displacement could last just a few days, but also that in some cases permanent migration is the final outcome. Out of 353,000 evacuees, 40,000 people decided to leave Dade County (FL) after Hurricane Andrew. Among other variables, strength of the hurricane and general housing vulnerability were significant predictors of evacuation behavior, while the extent of the damage to housing influenced the decision to leave permanently.

Overall, outmigration from US Gulf Coast following hurricanes Katrina and Rita varied by a county's proportion of disadvantaged populations, housing damage, and built environment density (Myers et al. 2008). In the case of the city of New Orleans, flood damage was extensive and most of New Orleans' population was evacuated or displaced (Sastry 2009), with black and low-income neighborhoods being deadly impacted as they were more likely to be located in lower-lying areas (Fussell et al. 2010). Residents were allowed back in the city at the end of September 2005, and by mid-2008 the population was estimated to be about 312,000.[7] The duration of displacement and the dynamics of return showed differences by age, race, education and socioeconomic status, with black residents staying away longer. Similarly to the hurricane's studies mentioned before, degree of housing damage was a key element affecting return (Fussell et al. 2010; Sastry 2009; Stringfield 2010). The selective return was also reflected in the selective recovery of neighborhoods across the city, feeding the hypothesis the hypothesis of larger segregation after major environmental disasters (Pais and Elliott 2008).

[7]Even 5 years after Katrina, in 2010, New Orleans' population was just 343,829, a 29.1 % decline from 2000 (US Census Bureau http://quickfacts.census.gov/qfd/states/22/2255000.html).

The category of environmental hardship is well represented by drought-induced displacement and migration. This category is heavily influenced by the effects of the droughts of the beginning of the twentieth century in the US Great Plains (Gutmann et al. 2005; McLeman 2006; McLeman and Smit 2006; McLeman et al. 2008) and in Canada (McLeman et al. 2010; McLeman and Ploeger 2011). The dust storms of the 1930s Dust Bowl affected Nebraska, Kansas, Oklahoma, Texas, Colorado and New Mexico. The authors coincide in highlighting the intricate web of factors that influence migration and displacement, including several socioeconomic drivers in addition to the prolonged and recurrent droughts that hit the US Great Plains and the Canada Prairies.

In the Great Plains, structural variables linked to a county unemployment level and degree of agricultural employment largely determined net migration sign, with environmental variables (temperature and precipitation) having an effect on net migration on those decades with very extreme values (Gutmann et al. 2005). In Eastern Oklahoma in the 1930s, harsh environmental conditions (years of drought followed by years of excessive rain and flooding) resulted in extensive crop failure. Migrant and non-migrant families showed differences while areas with strong local social networks were less likely to sustain outmigration (McLeman 2006).

Similar results were found for rural households impacted by both droughts and economic recession (including falling commodity process and employment) during the 1930s in Alberta (Canada) (Gilbert and McLeman 2010). Differences in soil quality across farms in Saskatchewan provided another clue to understand the different migration outcomes of drought occurrence (McLeman and Ploeger 2012). At the scale of the entire Canadian Prairies region and for the 1926–1936 period, areas experimenting severe summer drought conditions showed higher levels of population decline (even considering the general depopulation trend of the prairies at the time), although certain heterogeneity remains (McLeman et al. 2010).

Settlement abandonment and depopulation have been the aftermath of environmental crises. For example, the collapse of the cod fisheries in Canada's Outer Banks resulted in a fishing moratorium that contributed to economic stagnation and out migration from the maritime provinces (Hamilton and Butler 2001; Higgins 2008). Other examples are the relocation of Arctic communities due to changes in coastal ecosystems in the case of the Inuit' eastward journey in the Canadian Arctic (Friesen 2010), and the small scale antecedent of the effects of sea level rise on human settlement exemplified by the abandonment of the islands in the Chesapeake Bay that experimented net sea-level rise since mid-nineteenth century (Arenstam Gibbons and Nicholls 2006).

Finally, environment characteristics can also being pulling factors for migration movements, the so-called amenity driven migration. In the United States, basic climate characteristics of counties such as temperature, humidity and wind were found to be correlated with migration rates (in-, out- and return) (Poston et al. 2009). Elderly mobility is another example of population moving into sun-belt states of the South and Southwest (Florida and Arizona, to mention just two), either permanently or just during the winter months (the "snowbirds") (McLeman and Hunter 2010). Amenities are also the base of the tourism industry (Chen et al. 2009), and a typical example are the Mountain West amenity-rich counties combining proximity to high altitudes and water bodies (Gutmann et al. 2005; Shumway and Otterstrom 2001).

Climate Change in North America

It is expected that the North America region will be affected by climate change impacts in the form of extreme weather events (e.g., higher intensity hurricanes along the Gulf Coast), changes in water availability (e.g., decline in fresh water resources in the US Midwest and Southwest and the Canadian prairies), the reduction of arctic sea ice (which has already prompted resettlement of some coastal Alaskan communities), and seasonal droughts (which contribute to forest fires in all regions). This section begins with a consideration of the evidence for recent climatic changes, and then continues to a discussion of future scenarios.

In reality the climate has already changed in North America. Analyses by the National Climatic Data Center and represented in maps available from the U.S. Environmental Protection Agency (EPA undated) show that the Northeast and Northwest have seen increases in temperature of between 1 and 2 °C, and the Southwest has seen warming of up to 3 °C. Winter temperatures have risen even more quickly, with average winter temperatures in the Midwest and northern Great Plains increasing more than 3.8 °C (Karl et al. 2009). Precipitation patterns are more patchy but generally show secular increases over the south central U.S. of between 10 and 20 %, together with the great lakes region and New England. Only spotty areas of drying are found.

The IPCC Special Report on Extremes (henceforth "SREX report") (Field et al. 2012) has found an increasing trend in total precipitation over the past half century. Station data from Canada, the United States, and Mexico also shows that heavy precipitation and the average amount of precipitation falling on days with precipitation has been increasing over 1950–2004. There has been an increasing area experiencing above-normal proportion of heavy daily precipitation from 1950 to 2006, with much of that focused on the Great Plains and northwestern Midwest. According to the US National Climate Assessment (Karl et al. 2009), the top 1 % of precipitation events increased by 20 %, while total precipitation increased by only 7 % over the past century. The SREX report (p. 142) concludes, "Overall, the evidence indicates a *likely* increase in observed heavy precipitation in many regions in North America, despite statistically non-significant trends and some decreases in some subregions."

Although the media are quick to attribute (or question the attribution of) individual climate-related extreme event to climate change, as Trenberth (2012) and the SREX report, many current patterns of variability and extremes are consistent with what is anticipated on a warming planet (Field et al. 2012). As Trenberth points out, "all weather events are affected by climate change because the environment in which they occur is warmer and moister than it used to be." It is also known that severity of natural disasters can be a function of greater extremes, changes in the patterns of human settlement and activities, or both. For example, flooding can result both from a higher percentage of rainfall occurring in shorter periods, and from land cover changes such as wetlands loss and increasing impervious surface cover owing to urbanization patterns. The floods, in turn, may have a greater economic or human impact than in the past because of settlement in flood plains.

The 2011 heat wave and drought in Texas and Oklahoma is consistent with projected climate change impacts in the southwestern US (Weiss et al. 2012). In the Central US, on the other hand, the SREX report finds that drought periods have become less frequent but more intense. Flooding along Mississippi River and its tributaries has become more frequent and damaging (Trenberth 2012). According to the SREX report, there has been an increase in the numbers of cyclones over the last 50 years, with no statistically significant change in cyclone intensity. Finally, there has been some speculation that tornados have increased in frequency and intensity as a result of changes in climate. In 2011 the U.S. saw 1,612 tornados, with major outbreaks near Huntsville, Alabama and in the mid-West (with 158 deaths in Joplin, Missouri), putting it in the 80+ percentile in terms of annual numbers (Blunden and Arndt, 2012: 184). According to Trenberth (2012), "Global warming does not contribute directly to tornadoes themselves, but it does contribute to the vigor of the thunderstorms that host them through the increased warmth and moisture content (moist static energy) of the low level air flow."

So what does the future hold? Based on climate modeling conducted for the IPPC's AR4 (Field et al. 2007), average temperatures will increase in North America in the next 25 years by 1–3 °C, and by the end of the century they will increase by 2–3 °C and up to 5 °C in the high latitudes. According to the AR4's Working Group 1 report (Christensen et al. 2007:887), annual mean precipitation is very likely to increase in Canada and the Northeast USA, and likely to decrease in the Southwest USA. In southern Canada, precipitation is likely to increase in winter and spring, but decrease in summer.

In terms of climate extremes, according to the SREX report there is medium confidence that droughts will increase in frequency and intensity in the twenty-first century in Central North America due to reduced precipitation and/or increased evapotranspiration. There will be a decrease in the return period of a 1 in 20 year flood over most of North America to roughly 1 in 10 years in the 2050 time period and to roughly 1 in 8 years by 2080–2100. Wang and Zhang (2008) used statistical downscaling to investigate possible changes in North American extreme precipitation probability during winter from the last half of the twentieth century to the last half of the twenty-first century. Downscaled results suggested that there will be a strong increase in extreme precipitation over the south and central United States but decreases over the Canadian prairies (Wang and Zhang 2008).

The heavier precipitation will likely contribute to flooding, both inland and coastal. According to the SREX report, beyond direct destruction of property, flooding has important negative impacts on a variety of economic sectors including transportation and agriculture (e.g., soil compaction and crop losses). If losses exceed a certain threshold, or the return period is shorted dramatically, as suggested by the SREX report, it is conceivable that this could precipitate out-migration from vulnerable areas.

Scenarios suggest that the southwestern U.S. will become hotter and drier (Field et al. 2007; Karl et al. 2009; Weiss et al. 2012), caused by a weakening of the North American monsoon as a result of the poleward expansion of the Hadley Cell over dry subtropical areas (Lu et al. 2007). Water resources may become more highly contested, and as a result, this could exert pressure on communities that depend on scarce water resources for domestic or agricultural supply. In the history of contested water resources, it is usually the wealthier cities that ultimately gain control of water resources, which means that in agricultural areas that have reduced water supplies (potentially including the Imperial Valley of southern California), farming communities may be depopulated. According to Christensen et al. (2004), climate changes may reduce states' abilities to meet water sharing agreements in contested regions such as the Colorado River Basin (Christensen et al. 2004). This is compounded by likely reductions in snowpack in the Rocky mountains that feed the Colorado (Field et al. 2007). Ironically, the Southwest has also been an area of heavy in-migration over the past two decades (as was mentioned in the previous sections, see also Table 6.2), so it will make adaptation to the reduced water supplies more challenging (de Sherbinin et al. 2012). This issue was underscored in the 2009 U.S. National Climate Assessment (Karl et al. 2009:129–134).

Future sea level rise and potential increases in storm surge are likely to increase inundation and property damage in coastal areas. Frey et al. (2010) simulated the combined effects of sea level rise and more powerful hurricanes on storm surge in southern Texas in the 2080s, finding that the inundated area could increase by 60–230 % across scenarios evaluated (Frey et al. 2010). The SREX report finds that the uncertainties surrounding changes in tropical and extratropical cyclones mean that it is not possible to assess future storm surge with accuracy. Yet the earlier IPCC AR4 report (Field et al. 2007) finds that extra-tropical storms such as those that affect the US and Canada are likely to become more intense (but perhaps less frequent), leading to increased extreme wave heights in the mid-latitudes. The AR4 report assigns very high confidence to higher flood levels as a result of high rainfall and storm-surge.

In summary, the principal climate impacts that will affect migration in North America will be determined by the interplay of higher temperatures, which result in higher levels of evapotranspiration, and regional changes in precipitation. Additional factors will be decreased winter snow pack, which is an important source of water supply in the Western U.S., and the impacts of sea level rise and storm surge on coastal communities.

A Synthesis of Existing Case Studies on the Links Between Migration and Climate-Related Events

Population displacement is likely to increase as a result of fast-onset extreme events – cyclones, flooding, and tornados – all of which are increasing in severity and, with the exception of cyclones, also in frequency. There are uncertainties, however, in the degree to which climate change will contribute to each of these events (covered above), and there are further uncertainties concerning the degree to which this displacement will result in permanent relocation (Bakewell 2011), as happened in the case of Hurricane Katrina (Cutter 2011; Zaninetti and Colten 2012). Continued patterns of coastal development, when combined with sea level rise, issues of land subsidence in some regions, and storm surges, suggest that U.S. coastal areas, particularly flat and low lying areas on the East and Gulf Coasts, may experience significant population displacement from storm-related activities.

Increase in the strength of hurricanes ranks high in the list of concerns. Esnard et al. (2011:845) developed an index of relative displacement risk[8] (DRI), which combines measures of physical vulnerability and community resilience with hurricane exposure probabilities. The index was applied to the study of the effects of the 2007 hurricane season in an area defined by 158 coastal counties from North Carolina to Texas. The authors found that the overall DRI was higher in the coastal counties of the Gulf of Mexico than in the counties located on the Atlantic Ocean coast, with clusters of high risk in South Florida, Louisiana and Texas. Their results also showed that the displacement risk was not strictly limited to coastal counties, as some inland counties had risk scores associated with moderate risk of displacement, linked to higher scores in vulnerability that coastal counties (Esnard et al. 2011).

The elaboration of indices brings to the front the issue of information sources. There has been some research on data development looking for alternative and accurate data sources, for example administrative records, in addition to the usual sources (census and surveys) in order to estimate displacement and population redistribution after hurricanes and other catastrophic disasters (Plyer et al. 2010). More specifically, the authors suggest to combine census population data, administrative data collected on an ongoing basis (utility accounts, residences receiving mail and change of address data, driver licenses, voter registration, traffic, school enrollment), and data collected after the disaster by FEMA and others (housing damage estimates, FEMA's individuals and households program). They conclude by stating that "the appropriate administrative data sets to be used as a basis for monthly population estimates may differ by point in time relative to the disaster and by the impact of the disaster on each county" (Plyer et al. 2010:170).

Populations in low elevation coastal zones (LECZs) are another concern. By 2000, 24 million people (8 % of the total population) lived in North America's LECZ, estimated in 553 thousand square kilometers, and 21 million of them were located in urban areas. United States ranked 8th in terms of total population in LECZ, while Canada and Unites States ranked 2nd and 3rd respectively in terms of total land area of the country in LECZ. These numbers make for very large area and population at risk (McGranahan et al. 2007).

---

[8]The displacement risk index (DRI) combines measures of socioeconomic vulnerability, physical vulnerability, community resilience, hurricane return periods, and probability of strikes. The authors defined displacement as "the uprooting of people from their homes resulting from a hurricane disaster for periods of time that exceed the typical temporary shelter timeframe of 3 months". Raw scores where converted to "a standard normal percentile rank where a percentile of 100 denotes maximum hurricane-related displacement risk for a county" (Esnard et al. 2011:834, 849).

Curtis and Schneider (2011) estimated that about 20 million people could be affected by sea-level-rise (SLR) by 2030 in their study area of California, Florida, New Jersey and South Carolina. The effects of SLR will be both direct (through immediate displacement) and indirect (through disruption of livelihoods), but the impacts on displacement will have considerable spatial heterogeneity depending on socio-demographic characteristics and associated degree of vulnerability. Very importantly, impacts will very likely extend *beyond* the inundated counties (the ripple effect) towards other counties through existing, modified and new migration flows, and effectively restructuring existing migration patterns (Curtis and Schneider 2011).

According to Field et al. (2007:632), "residents of northern Canada and Alaska are likely to experience the most disruptive impacts of climate change, including shifts in the range or abundance of wild species crucial to livelihoods and well-being of indigenous peoples (high confidence)." Changing weather patterns and reduction in ice pack has already affected the subsistence activities of indigenous groups, and this may have impacts on migration, although the numbers of migrants would be small. The cultural dislocation is likely to be the most significant impact. Several papers address the displacement and planned relocation of Arctic communities in Alaska and Canada, as ice decreases, warmer winters and permafrost thaw threatened villages' infrastructure, housing and livelihoods (Turner and Clifton 2009). In the Arctic, melting permafrost and coastal flooding and erosion have resulted in the relocation of some coastal indigenous communities (Bronen 2011; Marino 2012; Oliver-Smith 2009), and there are plans for further relocation despite the uncertain results.

Mobility and displacement could also be the direct or indirect result of mitigation measures (de Sherbinin et al. 2011). For example, low carbon policies might affect migration in several ways: in receiving areas, by changing the characteristics of labor markets (fewer low-skilled jobs in a low-carbon economy) and increasing the cost of services; in sending areas, by improving access to energy sources, and improving income diversification with biofuel production (Reilly and Hossain 2011). However, despite hopeful expectations, no association has been found so far between the expansion of biofuel plants in west North Central United States and demographic trajectories of rural counties (Kulcsár and Bolender 2011).

Finally, changes in water availability are also likely to have effects on internal population mobility. For example, a study by Feng and others examined the sensitivity of US internal migration to weather-induced crop yields, for the 1970–2009 period. The authors found a non-trivial effect on future movements, concluding that in the medium term (2020–2049), about 3.7 % of the adult population could leave rural the Corn Belt counties under the yield changes predicted by the B2 scenario.[9] The effect could be even larger for young adults (Feng et al. 2012).

[9]"The B2 storyline and scenario family describes a world in which the emphasis is on local solutions to economic, social, and environmental sustainability. It is a world with continuously increasing global population at a rate lower than A2, intermediate levels of economic development,

International migration flows to the United States could be affected by the effects of climate change in sending countries (WBGU (German Advisory Council on Global Environmental Change) 2007). Declines in water availability and increase frequency of droughts could led to the intensification of Mexico-US migration flows (Schwartz Leighton and Notini 1994). A recent study indicates that water scarcity in Mexico have the potential to increase Mexico-US migration through the impacts on agriculture productivity in the sending areas (Feng et al. 2010). The authors estimated that each 10 % reduction in crop yields might increase Mexican emigration to USA by 2 %.

On the same topic, Schmidt-Verkerk (2011) analyzes the sensitivity of legal and undocumented Mexico-US migration flows to climate change impacts. CC impacts in Mexico's rural sending areas may increase the pressure to migrate (through reducing employment opportunities, higher commodity prices, changes in financial resources) while at the same time reducing the means to do so. On the other hand, impacts in the United States' receiving areas may reduce labor needs in commercial farming and have a chilling effect on immigration (Schmidt-Verkerk 2011:237–239).

To the extent that North America is a region of immigration, changes in international migration flows due to CC impacts are bound to affect it. But there is a lack of international legal and normative frameworks for addressing environmental migration, combined with the fact that immigration laws in most receiving countries do not contemplate the case of environmental immigrants, unless they agree with some already existing category (Martin 2010b:405). As examples, Canada (net immigration country) agreed to suspend deportations of citizens of countries affected by the 2004 tsunami; Mexico's National Strategy on Climate Change does not address migration issues at all despite this being a net emigration country (Martin 2009:368). In the particular case of the United States, the 1990 temporary protected status legislation does contemplate the case of environmental disaster (earthquake, flood, drought, epidemic or other), but only for persons already in the country and on a temporary basis (Martin 2010a). This legislation has been applied in the cases of earthquakes in El Salvador and Haiti, and countries affected by Hurricane Mitch (Honduras y Nicaragua).

Final Remarks

This brief review of past events and future scenarios of climate change and migration in North America illustrate once again the complexity of the issues at hand. It is difficult to avoid the conclusion that, at the national and state or province levels, the major driver of migration in North America is still economics. It is past and present

and less rapid and more diverse technological change than in the B1 and A1 storylines. While the scenario is also oriented toward environmental protection and social equity, it focuses on local and regional levels" (IPCC 2000).

patterns of, and differences in, economic growth – and therefore in the development of housing and residential areas – that have often led to environmental issues that may contribute to environmental displacement and migration today.

However, the examination of relatively new research on possible scenarios of climate change-induced migration suggests that the direct and indirect contribution of environmental factors at the local level is not trivial, and that it could escalate and extend to other geographic areas through the ripple effects of existing migration networks. It also indicates that the impacts are likely to show heterogeneities across regions (for example, coastal areas and drylands would be more exposed) and social groups (e.g. indigenous communities would be more vulnerable). These results need to be incorporated into existing adaptation and mitigation policies.

References

Adamo, S. B., & de Sherbinin, A. (2011). The impact of climate change on the spatial distribution of the population and on migration. In United Nations. Population Division (Ed.), *Population distribution, urbanization, internal migration and development: An international perspective* (pp. 161–195). New York: UN. Population Division.

Arenstam Gibbons, S. J., & Nicholls, R. J. (2006). Island abandonment and sea-level rise: An historical analog from the Chesapeake Bay, USA. *Global Environmental Change, 16*(1), 40–47.

Bakewell, O. (2011). Conceptualising displacement and migration: Processes, conditions, and categories. In K. Koser & S. Martin (Eds.), *The migration-displacement nexus: Patterns, processes, and policies*. Oxford: Berghahn Books.

Black, R., Bennett, S. R. G., Thomas, S. M., & Beddington, J. R. (2011a). Climate change: Migration as adaptation. *Nature, 478*(7370), 447–49.

Black, R., Neil Adger, W., Arnell, N. W., Dercon, S., Geddes, A., & Thomas, D. (2011b). The effect of environmental change on human migration. *Global Environmental Change, 21*(Supplement 1(0)), S3–S11.

Blunden, J., & Arndt, D. S. (Eds.) (2012). State of the climate in 2011. *Bulletin of the American Meteorological Society, 93*(7), S1–S264.

Boyd, M. (1989). Family and personal networks in international migration: Recent developments and new agendas. *International Migration Review, 23*(3), 638–670.

Bronen, R. (2011). Climate-induced community relocations: Creating an adaptative governance framework based in human rights doctrine. *NYU Review of Law & Social Change, 35*, 356–406.

Chen, Y., Irwin, E. G., & Jayaprakash, C. (2009). Dynamic modeling of environmental amenity-driven migration with ecological feedbacks. *Ecological Economics, 68*(10), 2498–2510.

Christensen, N. S., Wood, A. W., Voisin, N., Lettenmaier, D. P., & Palmer, R. N. (2004). The effects of climate change on the hydrology and water resources of the Colorado River basin. *Climatic Change, 62*(1), 337–363.

Christensen, J. H., Hewitson, B., Busuioc, A., Chen, A., Gao, X., Held, I., Jones, R., Kolli, R. K., Kwon, W.-T., Laprise, R., Magaña Rueda, V., Mearns, L., Menéndez, C. G., Räisänen, J., Rinke, A., Sarr, A., & Whetton, P. (2007). Regional climate projections. In S. Solomon, D. Qin, M. Manning, Z. Chen, M. Marquis, K. B. Averyt, M. Tignor, & H. L. Miller (Eds.), *Climate change 2007: The physical science basis. Contribution of Working Group I to the fourth assessment report of the Intergovernmental Panel on Climate Change*. Cambridge/New York: Cambridge University Press.

Curtis, K., & Schneider, A. (2011). Understanding the demographic implications of climate change: Estimates of localized population predictions under future scenarios of sea-level rise. *Population and Environment, 33*(1), 28–54.

Cutter, S. (2011). *The Katrina exodus: Internal displacements and unequal outcomes*. London, UK Government's Foresight Project, Migration and Global Environmental Change. http://www.bis.gov.uk/assets/foresight/docs/migration/case-studies/11-1156-cs1-katrina-exodus-internal-displacements.pdf

de Sherbinin, A., Castro, M., Gemenne, F., Cernea, M. M., Adamo, S., Fearnside, P. M., Krieger, G., Lahmani, S., Oliver-Smith, A., Pankhurst, A., Scudder, T., Singer, B., Tan, Y., Wannier, G., Boncour, P., Ehrhart, C., Hugo, G., Pandey, B., & Shi, G. (2011). Preparing for resettlement associated with climate change. *Science, 334*(6055), 456–457.

de Sherbinin, A., Levy, M., Adamo, S. B., MacManus, K., Yetman, G., Mara, V., Razafindrazay, L., Goodrich, B., Srebotnjak, T., Aichele, C., & Pistolesi, L. (2012). Migration and risk: Net migration in marginal ecosystems and hazardous areas. *Environmental Research Letters* 7(4):045602. http://stacks.iop.org/1748-9326/7/i=4/a=045602.

Donner, W., & Rodríguez, H. (2008). Population composition, migration and inequality: The influence of demographic changes on disaster risk and vulnerability. *Social Forces, 87*(2), 1089–1114.

Durand, J., & Massey, D. S. (2010). New world orders: Continuities and changes in Latin American migration. *Annals of the American Academy of Political and Social Science, 630*, 20–52.

Emerson, J. W., Hsu, A., Levy, M. A., de Sherbinin, A., Mara, V., Esty, D. C., & Jaiteh, M. (2012). *2012 Environmental performance index and pilot trend environmental performance index*. New Haven: Yale Center for Environmental Law and Policy.

Esnard, A. M., Sapat, A., & Mitsova, D. (2011). An index of relative displacement risk to hurricanes. *Natural Hazards, 59*(2), 833–859.

Faist, T. (2004). *The volume and dynamics of international migration and transnational social spaces*. Oxford: Clarendon.

Fawcett, J. T. (1989). Networks, linkages and migration systems. *International Migration Review, 23*(3), 671–680.

Feng, S., Krueger, A. B., & Oppenheimer, M. (2010). Linkages among climate change, crop yields and Mexico–US cross-border migration. *Proceedings of the National Academy of Sciences, 107*(32), 14257–14262.

Feng, S., Oppenheimer, M., & Schlenker, W. (2012). Climate change, crop yields, and internal migration in the United States (NBER Working Paper).

Field, C. B., Mortsch, L. D., Brklacich, M., Forbes, D. L., Kovacs, P., Patz, J. A., Running, S. W., & Scott, M. J. (2007). North America. In M. L. Parry, O. F. Canziani, J. P. Palutikof, P. J. van der Linden, & C. E. Hanson (Eds.), *Climate change 2007: Impacts, adaptation and vulnerability. Contribution of Working Group II to the fourth assessment report of the Intergovernmental Panel on Climate Change* (pp. 617–52). Cambridge: Cambridge University Press.

Field, C. B., Barros, V., Stocker, T. F., Qin, D., Dokken, D. J., Ebi, K. L., Mastrandrea, M. D., Mach, K. J., Plattner, G.-K., Allen, S. K., Tignor, M., & Midgley, P. M. (Eds.). (2012). *Managing the risks of extreme events and disasters to advance climate change adaptation. A special report of Working Groups I and II of the Intergovernmental Panel on Climate Change*. Cambridge/New York: Cambridge University Press.

Findlay, A. M. (2011). Migrant destinations in an era of environmental change. *Global Environmental Change, 21*(Supplement 1(0)), S50–S58.

Frey, A. E., Olivera, F., Irish, J. L., Dunkin, L. M., Kaihatu, J. M., Ferreira, C. M., & Edge, B. L. (2010). The impact of climate change on hurricane flooding inundation, population affected, and property damages. *Journal of the American Water Resources Association, 46*(5), 1049–1059.

Friesen, T. M. (2010). Dynamic Inuit social strategies in changing environments: A long-term perspective. *Geografisk Tidsskrift-Danish Journal of Geography, 110*(2), 215–225.

Fussell, E., Sastry, N., & VanLandingham, M. (2010). Race, socioeconomic status, and return migration to New Orleans after Hurricane Katrina. *Population and Environment, 31*(1), 20–42.

German Advisory Council on Global Environmental Change (WBGU). (2007). *Climate change as security risk*. Berlin: WBGU.

Gilbert, G., & McLeman, R. (2010). Household access to capital and its effects on drought adaptation and migration: A case study of rural Alberta in the 1930s. *Population and Environment, 32*(1), 3–26.

Goldring, L., Berinstein, C., & Bernhard, J. (2009). Institutionalizing precarious migratory status in Canada. *Citizenship Studies, 13*(3), 239–265.

Gutmann, M., & Field, V. (2010). Katrina in historical context: Environment and migration in the U.S. *Population and Environment, 31*(1), 3–19.

Gutmann, M., Deane, G., Lauster, N., & Peri, A. (2005). Two population-environment regimes in the Great Plains of the United States, 1930–1990. *Population and Environment, 27*(2), 191–225.

Hamilton, L. C., & Butler, M. J. (2001). Outport adaptations: Social indicators through Newfoundland's cod crisis. *Human Ecology Review, 8*(2), 1–11.

Higgins, J. (2008). *Rural depopulation*. Newfoundland/Labrador Heritage: Memorial University of Newfoundland.

IOM. (2010). *World migration report 2010: The future of migration: Building capacities for change*. Geneva: IOM.

IPCC. (2000). *Special report on emissions scenarios. Summary for policy makers*. http://www.ipcc.ch/pdf/special-reports/spm/sres-en.pdf

Karl, T. R., Melillo, J. M., & Peterson, T. C. (Eds.). (2009). *Global climate change impacts in the United States*. Cambridge: Cambridge University Press.

Kritz, M., & Zlotnik, H. (1992). Global interactions: Migration systems, processes and policies. In M. Kritz, L. Lin Lean, & H. Zlotnik (Eds.), *International migration systems. A global approach* (pp. 1–16). New York: Oxford University Press.

Kulcsár, L., & Bolender, B. (2011). If you build it, will they come? Biofuel plants and demographic trends in the Midwest. *Population and Environment, 32*(4), 318–331.

Lu, J., Vecchi, G. A., & Reichler, T. (2007). Expansion of the Hadley cell under global warming. *Geophysical Research Letters, 34*, L06805.

Magalhaes, L., Carrasco, C., & Gastaldo, D. (2010). Undocumented migrants in Canada: A scope literature review on health, access to services, and working conditions. *Journal of Immigrant and Minority Health, 12*, 132–151.

Marino, E. (2012). The long history of environmental migration: Assessing vulnerability construction and obstacles to successful relocation in Shishmaref, Alaska. *Global Environmental Change, 22*, 374–381.

Martin, S. (2009). Managing environmentally induced migration. In F. Laczko & C. Aghazarm (Eds.), *Migration, environment and climate change: Assessing the evidence*. Geneva: IOM.

Martin, S. (2010a). *Climate change and international migration*. Washington, DC: The German Marshall Fund of the United States.

Martin, S. (2010b). Climate change, migration and governance. *Global Governance, 16*, 397–414.

McGranahan, G., Balk, D., & Anderson, B. (2007). The rising tide: Assessing the risks of climate change and human settlements in low elevation coastal zones. *Environment and Urbanization, 19*(1), 17–37.

McLeman, R. (2006). Migration out of 1930s rural eastern Oklahoma: Insights for climate change research. *Great Plains Quarterly 26*(1), 27–40.

McLeman, R. A., & Hunter, L. M. (2010). Migration in the context of vulnerability and adaptation to climate change: Insights from analogues. *Wiley Interdisciplinary Reviews: Climate Change, 1*(3), 450–461.

McLeman, R., & Ploeger, S. (2012). Soil and its influence on rural drought migration: Insights from Depression-era Southwestern Saskatchewan, Canada. *Population & Environment, 33*(4), 304–333.

McLeman, R., & Smit, B. (2006). Migration as an adaptation to climate change. *Climatic Change, 76*(1), 31–53.

McLeman, R., Mayo, D., Strebeck, E., & Smit, B. (2008). Drought adaptation in rural eastern Oklahoma in the 1930s: Lessons for climate change adaptation research. *Mitigation and Adaptation Strategies for Global Change, 13*(4), 379–400.

McLeman, R., Herold, S., Reljic, Z., Sawada, M., & McKenney, D. (2010). GIS-based modeling of drought and historical population change on the Canadian Prairies. *Journal of Historical Geography, 36*(1), 43–56.

Myers, C., Slack, T., & Singelmann, J. (2008). Social vulnerability and migration in the wake of disaster: The case of Hurricanes Katrina and Rita. *Population and Environment, 29*, 271–291.

Oliver-Smith, A. (2009). Sea level rise and the vulnerability of coastal peoples: Responding to the local challenges of global climate change in the 21st century. *InterSecTions, No. 7*.

Orlove, B. (2005). Human adaptation to climate change: A review of three historical cases and some general perspectives. *Environmental Science & Policy, 8*, 589–600.

Pais, J. F., & Elliott, J. R. (2008). Places as recovery machines: Vulnerability and neighborhood change after major hurricanes. *Social Forces, 86*(4).

Passel, J., & Cohn, D. (2011). Unauthorized immigrant population: National and state trends, 2010. Washington, DC, Pew Research Center. http://www.pewhispanic.org/files/reports/133.pdf.

Passel, J., Cohn, D., & Gonzalez-Barrera, A. (2012). *Net migration from Mexico falls to zero – and perhaps less*. Washington, DC: Pew Hispanic Center.

Plyer, A., Bonaguro, J., & Hodges, K. (2010). Using administrative data to estimate population displacement and resettlement following a catastrophic U.S. disaster. *Population and Environment, 31*(1), 150–175.

Poston, D., Zhang, L., Gotcher, D., & Gu, Y. (2009). The effect of climate change on migration: United States, 1995–2000. *Social Science Research, 38*, 743–753.

Ratha, D., Mohapatra, S., & Silwal, A. (2011). Migration and remittances factbook 2011. Washington, DC, IBRD and World Bank. http://data.worldbank.org/data-catalog/migration-and-remittances.

Reilly, M., & Hossain, Y. (2011). The impact of low-carbon policy on migration. In Foresight (Ed.), *Migration and global environmental change*. London: Government Office for Science UK.

Sastry, N. (2009). Displaced New Orleans residents in the aftermath of Hurricane Katrina: Results from a pilot survey. *Organization & Environment, 22*(4), 395–409.

Schmidt-Verkerk, K. (2011). *The potential influence of climate change on migratory behaviour – A study of drought, Hurricanes and migration in Mexico*. Doctoral thesis. Brighton: University of Sussex.

Schwartz Leighton, M., & Notini, J. (1994). *Desertification and migration: Mexico and the United States*. San Francisco: U.S. Commission on Immigration Reform.

Sharpe, A., Arsenault, J.-F., & Ershov, D. (2007). *The impact of interprovincial migration on aggregate output and labour productivity in Canada, 1987–2006*. Ontario: Center for the Study of Living Standards.

Shumway, J. M., & Otterstrom, S. M. (2001). Spatial patterns of migration and income change in the mountain west: The dominance of service-based, amenity-rich counties. *The Professional Geographer, 53*(4), 492–502.

Smith, S., & McCarty, C. (1996). Demographic effects of natural disasters: A case study of Hurricane Andrew. *Demography, 33*(2), 265–275.

Smith, S., & McCarty, C. (2009). Fleeing the storm(s): An examination of evacuation behavior during Florida's 2004 hurricane season. *Demography, 46*(1), 127–145.

Statistics Canada. (2008). *Report on the demographic situation in Canada*. Ottawa: Statistics Canada.

Stringfield, J. (2010). Higher ground: An exploratory analysis of characteristics affecting returning populations after Hurricane Katrina. *Population and Environment, 31*(1), 43–63.

Trenberth, K. (2012). Framing the way to relate climate extremes to climate change. *Climatic Change, 115*(2), 283–290.

Turner, N. J., & Clifton, H. (2009). "It's so different today": Climate change and indigenous lifeways in British Columbia, Canada. *Global Environmental Change, 19*(2), 180–190.

U.S. Environmental Protection Agency (EPA). (Undated). Recent climate change. Available at http://epa.gov/climatechange/science/recenttc_tempanom.html and http://epa.gov/climatechange/science/recentpsc_precipanom.html. Accessed 30 June 2012.

UNHCR. (2012). *Global trends 2011*. A year of crises. Geneva, UNHCR. http://www.unhcr.org/4fd6f87f9.html.

United States Census Bureau. (2011). *Statistical abstract of the United States: 2012. Historical statistics*. Washington, DC: US Census Bureau.

Walters, N., & Trevelyan, E. (2011). *The newly arrived foreign born population of the United States: 2010*. Washington, DC: United States Census Bureau.

Wang, J., & Zhang, X. (2008). Downscaling and projection of winter extreme daily precipitation over North America. *Journal of Climate, 21*(5), 923–937.

Weiss, J. L., Overpeck, J. T., & Cole, J. E. (2012, March). Warmer led to drier: Dissecting the 2011 drought in the southern U.S. *Southwest Climate Outlook, 11*(3), 3–4.

Westerling, A. L., Hidalgo, H. G., Cayan, D. R., & Swetnam, T. W. (2006). Warming and earlier Spring increase Western U.S. forest wildfire activity. *Science, 313*, 940–943.

World Bank. (2012). *World Bank data: Global bilateral migration*. http://databank.worldbank.org/data/views/variableselection/selectvariables.aspx?source=global-bilateral-migration

Zaninetti, J.-M., & Colten, C. E. (2012). Shrinking New Orleans: Post-Katrina population adjustments. *Urban Geography, 33*(5), 675–699.

Migration and Climate Change in Latin America and the Caribbean

7

Raoul Kaenzig and Etienne Piguet

Introduction

This chapter is based on existing empirical studies related to climate change and migration in Latin America and the Caribbean.[1] It looks at the situation in the region as compared with that in other regions of the world, while underscoring certain aspects that are specific to Latin America and the Caribbean. Historical analogy is used; a summary of the past consequences of environmental degradation for migration facilitates an evaluation of the future consequences of climate change. In addition, this compilation of existing studies also makes it possible to reflect critically on the geographic and theoretical distribution of the case studies and to identify the regions for which additional and complementary studies would be desirable, given the vulnerability of those regions. Based on the existing literature, three kinds of climate evolution are expected to have the greatest impact in terms of population displacement: natural hazards (tropical cyclones,[2] heavy rains and floods), droughts and sea level rise (Piguet 2008; Intergovernmental Panel on Climate Change 2007). Added to this list is the melting of glaciers, which is a particularly sensitive issue in some Andean countries. The present chapter includes an evaluation of the impact on migration of each of these phenomena based either on historical experience or on projections.

[1]The present chapter builds partly on the observations made in a previous article by the same authors on Central America and South America Kaenzig and Piguet (2011). It includes an updated look at the region and the analysis is extended to include the Caribbean.

[2]The generic term "tropical cyclone" includes hurricanes (in the western Atlantic and eastern Pacific), typhoons (in the western Pacific), cyclones (in the southern Pacific and the Indian Ocean), tropical storms, etc. For this chapter, in the particular case of Latin America and the Caribbean, the term "tropical storms and hurricanes" is used.

R. Kaenzig (✉) • E. Piguet
Institute of Geography, University of Neuchâtel, Neuchâtel, Switzerland
e-mail: Raoul.kaenzig@unine.ch; etienne.piguet@unine.ch

E. Piguet and F. Laczko (eds.), *People on the Move in a Changing Climate*, Global Migration Issues 2, DOI 10.1007/978-94-007-6985-4_7,

The case studies that have been drawn on for this overview were chosen for their primary focus on the relationship between migration and environmental changes linked to climate. In some cases, however, studies that relate to the migratory consequences of environmental degradations that are not linked to climate change are discussed, such as earthquakes (Belcher and Bates 1983; Halliday 2006), or where the causal link with global warming is not completely established, such as El Niño events. Proceeding by analogy, the lessons that can be drawn from these studies can also shed light on the future consequences of climate change for migration. This constitutes a widely accepted hypothesis in the field of migration and in environment studies, but it should be kept in mind that climate change could lead to environmental disruptions without historical precedent and that the current trends in migration also often differ from the migrations of the past. For those reasons, historical analogies should be completed whenever possible with other methods such as scenario building or exploratory modelling. In this chapter, when the findings of the studies permit, population mobility is described according to its probable duration and the distances covered.

This chapter does not include a discussion on the consequences for migration resulting from deforestation and large-scale development and construction projects. Although these issues are at the heart of the most important challenges for Latin America and the Caribbean, the authors consider them to be distinct from the issue addressed here,[3] namely the relationship between migration and the direct impacts of climate change (natural hazards, water shortages and rising sea levels).

Current Effects of Climate Change in Latin America and the Caribbean

The large variations across the region in terms of climate, ecosystems, population distribution and lifestyles (Intergovernmental Panel on Climate Change 2007; Nagy et al. 2006) mean that climate change manifests itself in very different ways depending on the area and the type of physical phenomenon. This section includes a brief look at each of these manifestations.

Precipitation

Over the last decade, an increase in precipitation has been observed in south-eastern Brazil, in Paraguay, in Uruguay, in the Argentine Pampas and in parts of Bolivia. Meanwhile, other regions, such as southern Chile, southern Peru and western Central America, have experienced a sharp decline. This increased variability in

[3]Research on the stakes for migration related to issues of deforestation and of development projects has focused primarily on the area of the Amazon Basin, with case studies on Brazil Capellini et al. (2011), Rodrigues et al. (2009), Fearnside (2008), and on Ecuador Barbieri et al. (2008), Carr (2009). in particular.

precipitation, exacerbated by the El Niño and La Niña cycles,[4] can intensify the periods of drought,[5] leading to severe shortages in the water supply or, alternatively, can increase the intensity and/or the frequency of floods (Warner et al. 2009; Intergovernmental Panel on Climate Change 2007). These changes are particularly worrisome in Central America and in Mexico. Countries such as Guatemala, Nicaragua and Mexico are already frequently affected by periods of drought. A special report on extreme events published by the Intergovernmental Panel on Climate Change (IPCC) points out that, in addition, Central America, Mexico and north-eastern Brazil will (with a moderate amount of certainty) experience an intensification of their drought phases by the end of the twenty-first century (IPCC 2012). These episodes of drought have a major impact on the economy and on development in general because households rely almost entirely on rain-fed agriculture for their livelihoods (Adger 2006; Eakin 2005). According to the Stern Review on the Economics of Climate Change, in Latin America and the Caribbean, agriculture is the sector most affected by the increase in temperatures and in the variability of precipitation (Nagy et al. 2006).

Temperature

Latin America and the Caribbean have seen an average increase in temperatures of between 0.5 and 1 °C over the last 15 years.[6] The acceleration of the melting of glaciers is one of the most tangible consequences of this rise in temperature. The tropical glaciers have lost more than one third of their surface area in recent decades (Coudrain et al. 2005). This phenomenon is particularly worrisome in the Andean countries, such as Bolivia, Colombia, Ecuador and Peru, where access to water, whether for consumption or for hydroelectric production, already represents a serious challenge. Taking the warming into account, these problems with water supply will most likely become more severe over the years to come. According to the various climatic scenarios developed by IPCC, the region will see an increase in average temperatures of anywhere between 1 and 6 °C by 2100 (scenarios B2 and A2, respectively), with the greatest amount of warming affecting the tropical areas of South America (Intergovernmental Panel on Climate Change 2007).

[4]In an El Niño phase, the surface water temperature in the Pacific Ocean increases considerably, modifying the pressure system and causing major changes in precipitation patterns. The coastlines of South America are particularly exposed to the changes that result from the cycles of El Niño Intergovernmental Panel on Climate Change (2007).

[5]In the Amazon Basin, as well as in Central America, there has been an observable increase in changes to the patterns of the seasons. This is seen most obviously in the later start to the rainy season, which therefore also means a lengthening of the dry seasons Aguilar et al. (2005), Marengo et al. (2011).

[6]In Latin America and the Caribbean, it has not been possible to establish any long-term tendency in temperature averages. Nevertheless, one can see a significant trend toward warming in certain areas (Amazon Basin and the north-western part of South America) and, in certain rare cases, toward cooling (Chile) Intergovernmental Panel on Climate Change (2007).

Natural Hazards

The following map (Fig. 7.1) shows the number of people (in millions) affected by tropical storms and hurricanes (shown in grey) and by floods (in black) over the last 40 years. Central America and the Caribbean are clearly the regions most affected by tropical storms and hurricanes, while South America was particularly affected by floods.

Over recent decades, Central America and the Caribbean have been visited by a large number of tropical storms and hurricanes, with a record number of extreme events occurring during the 2005 season; the National Oceanic and Atmospheric

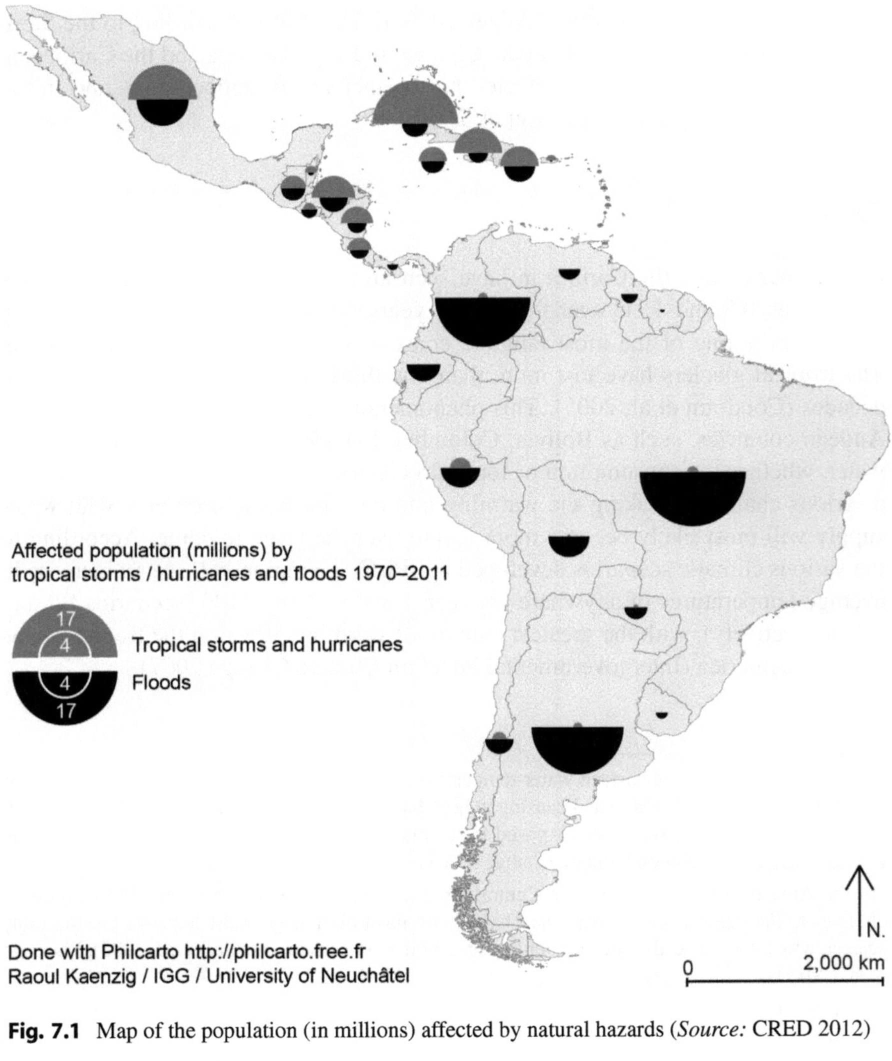

Fig. 7.1 Map of the population (in millions) affected by natural hazards (*Source:* CRED 2012)

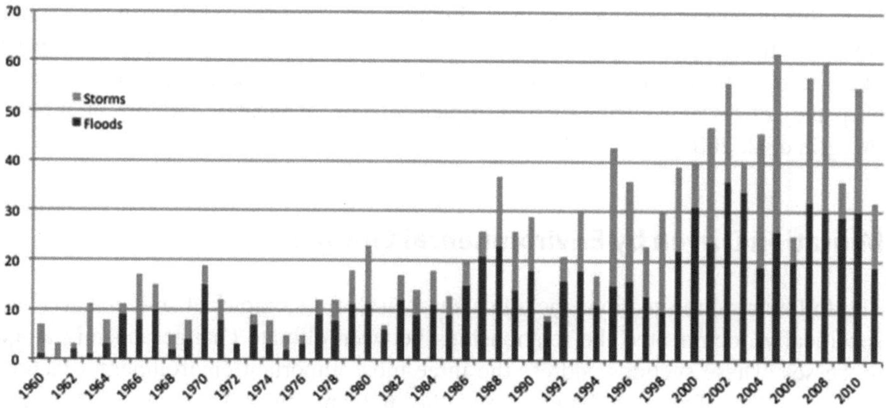

Fig. 7.2 Population (in millions) affected by tropical storms or hurricanes and by flooding in Latin America and the Caribbean, by year (*Source*: CRED 2012)

Administration[7] recorded no fewer than 31 such storms that season, of which 4 were Category 5 hurricanes. This highly unusual number of storm activities resulted in dramatic material and human losses: Hurricane Stan, for example, struck Mexico and four Central American countries (Costa Rica, El Salvador, Guatemala and Nicaragua), killing 1,500 people and causing 3 billion dollars' worth of losses (Magrin et al. 2007).

Floods and mudslides often accompany hurricanes, and this was certainly the case for Hurricane Mitch, which struck Honduras and Nicaragua in 1998 with devastating consequences, killing almost 10,000 Hondurans (Simms and Hannah 2006; Warner et al. 2009; Wrathall 2008). More recently, in September 2010, Hurricane Karl reached the coasts of Mexico and the heavy rains that it caused resulted in major flooding throughout the Veracruz region. Along the same lines, the Bahamas, Belize, El Salvador and Puerto Rico[8] are all among the top 10 countries in the world in terms of the population's exposure to coastal storms, according to Dasgupta et al. (2009). It should be noted that the first-ever hurricane to have been observed in the southern Atlantic hit southern Brazil in 2004 and was followed by major flooding in the eastern Amazon Basin (Pezza and Simmonds 2005). There were also serious floods in Argentina in 2000 and 2002, and torrential rainfall affected the Bolivarian Republic of Venezuela in 2005.

Figure 7.2 shows the steep increase in the number of people affected by tropical storms or hurricanes and by flooding over the last four decades. In fact, from 2001 to 2011, five times as many people were affected by natural catastrophes as was the case from 1960 to 1970.

[7] www.nhc.noaa.gov/2005atlan.shtml

[8] The Commonwealth of Puerto Rico is an unincorporated territory of the United States.

In this overview, one can see that, although environmental disturbances are not new to Latin America and the Caribbean, they are rapidly increasing and are very likely to become worse in the future, and this is especially so for the intensity and frequency of extreme events such as tropical storms, hurricanes and torrential rains (Magrin et al. 2007; IPCC 2012).

Migration Driven by Environmental Change

The following three sections include a summary of the empirical studies that have specifically documented the relationships between climate change and migratory processes. These sections address the three most important environmental factors that, as a result of climate change, could play an increasing role in the coming years and that may affect migration.

In the first section, the connection between environmental hazards and population displacements is analysed. As previously mentioned, the authors' primary method here is historical analogy, which draws on lessons from the past, as there are no precise forecasts available for future hydrological, meteorological or climatological catastrophes. The second section addresses the question of water shortages. Because droughts occur less abruptly than do other natural hazards, they have resulted in very different outcomes, and whether mobility is used as a response depends heavily on the context in which the exposed populations live. Particular attention is paid to the Andean regions, where retreating glaciers present a major challenge. And finally, in the third section, the occurrence of rising sea levels is addressed. In this section, there are almost no historical analogies available, but it is possible to use fairly precise projections about the populations that are threatened by this phenomenon.[9]

From Natural Catastrophes to Migration?

Tropical storms, hurricanes and floods are typical examples of phenomena that appear with very little warning and can result in the displacement of populations in search of shelter, care or food. Estimates of the number of people affected each year in Latin America and the Caribbean by floods (330 million between 2000 and 2010) and by tropical storms and hurricanes (226 million in the same period) give an idea of the magnitude of the threat (EMDAT 2012),[10] but it is very difficult to estimate how many people would be affected if there were to be more of these catastrophes due to climate change. No climatological model is capable of predicting precisely

[9]For a methodological discussion, cf. Piguet (2010).

[10]The Emergency Events Database (EM-DAT) is an international catastrophe database used as a classification system (see www.emdat.be/classification). Floods are listed as hydrological catastrophes, cyclones as meteorological catastrophes, droughts as climatological catastrophes and earthquakes as geophysical catastrophes.

where or when catastrophes will occur, and it is therefore not possible to know whether the areas affected are densely populated or not. Past experience, however, does offer a wealth of information on the kinds of displacement caused by these phenomena and on the specifics of the roles played by the phenomena themselves.

Most of the existing research on meteorological catastrophes in Latin America and the Caribbean focuses on hurricanes and tropical storms in Mexico, Central America (El Salvador, Guatemala, Honduras and Nicaragua) and the Caribbean (Dominican Republic and Haiti). There appears to be only one study that was carried out further south; and indeed such catastrophes take place much less often there. This qualitative research undertaken in Argentina as part of the Environmental Change and Forced Migration Scenarios (EACH-FOR) project,[11] comes to the conclusion that environmental risks[12] do not lead to substantial population movements, that economic reasons appear to be the main drivers of such movements (Irianni et al. 2008).

Saldaña-Zorrilla and Sandberg (2009) measured the influence of floods, storms, hurricanes, droughts and frosts on migratory phenomena in Mexico from 1990 to 2000. The results of their multivariable analyses show that, all other things being equal, an increase of 10 % in the frequency of natural disasters results in an increase of 5–13 % in migration, depending on the affected region. Studies by Alscher and Faist (2008) and Escobar et al. (2006) confirm that, in the Chiapas region, Hurricane Mitch (1998) and Hurricane Stan (2005) were triggers among economically vulnerable populations for the decision to migrate. Wrathall's study (2008) offers a refinement of this result, affirming that, among vulnerable populations, the most economically disadvantaged often resort to migrating within their country.

Research carried out in Nicaragua and in Honduras, also on the consequences of Hurricane Mitch for migration, reach similar conclusions. In Nicaragua, the results of a study carried out before and after the hurricane indicate that the households that were most exposed to torrential rainfall during the hurricane had a greater inclination to migrate abroad than did households with similar adaptive capacities but living in an unexposed region (Carvajal and Pereira 2008). In Honduras, Oliver-Smith (2009) significant increase in the number of arrests of migrants along the American-Mexican border in the months following the hurricane. These studies, however, also highlight the crucial importance of the social context, because these population displacements are occurring in regions that have been rendered fragile by their great social, environmental and economic vulnerability. As Comfort states in Oliver-Smith: "The uprooted people of Honduras were not refugees of nature, but were displaced by socially inscribed and enacted changes in their environment that combined with a naturally formed agent called a hurricane" (2009:19). The hurricane, then, in aggravating the precariousness of the situation, acts as a catalysing element that can result in the decision to migrate.

[11] See www.each-for.eu

[12] The study considers floods, cyclones, earthquakes and soil erosion.

Though Halliday's study (2006) focuses not on climate deterioration but on migration between El Salvador and the United States of America as a result of earthquakes[13] and agricultural misfortunes,[14] is of great interest as an analogy because it provides some nuance for the results discussed immediately above. This study is based on information gathered over several different periods (longitudinal research undertaken with panel studies). The statistical analyses show that, while unfavourable agricultural conditions significantly increase migration to the United States, the sudden events studied, namely the earthquakes, have the opposite effect, with a 40 % decrease in departures. Belcher and Bates (1983) obtained similar results in their study of two earthquake episodes, one in 1972 in Nicaragua and one in 1976 in Guatemala. They also observed that emigration from the affected communities was no greater than emigration from unaffected areas. From this, the authors conclude that the incidences of migration cannot be directly attributed to the earthquakes but that they must be seen within their specific local contexts. In Chile, for example, in spite of the dire material consequences of the 2010 earthquake, the resulting population displacements were of short duration, essentially for the purpose of dealing with the immediate needs of the situation (André 2011). In some cases, net migration can even turn out to be positive (more immigration than emigration) after such an event because of the increased need for human resources to help with reconstruction in the affected areas. This phenomenon was observed in Brazil in early 2011, when violent rains resulted in landslides and flooding in the state of Rio de Janeiro (Capellini et al. 2011).

In the Caribbean, Haiti is often cited as a case in point in studies on the vulnerability of populations to sudden environmental damage. In Haiti, almost all the forests have disappeared and a large proportion of the country is subject to serious soil erosion, which exacerbates the consequences of the tropical storms or hurricanes and the floods that strike the island, threatening the population's means of subsistence (Alscher 2011; Simms and Hannah 2006). According to Jacobson (1988), the first references to Haitian boat people as "environmental refugees" go back to 1984. The environmentalist Myers evokes the fundamental role that environmental damage plays on the island, stating that Haitians "are abandoning their homeland in part because their country has become an environmental basket case" (Myers 1993). Nevertheless, the existing empirical studies (using qualitative methods and/or literature reviews) allow for the conclusion that migratory processes continue to be deeply motivated by factors related to the political and economic order. At the same time, the studies also stress that environmental factors also significantly weaken the sometimes precarious situation in Haiti (Alscher 2011; Leighton-Schwartz and Notini 1995; Catanese 1999; Barker 1989). Along these lines, after the 2010 earthquake in Haiti, the humanitarian conditions and the

[13]Earthquakes are not a result of climate warming, but some of the consequences of earthquakes are similar to the consequences of hurricanes and other climate-related catastrophes and some analogies can therefore be drawn with them.

[14]Halliday defines agricultural misfortunes as "agricultural conditions that caused livestock loss and/or harvest loss" Halliday (2006).

authorities' inability to respond to the urban population's basic needs, in particular, drove city dwellers to return to the country's rural areas, thereby endangering the resident population's food security along with that of the new arrivals. The overexploitation of natural resources had previously led the inhabitants of the rural areas to leave those regions. The International Organization for Migration (IOM) (2012) estimates that 1.5 million people were subject to internal displacement after the earthquake.[15] Others migrated abroad, mostly to the countries of Latin America or the Caribbean, while attempts to emigrate to the United States were severely limited by the increased enforcement of migration policies and the high cost of the trip (Gütermann and Schneider 2011).

It becomes clear that Central America has been the object of much more research than has the Caribbean region, which, despite having had to face intense and repeated natural catastrophes, has not been very well documented. There is research on environmental degradation processes, but most of these studies do not include information about the stakes for migration. Looking at the island of Hispaniola (Dominican Republic and Haiti), however, it is very clear that the effects of tropical storms or hurricanes and of floods are greatly exacerbated by the high degree of deforestation and soil erosion. This fragile environmental context makes the region very vulnerable to natural hazards, thereby exposing the population to greater risks than would otherwise be the case.

The tropical storms and hurricanes that have hit Central America have had a somewhat different impact on migration from that which can be observed elsewhere in the world. In Central America, migration appears to be more often international than what has been underlined in studies about other regions of the world. It is possible to assume that the proximity of the United States as a migration destination, along with migrants' greater economic resources (partly thanks to remittances from earlier migrants), may explain this difference. In addition, migration is most often seen in contexts in which there are pre-existing relationships of migration between the sending and receiving regions (Deprez 2010). However, it should be kept in mind that local and temporary displacements are still a common characteristic. As in Asia, a large part of the affected populations in Central America do not have the means to move far and soon return to the disaster area to rebuild their homes (Paul 2005).

Nevertheless, the situation in Latin America and the Caribbean shares many features with what studies have shown in the rest of the world. For example, as emphasized by Kniveton et al. (2008), the level of vulnerability differs greatly from one region to another, and social and economic contexts also play important roles. In Latin America and the Caribbean as anywhere else, in order for significant and long-lasting migration to occur, the society must be largely dependent on the environment for its survival and social factors must also exacerbate the impact of the catastrophe.

[15] At the end of 2011, half of a million people were still staying in temporary camps. The population that remains in the camps is the most vulnerable. When the camps were established, 60 % of their residents were homeowners (of their previous residence) and 40 % were not. By the end of 2011, the non-homeowners made up 80 % of the population that were still living in the camps (IOM 2012).

Migration and Water Shortages

Existing research on the global scale shows that shortages of water for human consumption and for irrigation occur much less suddenly than do tropical storms or hurricanes and floods. As a result, the kinds of mobility caused by water shortages are more gradual, especially in rural regions (Leighton 2011). Far from always representing a tragedy in itself, the resulting migration, in such a context, can be the basis of a diversification of revenue sources for coping with decreased agricultural productivity, revenues and means of subsistence (Bilsborrow 1992).

The existing empirical studies in this field are very mixed, and it is difficult to weigh the importance of specifically environmental factors. Many researchers argue against a direct link between drought and emigration, insisting instead on the multiplicity of causes, on the variety of survival strategies adopted by the affected populations, and on the fact that the displacements caused by water scarcity and desertification are essentially moves over short distances (Kniveton et al. 2008; Leighton 2011; Meze-Hausken 2004).

In Brazil, the north-eastern part of the country is particularly affected by a high incidence of droughts.[16] This semi-arid region relies mainly on agriculture made up of small subsistence farms, and historical studies have shown that, during periods of drought, the region can suffer agricultural losses that are as high as 80 % of production, resulting in waves of emigration to more favourable regions, in the south of the country (Kahn and Campus 1992; Capellini et al. 2011). Leighton estimates that droughts were contributing factors in the emigration of 3.4 million people between 1960 and 1980 (Leighton 2006). More recently, Franke et al. (2002) show the effect of the El Niño fluctuations of the early 1980s and early 1990s on migration from rural regions to the cities of São Luís and Teresina (the capitals of their respective states). Confalonieri (2003) also establishes a correlation between the 1982–1983 El Niño period and the peaks that were observed in migration between the states of Maranhão and Pará. However, these studies do not allow a determination about whether these migrants were part of particularly vulnerable populations, with little financial or social capital, who were forced to move, or whether, instead, this was the migration of a relatively well-off population for whom migration was a strategic choice (Barbieri et al. 2010; Barbieri and Confalonieri 2011).[17]

[16]Capellini et al. (2011), who are critical of the determinist point of view, point out that drought episodes in north-eastern Brazil are exacerbated, maybe sometimes even caused, by the concentration of property and by the fact that water resources are managed by powerful landowners.

[17]It is also worth noting that these same authors developed an original approach for the purpose of estimating the impact of climate change on migration using a comprehensive economic model based on the warming scenarios of IPCC and applied to north-eastern Brazil. Their very interesting findings are, however, still exploratory. They seem to indicate a weak impact of climate change on migration, with, however, an increase in migration beginning in the 2030s. Barbieri et al. (2010), Barbieri and Confalonieri (2011).

There are two studies focusing on Ecuador. The research of Alvarez Gila et al. (2008) addresses Ecuadorian immigrants living in Spain. The interviews that they carried out show that the immigrants considered that mostly economic factors had caused their departure, but that environmental degradation (mainly soil erosion and desertification, often attributed to El Niño episodes) accentuated the difficulties. Therefore, the authors of those studies see these forms of mobility as alternatives that make it possible for the population to deal with environmental hazards. Gray (2009) carried out a study of 300 households in Ecuador between 1995 and 2006, examining the explanatory factors at work in migratory processes at the local, regional and international levels. Some of the environmental variables that were used included precipitation and changes in the volume of agricultural production.[18] The results show that environmental conditions play a role in migration at all three geographic levels, but that they are more significant for local and regional mobility.

Balderrama Mariscal et al. (2011) look at the characteristics of migration in the Bolivian Andes,[19] which is an area that has been particularly affected by gradual environmental degradations such as drought and desertification. Internal migration (to other rural regions or to neighbouring cities, which constitute major population magnets) has long been a revenue diversification strategy for communities that are dependent on agriculture. In the northern part of the administrative region of Potosi, households with agricultural holdings above 3,500 m adopt temporary or seasonal forms of migration, while the communities in the valleys, between 1,650 and 3,000 m, who are severely affected by processes of degradation and of soil erosion, tend to leave the area for good. Nevertheless, it is still difficult to weigh the influence of environmental factors here, since they are strongly connected to the general economic situation of the region, which suffers from a very high unemployment rate due to the decline of some of the mining sectors. This case study, however, does make it possible to note the way in which environmental changes affect the duration of internal migration.

Due to its sheer volume, migration between Mexico and the United States occupies a unique place in the field of Latin American migration.[20] From an environmental standpoint, 70 % of the surface area of Mexico is considered arid, and it suffers from advanced desertification, along with repeated periods of drought, especially in the northern and central regions of the country (Leighton 2006). For rural communities that are mainly dependent on rain-fed agriculture, internal or cross-border migration is therefore one of the options for dealing with adverse climatic conditions (Cohen et al. 2012).

[18] Gray refers to these changes as "unusual harvests" (Gray 2009).

[19] The study examines two regions in Bolivia: one in the Andes, in the northern zone of the Potosi Department, made up of the provinces of Ibañez, Bustillo, Bilbao, Charcas and Chayanta, and one on lower ground, in the municipality of San Julían, north-east of Santa Cruz, in the Santa Cruz Department. For the purposes of this chapter, the first of these two cases is of particular interest because in the study of the second, the environmental component is only weakly represented.

[20] Migrants originally from Latin America and the Caribbean currently make up 53 % of the total foreign-born population of the United States. Mexicans alone make up 30 % of this group (Hanson and McIntosh 2009; IOM 2011; Durand and Massey 2010).

The first set of results from the studies outlined in this chapter confirms the link among episodes of drought, processes of desertification and cross-border migration. Feng et al. (2010) show that a reduction of 10 % in crop yields that is attributable mainly to periods of drought leads to an increase of 2 % in Mexican emigration to the United States. Although they are methodologically controversial (Auffhammer and Vincent 2012), these results confirm Munshi's analyses, which also measure a negative correlation between the inclination to migrate to the United States and the rainfall levels in the region of origin in Mexico (2003). Likewise, according to the research of Alscher and Faist (2008), environmental degradations such as soil erosion and changes in precipitation patterns represent explanatory factors for heavy internal migration and also for emigration to the United States. These degradations come on top of economic and structural difficulties such as the liberalization of markets, increased international competition and reductions in state subsidies (Jungehülsing 2010; Alscher and Faist 2008) and are exacerbated by poor soil management (de Janvry et al. 1997; Schwartz Leighton and Notini 1994; Escobar et al. 2006).

This result, showing a relatively clear link between drought and emigration to the United States, is nuanced by the studies of Kniveton et al. (2008) and Schmidt-Verkerk (2009). The work of Kniveton et al. on the Mexican states of Zacatecas and Durango, which suffered periods of drought between 1951 and 1991, shows no significant correlation between drought and emigration in Zacatecas. On the contrary, a correlation between an increase in emigration and periods of increased rainfall can be seen in Durango (Kniveton et al. 2008).[21] Meanwhile, Schmidt-Verkerk's 2009 qualitative study, also on Zacatecas, shows that emigration is considered an option only by populations living in regions with an already-established tradition of migration. Environmental degradation, therefore, does not overturn existing practices but rather reinforces them. Migration is not an option for certain particularly vulnerable populations that have no tradition of migration, and those populations are forced to cope with environmental hazards where they are.

Andean Region and Retreating Glaciers

The retreating of the glaciers is one of the most tangible manifestations of global warming in South America. Although the physical dynamics behind the melting of the ice are quite well understood, and the media are keenly interested in the phenomenon of the melting, it has been only marginally addressed in studies that deal with migration. This subsection will therefore provide a brief summary of this specific topic.

[21] In order to explain this inverted relationship, the authors refer to the work of Findley (1994) and her case study on Mali, where she shows that long-distance migration tends to decrease during periods of drought. When experiencing water shortages, the affected households cannot afford the investment required for a long-distance migration because basic needs, such as securing a food supply, must be met first.

Since 1970, the Andean glaciers have lost almost half of their area, and the melting process has been significantly accelerating over time (Coudrain et al. 2005; Soruco et al. 2009). In Bolivia, for example, many small low- and medium-altitude tropical glaciers have completely disappeared over this period, and projections indicate that many others are likely to disappear over the coming decades (Vergara 2005; Ramirez 2006; Inter-American Institute 2010). The decreasing water supply is one of the major effects that have already been observed in the region; run-off from glacier basins is an essential element in the water supply for agriculture, drinking water, energy production and ecosystem integrity (Hoffmann 2008; Vergara 2005; Paz Rada 2007; Viviroli et al. 2007). The disruption of this supply due to global warming presents serious challenges in terms of adaptation. Migration to nearby urban centres, such as La Paz and El Alto, constitutes a common revenue-diversification strategy in the mountain regions' rural communities. While this form of mobility, which is often temporary, allows subsistence-farming households to be less dependent on their unpredictable environmental context, it also exposes them to the market's price fluctuations (Kaenzig 2011; McDowell and Hess 2012).

Other ethnographic studies have also addressed this issue (Rhoades 2008; Orlove 2009; Carey 2010; Young and Lipton 2006), providing valuable insights into the ways in which communities living along the edges of the retreating glaciers, especially in the mountainous areas of Ecuador and Peru, have adapted. However, even though these studies sometimes refer to migration as an inescapable consequence of the melting of the glaciers, they do not specifically analyse the relationship. At the same time, this very relationship – between the melting of the glaciers and migration – has been the basis for numerous discussions, many of them alarmist, in the international media and in reports[22] by non-governmental organizations, international organizations and governments, although no scientific study has rigorously documented the mechanisms connecting the retreat of the glaciers with migration in South America. At present, then, research on the implications of glacier melting for migration, and more generally of climate change in the Andes, would appear to be highly desirable. In particular, the focus should be put on urban areas (for example, La Paz, El Alto, Lima and Quito), as they are often important destinations, but also extremely vulnerable to climate change.

The present overview of the studies that examine droughts and, to a lesser extent, glacier retreat, confirms that these phenomena are connected, sometimes strongly, with migration phenomena. Yet the results are inconsistent from one study to the next, which can be explained by the lack of reliable and consistent data on migration and precipitation but, even more so, by the variability of local contexts and of the adaptation strategies of different populations. In conclusion, migration represents one possible form of adaptation among several, while at the same time it is not available to everyone. Therefore, it would be going too far to predict that, as water resources decrease, there will be an inevitable increase in migration.

[22] See especially Vergara (2005), OXFAM (2009) and Ross (2010).

Table 7.1 Population (in millions) living at or below an altitude of 1, 5 and 10 m above sea level

Sea-level rise: Authors and regions studied	1 m	5 m	10 m
Nicholls (2004), Latin America and the Caribbean	9	–	–
Anthoff et al. (2006), Latin America and the Caribbean	9.9	17	24.7
Dasgupta et al. (2007), Latin America and the Caribbean	2.9	13.5	–
McGranahan et al. (2007), Latin America and small island States, including the Caribbean	–	–	35

Sea-Level Rise: A Long-Term Factor for Migration

Unlike natural catastrophes and droughts, which are context-specific and sometimes temporary, the rise in sea levels is unequivocally linked with migration. It is a practically irreversible phenomenon that manifests itself in a more-or-less linear way over a long period of time. In the absence of new infrastructure, such as dikes and raised structures, sea-level rise will make migration unavoidable, while allowing for gradual, planned departures. The consequences of rising sea levels can be predicted and localized with some degree of confidence, given that the configuration, altitude and population of the coasts are known. As a result, it is possible to calculate on the global scale the number of people living in areas threatened by rising sea levels, high tides, high-amplitude waves, increased salinity or coastal erosion. McGranahan et al. (2007) estimate that, worldwide, 602 million people live at altitudes of less than 10 m. Relatively speaking, the effect would not be as great in Latin America and the Caribbean as in other regions, because the number there is only 29 million, whereas in Asia it is 466 million. Nevertheless, as Table 7.1 shows, estimates of the population–altitude intersection are still quite variable. At an elevation of 1 m, estimates range from 2.9 million to 9.9 million people affected, and at 10 m, they range from 24.7 to 35 million people.[23] (On the other hand, estimates of the number of people living at or below 5 m above sea level are relatively consistent, coming out to an average of 15 million people affected.)

The areas along Latin America's Atlantic coast as well as the Caribbean are the most exposed to the risks of flooding and increased salinity. This is especially true for the Gulf of Mexico and the Caribbean Sea, as well as the mouths of large rivers such as the Amazon in Brazil, the Orinoco in the Bolivarian Republic of Venezuela and the Parana between Argentina and Uruguay. Even some inland areas, such as the Argentine Pampas, may also be affected by flooding as rising sea levels weaken natural drainage systems (Vergara 2005). On the Pacific coast, Ecuador is the only country that is significantly at risk (Anthoff et al. 2006).

[23]One explanation for this large degree of variation has to do with the data sources used; although all of the studies use the same population data, no two use the same data for altitude. According to Farr and Kobrick (2000), determining the areas that are at or below 10 m above sea level is still very difficult, especially in regions with steep coasts.

In terms of surface area, urban zones make up a very small part of the areas at risk of flooding (8 % of the total floodable area)[24]; the at-risk regions are mostly made up of agricultural land and humid and marshy areas. On the other hand, from the point of view of the affected population, the areas at risk are mostly in urban or densely populated regions (of people living in at-risk areas, 77 % live in cities).[25] In all likelihood, the population of the coastal urban zones will continue to grow in the coming years (Black et al. 2011).

Assuming a scenario in which sea levels rise by 1 m, the populations of Suriname, Guyana and the Bahamas, would feel the greatest impact, with somewhere between 5 and 7 % of the total population being affected, followed by French Guyana and Puerto Rico (Dasgupta et al. 2007; MacGranahan et al. 2007). IPCC also notes that island States are particularly vulnerable, with little ability to withstand the effects of climate change. Environmental damage caused by rising sea levels could also harm biodiversity, fisheries, tourism and infrastructure (Nagy et al. 2006). Jamaica's situation is similar to that of other small island States in that beach erosion, coral bleaching, salinization of drinking water and intense tropical storms can significantly hurt tourism. According to Berringer (2012), a decline in the tourism industry would have adverse effects on the country's poverty level, which may provide a further incentive to migrate abroad to places where the diaspora is already well established, such as the United Kingdom, the United States or Canada.

The habitable areas of the Caribbean islands are densely populated, with high growth rates, especially in urban areas (Intergovernmental Panel on Climate Change 2007). For the island countries of the Caribbean, rising sea levels put forward the sensitive issue of the possible disappearance of a State. In an article on the issue of climate-driven statelessness, however, Piguet (2012) points out that, among the Member States of the Alliance of Small Island States, not one of the three States with a maximum altitude of under 10 m[26] is in the Caribbean. In fact, among the Caribbean States, only the Bahamas has a highest point lower than 100 m (its highest point is at 63 m), but the country's overall topography, as well as its resources for establishing protective measures, means that disappearance would be very unlikely.

In conclusion, rising sea levels probably constitute the most clearly threatening aspect of climate change in terms of the global effect on forced long-term migration. Although rising sea levels affect Latin America and the Caribbean in a less significant way than they do other regions, there are nevertheless large areas that are at risk and, even beyond the issue of possible habitat loss, economic activities are also in danger, and their loss could force migration as well. Tourism and fishing are the two most vulnerable activities, especially in the Caribbean. In addition,

[24] Based on the data given in McGranahan et al. (2007) and Dasgupta et al. (2007); in both cases, the result is the same.

[25] Based on the data given in McGranahan et al. (2007).

[26] The three States are the Maldives (300,000 inhabitants; maximum altitude of 2 m), Tuvalu (10,000 inhabitants; maximum altitude of 2 m) and the Marshall Islands (63,000 inhabitants; maximum altitude of 6 m). Piguet (2012).

responses to rising sea levels involve more than just the simple abandonment of land, and migration can occur long before the living space is truly uninhabitable. The context, again, becomes highly relevant, as the affected populations may develop context-specific adaptive or mitigating strategies that could allow them to either significantly delay or, alternatively, significantly advance the need to leave.

Conclusions: A Region Highly at Risk Yet Very Little Studied

This overview shows, first of all, how few empirical studies there still are with respect to the consequences of climate change or of environmental catastrophes for migration in Latin America and the Caribbean (there appear to be just over 20 such studies) and how unevenly they are distributed across the region. There are many more studies on Central America, and particularly on Mexico, and a certain number that deal with Brazil. On the other hand, aside from Ecuador, the Andean countries (that is to say, Bolivia, the Bolivarian Republic of Venezuela, Chile, Colombia and Peru), along with South America's north-eastern countries (French Guyana, Guyana and Suriname) have remained relatively unexplored. The effects of climate change and of environmental damage in the Caribbean have certainly been noted, especially in the media, but there has been very little research that also includes the issue of migration. As far as tropical storms and hurricanes are concerned, the frequency with which they occur could partially explain the greater interest in Central America, but that does not apply to other phenomena, such as flooding, the increased variability in precipitation and the melting of the glaciers, which affect many countries.

There are three hypotheses about why it is that these areas have been studied so unequally. The first has to do with the existence, or lack thereof, of local research centres on migration and, where such centres do exist, how widely their findings are disseminated in international journals, especially English-language ones, which play a dominant role in the international scientific landscape. The second hypothesis has to do with pre-established relationships of migration with the United States, which have so far clearly dictated the scientific research agenda. Thus, Mexico, with its strong history of emigration to the United States, has been the object of a large volume of research carried out by North American researchers. And finally, interest in migrant-sending countries can be explained by political motives having to do with the politics of migration and security. The widespread fear surrounding uncontrollable flows of "climate refugees," fleeing their countries to seek shelter in northern countries, has prompted the rich countries to intensify research in potential migrant-sending areas. One extreme example is the report "Climate change and immigration: Warnings for America's southern border," which attempts to analyse the "migration risk" for the United States posed by climate change in Latin America (Ross 2010). The complexity of the phenomena involved, as illustrated in the present chapter, shows the degree to which such concerns are based on simplistic ideas about the relationship between migration and the environment.

Table 7.2 Summary of the identified consequences for migration based on the type of effect of climate change

Effects of climate change	Consequences for migration
Natural hazards (hurricanes, storms and floods)	Mostly in Central America and the Caribbean: the observed resulting mobility is generally local, and a tendency to return has often been observed. When there is international migration, the United States is the main destination.
Water shortages	Many areas are very vulnerable. So far, there has been little proven impact on mobility. There have been documented cases of migration from Mexico to the United States but the mechanisms are particularly complex and context-specific, with multiple causes.
Rising sea levels	No migration observed at present, but the areas identified as being at risk are highly populated and often urban. Without heavy protective measures, migration will in some cases be inevitable in the medium term.

A second conclusion is that the relationship between environmental change and migration that can be observed in Latin America and the Caribbean confirms the main tendencies that have been noted in other parts of the world: most displacements take place over short distances, with urban centres having a strong attraction; and in the case of sudden catastrophes, displacements are often short-term. The strategies used to cope with environmental changes include leaving and returning, temporary stays and multiple residences. Rather than speaking of migration, let alone of "refugees," it would make sense in these cases to discuss mobility and "displaced persons." In addition, the findings of these studies on Latin America and the Caribbean confirm that displacements related to the environment are multi-causal and therefore also very context-specific. The same kind of disturbance will have completely different consequences depending on the economic, social and political situation in which it occurs. And finally, environmentally driven displacements, which must be considered within their historical context, occur most frequently where there are pre-existing relationships of migration between the sending and receiving countries. Treating migration as an inevitable consequence of climate change, as well as trying to quantify the number of future migrants based on a simple count of the population living in the threatened areas, is therefore a misleading simplification. "Roots of flight: Environmental refugees in Latin America," an article published in 1996, already then pointed out that the forced displacements that were attributed to environmental factors should instead be seen as the result of a colonial past that had resulted in many conflicts over the unequal distribution of land and property (Gosine 1996). This analysis, rooted in a post-colonial perspective, is yet another reminder that environmental damage is often just one more factor added to social and political contexts that are sometimes steeped in deep inequality. The current legitimate interest in the consequences of climate change should not mean that the power dynamics that shape the social realities of the regions being studied are forgotten (Table 7.2).

In summary, Latin America and the Caribbean, although very vulnerable to climate change, should not be considered as a special case as far as past environmentally driven migration is concerned. The effects of climate change on population displacements in those areas should therefore not be radically different from its effects on population displacements as observed elsewhere. There should be relatively little long-distance migration in response to hurricanes and droughts but existing migration channels will most likely be strengthened, and there is a significant potential for long-term displacement due to rising sea levels. However, certain effects of climate change on migration specific to South America can be seen. This is the case with the melting of the tropical glaciers, which has not yet been greatly studied. These results provide a strong argument for increased research into the complex interactions between migration and the environment, especially in the most vulnerable countries.

Bibliography

Adger, W. N. (2006). Vulnerability. *Global Environmental Change, 16*, 268–281.

Aguilar, E., Peterson, T. C., Obando, P. R., Frutos, R., Retana, J. A., Solera, M., Soley, J., García, I. G., Araujo, R. M., Santos, A. R., Valle, V. E., Brunet, M., Aguilar, L., Álvarez, L., Bautista, M., Castañón, C., Herrera, L., Ruano, E., Sinay, J. J., Sánchez, E., Oviedo, G. I. H., Obed, F., Salgado, J. E., Vázquez, J. L., Baca, M., Gutiérrez, M., Centella, C., Espinosa, J., Martínez, D., Olmedo, B., Espinoza, C. E. O., Núñez, R., Haylock, M., Benavides, H., & Mayorga, R. (2005). Changes in precipitation and temperature extremes in Central America and northern South America, 1961–2003. *Journal of Geophysical Research, 110*, D23107.

Alscher, S. (2011). Environmental degradation and migration on Hispaniola Island. *International Migration, 4*, 16–18.

Alscher, S., & Faist, T. (2008). Environmental factors in Mexican migration: The cases of Chiapas and Tlaxcala.

Alvarez Gila, O., López de Maturana Diéguez, V., & Ugalde Zaratiegui, A. (2008). *Ecuador – Case study report. Each for case study report.* Bonn: United Nations University EHS.

André, M. (2011). The earthquake in Chile. In F. Gemenne, P. Brücker, & J. Glaser (Eds.), *The state of environmental migration 2010* (pp. 49–56). Paris: IDDRI Sciences Po and IOM (International Organization for Migration).

Anthoff, D., Nicholls, P., Tol, R. S., & Vafeidis, A. (2006). *Global and regional exposure to large rises in sea-level: A sensitivity analysis.* Working Paper. Norwich: Tyndall Centre for Climate Change Research.

Auffhammer, M., & Vincent, J. R. (2012, July). Unobserved time effects confound the identification of climate change impacts. *Proceedings of the National Academy of Sciences, 109*(30), 11973.

Balderrama Mariscal, C., Tassi, N., Rubena Miranda, A., Aramayo Caned, L., & Cazorla, I. (2011). *Rural migration in Bolivia: The impact of climate change, economic crisis and state policy* (International Institute for Environment and Development (IIED) Human Settlements Working Paper, 45). London: IIED.

Barbieri, A. F., & Confalonieri, U. E. C. (2011). Climate change, migration and health: Exploring potential scenarios of population vulnerability in Brazil. In E. Piguet, A. Pécoud, & P. de Guchteneire (Eds.), *Migration and climate change* (pp. 49–73). Cambridge: Cambridge University Press.

Barbieri, A., Carr, D. L., & Bilsborrow, R. E. (2008). Out-migration of second-generation frontier colonists from the Ecuadorian Amazon. *Population Research and Policy review.*

Barbieri, A., Domingues, E., Queiroz, B., Ruiz, R., Rigotti, J., Carvalho, J., & Resende, M. (2010). Climate change and population migration in Brazil's Northeast: Scenarios for 2025–2050. *Population and Environment, 31*, 344–370.

Barker, D. (1989). Environmental migrants. *Draft Paper for World Resources 1990–1991*.

Belcher, J. C., & Bates, F. L. (1983). Aftermath of natural disasters: Coping through residential mobility. *Demography, 7*, 118–128.

Berringer, A. C. S. (2012). Climate change and emigration: Comparing "sinking islands" and Jamaica. *Miradas en movimiento,* Special Volume Naturally Immigrants, pp. 106–120.

Bilsborrow, R. E. (1992). *Rural poverty, migration, and environment in developing countries: Three case studies* (World Bank Policy Research Working Paper 1017). Washington, DC: World Bank.

Black, R., Bennett, S. R. G., Thomas, S. M., & Beddington, J. R. (2011). Climate change: Migration as adaptation. *Nature, 478*, 447–449.

Capellini, N., Castro, M. C., & Gutjahr, E. (2011). Patterns of environmental migration in Brazil: Three case studies. In F. Gemenne, P. Brücker, & J. Glaser (Eds.), *The state of environmental migration 2010* (pp. 87–110). Paris: IDDRI Sciences Po and IOM (International Organization for Migration).

Carey, M. (2010). *In the shadow of melting glaciers*. Oxford: Oxford University Press.

Carr, D. (2009). Population and deforestation: Why rural migration matters. *Progress in Human Geography, 33*, 355–378.

Carvajal, L., & Pereira, I. (2008, October 9–11). *Evidence on the link between migration, climate disasters, and human development*. Paper presented at the international conference on environment, forced migration and social vulnerability, Bonn.

Catanese, A. (1999). *Haitians – Migration and diaspora*. Boulder: Westview Press.

Cohen, I. S., Spring, Ú. O., Padilla, G. D., Paredes, J. C., Inzunza Ibarra, M. A., López, R. L., & Díaz, J. V. (2012). Forced migration, climate change, mitigation and adaptive policies in Mexico: Some functional relationships. *International Migration, 51*(4), 1–20.

Confalonieri, U. (2003). Variabilidade climática, vulnerabilidade social e saúde no. Brasil. *Terra Livre, I*, 193–204.

Coudrain, A., Francou, B., & Kundzewicz, Z. (2005). Glacier shrinkage in the Andes and consequences for water resources. *Hydrological Sciences, 50*, 925–932.

Dasgupta, S., Laplante, B., Maisner, C., Wheeler, D., & Yan, J. (2007). *The impact of sea level rise on developing countries – A comparative analysis* (World Bank Policy Research Working Paper 4136, 51).

Dasgupta, S., Laplante, B., Meisner, C., Wheeler, D., & Yan, J. (2009). The impact of sea level rise on developing countries: A comparative analysis. *Climatic Change, 93*, 379–388.

de Janvry, A., Sadoulet, E., Davis, B., Seidel, K., & Winters, P. (1997). *Determinants of Mexico-U.S. migration: The role of household assets and environmental factors* (Department of Agricultural & Resource Economics, UC Berkeley, Working Paper Series N° 853). Berkeley: California Agricultural Experiment Station, Giannini Foundation of Agricultural Economics.

Deprez, A. (2010). *Climate migration in Latin America: A future "Flood of Refugees" to the North?* COHA (Council On Hemispheric Affairs).

Durand, J., & Massey, D. S. (2010). New world orders: Continuities and changes in Latin American migration. *Annals of the American Academy of Political and Social Science, 630*, 20–52.

Eakin, H. (2005). Institutional change, climate risk, and rural vulnerability: Cases from Central Mexico. *World Development, 33*, 1932–1938.

EMDAT. (2012). *The emergency events database*. http://www.emdat.be/

Escobar, H., Sovilla, B., & Arévalo, L. A. (2006). Pobreza, desastres naturales y migración en la regiones Istmo-Costa, Sierra y Soconusco de Chiapas. *Observatorio de la Economía Latinoamericana*.

Farr, T. G., & Kobrick, M. (2000). Shuttle radar topography mission produces a wealth of data. *Transactions of the American Geophysical Union, 81*, 583–585.

Fearnside, P. (2008). The roles and movements of actors in the deforestation of Brazilian Amazonia. *Ecology and Society, 13*, 23.

Feng, S., Krueger, A. B., & Oppenheimer, M. (2010). Linkages among climate change, crop yields and Mexico–US cross-border migration. *Proceedings of the National Academy of Sciences of the United States of America (PNAS)*. http://www.pnas.org/content/early/2012/07/10/1202049109.full.pdf+html. Last accessed.

Findley, S. E. (1994). Does drought increase migration? A study of migration from rural Mali during the 1983-85 drought. *International Migration Review, 28*, 539–553.

Franke, C. R., Ziller, M., Staubach, C., & Latif, M. (2002). Impact of the El Niño: Southern oscillation on visceral leishmaniasis, Brazil. *Emerging Infectious Diseases, 8*, 914–917.

Gosine, A. (1996). Roots of flight: Environmental refugees in Latin America – A response to analysis by Homer-Dixon. *Refuge, 15*, 27–32.

Gray, C. L. (2009). Environment, land, and rural out-migration in the Southern Ecuadorian Andes. *World Development, 37*, 457–468.

Gütermann, N., & Schneider, E. (2011). The earthquake in Haïti. In F. Gemenne, P. Brücker, & J. Glaser (Eds.), *The state of environmental migration 2010* (pp. 39–48). Paris: IDDRI Sciences Po and IOM (International Organization for Migration).

Halliday, T. J. (2006). Migration, risk and liquidity constraints in El Salvador. *Economic Development and Cultural Change, 54*, 893–925.

Hanson, G. H., & McIntosh, C. (2009). The demography of Mexican migration to the US. *American Economic Review, 99*, 22–27.

Hoffmann, D. (2008). Consecuencias Cel Retroceso Glaciar En La Cordillera Boliviana. *Pirineos, 163*, 77–84.

Inter-American Institute (IAI). (2010). *Melting the ice – Receding glaciers in the American Cordillera* (Communiqué 2). Inter-American Institute for Global Change Research (IAI).

Intergovernmental Panel on Climate Change. (2007). *Climate change 2007 – IPCC fourth assessment report on scientific aspects of climate change for researchers, students, and policymakers*. Cambridge: Cambridge University Press.

International Organization for Migration (IOM). (2011). *World migration report 2010*. Geneva: IOM.

International Organization for Migration (IOM). (2012). *Haiti: From emergency to sustainable recovery* (IOM Haiti Two-Year Report (2010–2011), p. 56). Port-au-Prince: International Organization for Migration Haiti.

IPCC. (2012). Summary for policymakers. In C. B. Field, V. Barros, T. F. Stocker, D. Qin, D. J. Dokken, K. L. Ebi, M. D. Mastrandrea, G.-K. P. K. J. Mach, S. K. Allen, M. Tignor, & P. M. Midgley (Eds.), *Managing the risks of extreme events and disasters to advance climate change adaptation. A special report of Working Groups I and II of the Intergovernmental Panel on Climate Change* (pp. 1–19). Cambridge/New York: Cambridge University Press.

Irianni, M., García, M. C., Velázquez, G. A., Fernández Equiza, A. M., & Alvarez, G. O. (2008). *Argentina case study report. Each for case study report*. Bonn: United Nations University EHS.

Jacobson, J. (1988) *Environmental refugees: A yardstick for habitability*. Worldwatch Paper 86). Washington, DC: Worldwatch Institute.

Jungehülsing, J. (2010). *Las que se van, las que se quedan: reacciones frente al cambio climático Un estudio de caso sobre migración y género en Chiapas*. (H. B. Stiftung, Ed.). México.

Kaenzig, R. (2011). Bolivian Andes: From climate change to human displacements? In A.Borsdorf, J. E. Stötter, & E. Veulliet (Eds.), *Managing Alpine Future II: "Inspire and drive sustainable mountain regions". Proceedings of the Innsbruck conference November 21–23, 2011* (pp. 47–57). Wien: Austrian Academy of Sciences.

Kaenzig, R., & Piguet, E. (2011). Migrations et changement climatique en Amérique Latine: quels enjeux ? *Vertigo – Revue Electronique en Sciences de l'Environnement, 11*.

Kahn, A. S., & Campus, R. T. (1992). Effects of drought on agricultural sector of Northeast Brazil. In *Presented at ICID*. Fortaleza, Brazil.

Kniveton, D., Schmidt-Verkerk, K., Smith, C., & Black, R. (2008). *Climate change and migration: Improving methodologies to estimate flows*. Geneva: IOM (International Organization for Migration).

Leighton, M. (2006). Desertification and migration. In P. M. Johnson, K. Mayrand, & M. Paquin (Eds.), *Governing global desertification* (pp. 43–58). London: Ashgate.
Leighton, M. (2011). Drought, desertification and migration: Past experiences, predicted impacts and human rights issues. In E. Piguet, A. Pécoud, & P. de Gutcheneire (Eds.), *Migration and climate change*. Cambridge: Cambridge University Press.
Leighton-Schwartz, M. L., & Notini, J. (1995). Preliminary report on desertification and migration case studies and evaluation. In J. Puigdefábrigas & T. Mendizábal (Eds.), *Desertification and migrations* (pp. 69–113). Logroño: Geoforma Ediciones.
MacGranahan, G., Balk, D., & Anderson, B. (2007). The rising tide: Assessing the risks of climate change and human settlements in low elevation coastal zones. *Environment and Urbanization, 19*, 17–37.
Magrin, G., Gay García, C., Cruz Choque, D., Giménez, J. C., Moreno, A. R., Nagy, G. J., Nobre, C., & Villamizar, A. (2007). Latin America. Climate change 2007: Impacts, adaptation and vulnerability. In M. L. Parry, O. F. Canziani, J. P. Palutikof, P. J. Van der Linden, & C. E. Hanson (Eds.), *Contribution of Working Group II to the fourth assessment report of the Intergovernmental Panel on Climate Change (IPCC)* (pp. 581–615). Cambridge: Cambridge University Press.
Marengo, J. A., Tomasella, J., Alves, L. M., Soares, W. R., & Rodriguez, D. A. (2011). The drought of 2010 in the context of historical droughts in the Amazon region. *Geophysical Research Letters, 38*, L12703.
McDowell, J. Z., & Hess, J. J. (2012). Accessing adaptation: Multiple stressors on livelihoods in the Bolivian highlands under a changing climate. *Global Environmental Change, 22*, 342–352.
Meze-Hausken, E. (2004). Migration caused by climate change: How vulnerable are people in dryland areas? *Mitigation and Adaptation Strategies for Global Change, 5*, 379–406.
Myers, N. (1993). *Ultimate security: The environmental basis of political stability*. New York: W. W. Nortan.
Nagy, G. J., Caffera, R. M., Aparicio, M., Barrenechea, P., Bidegain, M., Jimenez, J. C., Lentini, E., & Magrin, G. (2006). *Understanding the potential impact of climate change and variability in Latin America and the Caribbean*. Report prepared for the Stern Review on the Economics of Climate Change. In http://www.sternreview.org.uk, 34.
Nicholls, R. J. (2004). Coastal flooding and wetland loss in the 21st century: changes under the SRES climate and socio-economic scenarios. *Global Environmental Change, 14*, 69–86.
Oliver Smith, A. (2009). *Nature, society, and population displacement. Toward an understanding of environmental migration and social vulnerability* (InterSecTions 'Interdisciplinary Security ConnecTions' publication series of UNU-EHS, n°8). Bonn: UNU-EHS.
Orlove, B. (2009). The past, the present, and some possible futures of adaptation. In W. N. Adger, I. Lorenzoni, & K. O'Brien (Eds.), *Adapting to climate change: Thresholds, values, governance* (pp. 63–131). Cambridge: Cambridge University Press.
OXFAM. (2009). *Bolivia: Climate change, poverty and adaptation*. Bolivia: OXFAM International.
Paul, B. K. (2005). Evidence against disaster-induced migration: The 2004 tornado in north-central Bangladesh. *Disasters, 4*, 370–385.
Paz Rada, O. (2007). *Proyecto Estudios de Cambio Climático. Programa Nacional de Cambios Climáticos (PNCC)*. Bolivia: Gobierno de Bolivia.
Pezza, A. B., & Simmonds, I. (2005). The first South Atlantic hurricane: Unprecedented blocking, low shear and climate change. *Geophysical Research Letters, 32*, L15712.
Piguet, E. (2008). *Climate change and forced migration* (New Issues in Refugee Research – United Nations High Commissioner for Refugees Research Paper N°153). Geneva: United Nations High Commissioner for Refugees.
Piguet, E. (2010). Linking climate change, environmental degradation and migration: A methodological overview. *Wiley Interdisciplinary Reviews: Climate Change, 1*, 517–524.
Piguet, E. (2012). Des apatrides du climat? *Annales de Géographie, 683*, 86–100.

Ramirez, E. (2006). Impacto del Cambio Climático sobre la Disponibilidad de los Recursos Hídricos. In J. Choquehuanca, D. Hoffmann, & M. Frias (Eds.), *Retroceso de los Glaciares y Recursos Hídricos en Bolivia – De la Investigación a la Acción* (pp. 19–31). La Paz: Instituto Boliviano de la Montaña – UMSA.

Rhoades, R. (2008). Disappearance of the glacier on Mama Cotacachi: Ethnoecological research and climate change in the Ecuadorian Andes. *Pirineos, 163*, 37–50.

Rodrigues, A. S. L., Ewers, R. M., Parry, L., Souza, C., Veríssimo, A., & Balmford, A. (2009). Boom-and-bust development patterns across the Amazon deforestation frontier. *Science, 324*, 1435–1437.

Ross, L. R. (2010). *Climate change and immigration: Warnings for America's southern border. In Perspectives*. Washington, DC: American Security Project.

Saldaña-Zorrilla, S., & Sandberg, K. (2009). Impact of climate-related disasters on human migration in Mexico: A spatial model. *Climatic Change, 96*, 97–118.

Schmidt-Verkerk, K. (2009). The potential influence of changing precipitation and temperature patterns on migratory behaviour in the State of Zacatecas, Mexico. In A. Oliver-Smith & X. Shen (Eds.), *Linking environmental change, migration & social vulnerability* (pp. 51–60). Bonn: Source 12 – Publication Series of UNU-EHS.

Schwartz Leighton, M., & Notini, J. (1994). *Desertification and migration: Mexico and the United States*. San Francisco: U.S. Commission on Immigration Reform.

Simms, A., & Hannah, R. (2006). *Up in smoke? Latin America and the Caribbean*. 48. Working Group on Climate Change and Development.

Soruco, A., Vincent, C., Francou, B., & Gonzalez Javier, F. (2009). Glacier decline between 1963 and 2006 in the Cordillera Real, Bolivia. *Geophysical Research Letters, 36*, L03502.

Vergara, W. (2005). *Adapting to climate change. Lessons learned, work in progress and proposed next steps for the World Bank in Latin America* (Sustainable Development Working Paper 25). Washington, DC: The World Bank.

Viviroli, D., Dürr, H., Messerli, B., Meybeck, M., & Weingartner, R. (2007). Mountains of the world – Water towers for humanity: Typology, mapping and global significance. *Water Resources Research, 43*(7), W07447.

Warner, K., Ehrhart, C., de Sherbinin, A., Adamo, S., & Chai-Onn, T. (2009). *In search of shelter: Mapping the effects of climate change on human migration and displacement*. CARE/CIESIN/UNHCR/UNU-EHS/World Bank.

Wrathall, D. J. (2008). The role of entitlements in shaping the migration behaviours of vulnerable people: A conceptual model based on lessons from Hurricane Mitch in Honduras, UNU Conference Paper, 27 July–3 August 2008.

Young, K. R., & Lipton, J. K. (2006). Adaptive governance and climate change in the tropical highlands of western South America. *Climatic Change, 78*, 63–102.

Migration and Climate Change in Oceania 8

John Campbell and Richard Bedford

Introduction

Oceania, or the region comprising Pacific Island Countries (PICs) and Australia and New Zealand, had an estimated population of 36 million people in mid-2011 with 10 million living in the 22 Pacific Island nations and territories (see Table 8.1 and Fig. 8.1) (SPC 2011). The PICs have been identified as a cluster of countries that are most likely to be affected by climate change in international agreements, academic writings and the media. Several events have seen some Pacific island communities labelled as the world's first climate refugees, although thus far there is little published research that indicates climate change forced migration has occurred anywhere in the region. Nevertheless, it is likely that island environments (including New Zealand), as well as those in continental Australia, will be significantly affected by climate change, and these effects may well impinge on the lives and livelihoods of many of Oceania's residents. In extreme cases it is possible that forced migration, especially from some PICs, may eventuate as a result of changes associated with global warming. More likely, though, for the great majority of island residents, climate change will exacerbate environmental factors that have been influencing migration decision making for decades.

In this chapter we focus on migration and environment in the Pacific island countries. The impacts of climate change on environments and their inhabitants in the large continental landmass of Australia and the temperate hill country and mountains of New Zealand will be very different from the impacts on small tropical islands. Climate change, especially if it results in prolonged drought over large parts

J. Campbell (✉)
Geography Programme, University of Waikato, Hamilton, New Zealand
e-mail: jrc@waikato.ac.nz

R. Bedford
Pro Vice-Chancellor (Research), Auckland University of Technology, Auckland

Professor of Population Geography, National Institute of Demographic and Economic Analysis, University of Waikato, Hamilton, New Zealand

E. Piguet and F. Laczko (eds.), *People on the Move in a Changing Climate*, Global Migration Issues 2, DOI 10.1007/978-94-007-6985-4_8,

Table 8.1 Pacific Island Populations

Sub-region/country	Mid-year population estimates 2011	2050	Population increase (%) 2011–2050	Crude population density (persons/km^2) 2011	2050
Melanesia	8,797,410	16,339,285	86	16.2	30.1
Fiji Islands	851,745	1,060,706	25	46.6	58.0
New Caledonia	252,331	343,175	36	13.6	18.5
Papua New Guinea	6,888,297	13,271,057	93	14.9	28.7
Solomon Islands	553,254	1,181,299	114	18.2	38.8
Vanuatu	251,784	483,048	92	20.5	39.3
Micronesia	546,491	720,448	32	173.2	228.3
Federated States of Micronesia	102,360	109,265	7	146.0	155.9
Guam	192,090	267,820	39	355.1	495.0
Kiribati	102,697	163,266	59	126.6	201.3
Marshall islands	54,999	61,217	11	303.9	338.2
Nauru	10,185	16,283	60	485.0	775.4
Northern Mariana Islands	63,517	80,137	26	138.9	175.3
Palau	20,643	22,459	9	46.5	50.6
Polynesia	668,470	825,633	24	83.7	103.4
American Samoa	66,692	98,271	47	335.0	493.6
Cook Islands	15,576	15,977	3	65.7	67.4
French Polynesia	271,831	348,778	28	77.2	99.1
Niue	1,446	1,283	−11	5.6	5.0
Samoa	183,617	209,740	14	62.6	71.5
Tokelau	1,162	1,148	−1	96.9	95.6
Tonga	103,682	123,008	19	159.5	189.2
Tuvalu	11,206	13,858	24	431.0	533.0
Wallis and Futuna	13,193	13,570	3	92.9	95.6
Total	10,012,338	17,885,366	79		

Source: SPC (2011)

of southeast and western Australia may contribute to the already well-established net migration gains through internal migration to the wetter and warmer north-eastern and northern Australia, as well as some return migration of New Zealanders from Australia to New Zealand.[1] By comparison with many parts of the world, climate

[1] A useful overview of issues relating to Australia's population, including internal migration trends, recent immigration and the impact of climate change on the different regions of the country, can be found in Pincus and Hugo (2012). Migration between New Zealand and Australia is a prominent component of New Zealand's international migration system, as are flows between island countries in the eastern and central Pacific and New Zealand. These connections, together with long-standing flows between Europe and New Zealand, and the more recent flows from countries in the Americas, Asia, Africa and the Middle East, are examined in Spoonley and Bedford (2012). While successive governments in both New Zealand and Australia have encouraged immigration since the late nineteenth century, the two countries also restricted severely immigration from countries in Asia until the early 1970s (Australia) and mid-1980s (New Zealand). Jupp (2007) and McKinnon (1996) provide useful historical accounts of immigration policies in the two countries, while Bedford and

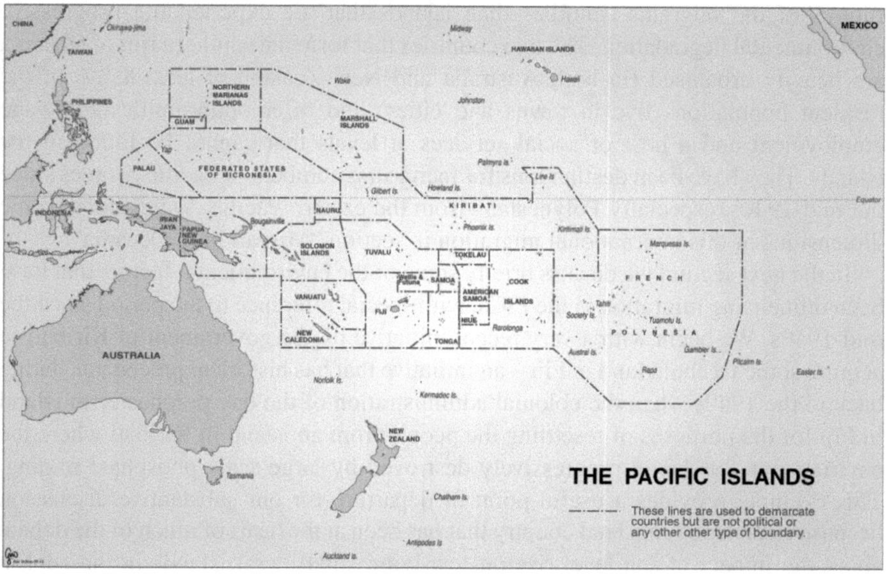

Fig. 8.1 Map of Oceania showing the 22 countries and territories

change may have less severe impacts on New Zealand, although sea level rise is likely to impact on the country's population which is concentrated in coastal locations.

In Australia and New Zealand relocation of people elsewhere in the two countries is not constrained to anything like the same extent as in the PICs by the customary land tenure systems of indigenous populations. In the PICs 90 % of land is still held under forms of customary title (see section "Overview of the main current environmental issues in the region" below). Most land in the islands is not treated as a commodity subject to market forces in the sense that it is in Australia and New Zealand. In the Pacific a great premium is placed on adjusting to environmental damage *in situ*; moving to another area, especially another rural area, is not straightforward given that all land is owned by some indigenous group. In these situations it can be easier to move into local towns or on to towns overseas where land might be available for leasing or purchase. Occasionally rural land which is no longer subject to customary title becomes available for purchase and, as we show in the next section, such land has been acquired in the past as well as currently for resettlement purposes.

In this chapter Australia and New Zealand enter the discussion mainly as destinations for Pacific migrants, some of whom are seeking greater livelihood security in countries which are perceived to offer better opportunities for the

Hugo (2012) deal more specifically with trends in and prospects for migration between the Pacific Islands and Australia and New Zealand.

futures of the migrants' families than islands that are experiencing progressive environmental degradation. The two countries that form the southern rim of Oceania are heavily urbanised (in both Australia and New Zealand at least 85 % of the resident populations live in towns and cities) and offer opportunities for wage employment and a host of social services at levels that cannot be found in the islands. They have been destinations for increasing numbers of Pacific peoples since the mid-1940s, especially Polynesians from the eastern Pacific. We examine some dimensions of this international migration in section "Migration in Oceania".

In the next section we discuss briefly some of the environmental factors that have been influencing migration in the PICs, with special reference to the period since the mid-1940s. We begin with a very recent initiative by the government of Kiribati to acquire some freehold land in Fiji – an initiative that has historical precedents dating back to the 1940s when the colonial administration of the day purchased an island in Fiji for the purposes of resettling the people from an island in Kiribati where the environment was being progressively destroyed by large-scale phosphate mining. This example provides a useful point of departure for our substantive discussion because it relates to an island country that has been at the heart of much of the debate about the impact of sea level rise on low-lying small coral islands in the central Pacific. It demonstrates the long-standing concerns about relationships between population growth, urbanisation, economic development and environment in many of the island states and territories in Oceania.

Historical Overview of Migration Related to Environmental Events

In April 2012 the Government of Kiribati approved the purchase of 2,282 ha of free-hold land located in the Savusavu area of Fiji's second largest island, Vanua Levu. The land block, known as Natoavatu, is owned by the Anglican Church and is one of the largest free-hold land areas remaining in Fiji. A report on a site inspection of Natoavatu late in 2011 placed emphasis on "the great opportunity this huge land mass would provide Kiribati, not only in terms of advancing its development aspirations as highlighted in the National Development Plan but, more importantly, in terms of providing long-term security for the people of Kiribati within the context of climate change and its impacts" (Anon 2012, p. 1). The block of land extends 10 km inland from the coast and has potential for supporting both agricultural as well as fisheries development, as well as having an extant mature pine plantation that is ready for milling. In the wider context of land available for development by I-Kiribati settlers, the area of Natoavatu is equivalent to around 10 % of the total land area of the Gilbert Group (around 28,300 ha) where 90 % of Kiribati's population was living in 2010. It is not a future home for all I-Kiribati, in the extreme event that their atolls and reef islands become progressively uninhabitable during this century, but it is certainly one of a number of options for future residence for a central Pacific people who have long had to grapple with environmental uncertainty in their low-lying coral islands (Bedford et al. 2007).

The Kiribati government's strategic investment in land in Fiji in 2012 mirrors a search for land for this purpose by the former British colonial administration of Kiribati and its southern neighbour, Tuvalu, in what was known as the Gilbert and Ellice Islands Colony (GEIC) until 1978 (Bedford 1967; Lambert 1975; Macdonald 1982). Finding opportunities for settlement elsewhere of I-Kiribati and Tuvaluans, who were perceived to be experiencing 'land hunger' on those islands they considered their ancestral homes, was a priority for the GEIC administration from as early as the 1930s (Maude 1938). There is nothing new about I-Kiribati or Tuvaluan land purchases overseas; indigenous peoples from the GEIC were involved in a range of resettlement schemes in Fiji and the Solomon Islands from the mid-1940s, commencing with the forced resettlement of Banabans on Rabi Island in Fiji to facilitate extraction of phosphate on their home island of Banaba (Ocean Island) by the British Phosphate Commission (Silverman 1971, 1977; Macdonald 1982; Williams and Macdonald 1985). This was followed by the voluntary purchase and subsequent settlement of Kioa in Fiji, an island close to Rabi, by the people of Vaitupu in Tuvalu in the late 1940s and early 1950s (Bedford 1967; Kock 1973; Butcher 2012).

In the late 1950s and early 1960s there was a managed resettlement programme in the western Solomon Islands (Wagina Island and Titiana, near Gizo) (Knudson 1965, 1977) for I-Kiribati who had earlier been resettled on uninhabited islands in the Phoneix Islands within the GEIC in the 1930s (Maude 1952) and who subsequently had to leave the Phoenix Islands because of water shortages. These forced, voluntary and managed resettlement schemes were all linked with perceived (current or future) pressures of people on fragile atoll environments in Kiribati and Tuvalu – the sorts of pressures that are assuming new dimensions with accelerated acidification of oceans, rising sea levels and changing patterns of precipitation and storm frequency that are often emphasized in the current debates about climate change.

This reference to a history of over 70 years of concern by governments about the ability of "fragile" atoll and reef island environments to support populations that are much larger than those that were there before sustained European intervention from the mid-nineteenth century in the lives of I-Kiribati and Tuvaluans (Bedford et al. 1980) provides a timely reminder of the long-standing interest in and debate about the impact of environmental factors on migration in Oceania. Campbell et al. (2007) documented over 500 cases of extreme events and migration in, from and between PICs: of these 86 involved permanent community relocation and 50 (58 %) of these were related to environmental variability and degradation. Movement in response to environmental change in association with other factors is not new in Oceania and indigenous strategies for adapting to both rapid-onset and slow-onset environmental events have tended to be ignored in the more sensationalist reporting of possible long-term impacts of climate change in the region.

The increased 'vulnerability' of many Pacific peoples to extreme environmental events is, in fact, an unintended consequence of, often, well-meaning state intervention at times of natural disaster and the devaluing and subsequent destruction of time-honoured indigenous systems of adaptation to environmental risk and uncertainty (Campbell 1985). Reclaiming indigenous capacities to respond to environmental change is a critical challenge for Pacific societies in the twenty-first century, especially given the unwillingness of the international community to adopt mitigation strategies to address some of the drivers of accelerated climate change. There is a preference instead for putting the onus on people in the places most affected by environmental degradation to adapt to adverse impacts of climate change.

One of the most profound changes that has occurred in Pacific populations that has resulted in increased vulnerability to several environment-related events is the increasing concentration of populations in low-lying coastal locations. While atoll and reef island populations by definition have always lived close to the sea, the great majority of Pacific peoples used to live inland from the coast for two main reasons: ease of security and protection from malaria.[2] The first major migration of Pacific peoples following sustained European contact was a movement from interior locations to the coast, partly for work on plantations and for easier access to churches, trading posts and later schools and medical centres, and partly because colonial governments everywhere encouraged greater concentration of dispersed rural populations in larger villages for administrative purposes.

From the 1950s this movement to coastal locations was stimulated by the development of towns – urbanisation in the Pacific is essentially a post-Second World War phenomenon (Brookfield 1972; Brookfield with Hart 1971; Connell and Lea 1993, 1995, 1998). In many Pacific colonies the indigenous populations were actively discouraged from living in towns; their livelihoods were to be obtained in rural communities from work on their lands. With the development of commercial coconut plantations on the low-lying coastal plains of most Pacific 'high islands', along with sugar farming (Fiji), cattle grazing and the cultivation of a range of horticultural crops in river valleys and on coastal plains, the Pacific's indigenous peoples progressively shifted from inland locations to be closer to the economic and social action associated with 'development'. The major exception was the Central Highlands of Papua New Guinea where sizeable populations remained on their lands in the interior of the largest Pacific country.

Many of the environmental events documented by Campbell et al. (2007), that were accompanied or followed by migration (often temporary rather than permanent), were tropical cyclones. Flooding, storm surges and widespread damage to crops and houses caused by high winds, heavy rain and storm surges is a regular part of life in several countries in the western Pacific, especially Fiji and Vanuatu. The increasing concentrations of people in villages and towns located in river valleys and along the coastal littoral strips of high islands has exposed higher proportions of Pacific peoples to risk of damage to their houses, cash crops and subsistence gardens from these events. Many are also more at risk to inundation from tsunami,

[2]There is an extensive literature on pre-colonial settlement in Pacific societies. Brookfield with Hart's (1971) detailed geographical interpretation of the islands in the western Pacific remains an authoritative analysis of people and places in the larger islands of the Pacific.

and these events, while much less frequent than cyclones, are not uncommon given the location of many islands along what is known as the 'Pacific rim of fire'. Earthquakes, land slips and volcanic activity are all environmental events which have generated migration flows in the past. Notwithstanding the label 'Pacific', these islands are anything but 'pacific' places when it comes to environmental processes and their impacts on human settlement and activity. Migration linked with environmental change long pre-dates the current concerns with global warming and associated rising sea levels.

Regional Projections Relating to Climate Change in Pacific Islands

The findings of the IPCC AR4 (2007) in relation to small islands states were hampered by the low levels of resolution of the GCM models used in the projection process. Overall it was found that small islands were likely to experience sea level rise (though not uniformly), and were very likely to experience atmospheric warming at rates lower than the global average. Increases in rainfall were projected to be likely in the equatorial Pacific (Christensen et al. 2007).

Since this most recent IPCC Assessment Report, the Pacific Climate Change Science Programme (PCCSP) has published more specific findings for the region (Australian Bureau of Meteorology and CSIRO 2011). Using updated records the PCCSP indicates that over the past 50 years there has been atmospheric warming (0.08–0.20 °C per decade), increasing sea surface temperatures, and sea level rise including increases in extreme high sea level events. Rainfall has tended to be highly variable, although increasing north-east of the South Pacific Convergence Zone[3] and decreasing to the south. Ocean acidification is also occurring with implications for coral growth. There were no trends detected in frequency or magnitude of tropical cyclones.

The PCCSP also produced some finer resolution projections for the region based on the IPCC SRES B1, A1B and A2 emissions scenarios. Among the projections are increases in temperature of between +0.5 and 1.0 °C by 2030 under all scenarios through to +1.5 to 3.0 °C by 2090 (depending on the emissions scenario). The study indicates that the number of extremely hot days and warm nights will also increase. Rainfall is projected to increase near the SPCZ and ITCZ and remain relatively unchanged elsewhere in the region. The projections include a decline in the frequency of droughts. This is in comparison with Working Group II of the IPCC AR4 which indicated with high confidence that water resources in SIDS are likely to be seriously compromised (Mimura et al. 2007, p. 689) although rainfall is not the only influence on water quality and quantity. Sea level is anticipated to rise by about 80 cm by century's end. There is still considerable uncertainty about the

[3]There are two convergence zones in the Pacific where moisture laden trade winds meet – they are characterised by a zone of cloudiness and increased rainfall. These are the Intertropical Convergence Zone (ITCZ) and the South Pacific Convergence Zone (SPCZ).

Table 8.2 Types of Islands in Oceania

Island type	Implications for climate change
Continental	
Large High elevations High biodiversity Well developed soils River flood plains Orographic rainfall	River flooding more likely to be a problem than in other island types. Deltas and flood plains at risk. Drought conditions often experienced during El Niño events. Frosts in the Papua new guinea Highlands, severe during El Niño conditions
Volcanic High Islands	
Steep slopes Different stages of erosion/subsidence Barrier reefs Relatively small land area Less well developed river systems Orographic rainfall	Because of size few areas not exposed to tropical cyclones. Streams and rivers subject to flash flooding. Barrier reefs may ameliorate storm surge and tsunami.
Atolls	
Very small land areas Very low elevations No or minimal soil Small islets surround a lagoon Shore platform of ocean side of islets Larger islets on windward side No surface (fresh) water Ghyben-Herzberg (fresh water) lens Convectional rainfall	Exposed to storm surge, 'king' tides and high waves can inundate islets, cause coastal erosion and salinise the freshwater lens. Narrow resource base and use of anthropogenic soils and freshwater lens to grow *cyrtosperma* or swamp taro. Exposed to drought and fresh water shortages. Water problems may lead to health hazards.
Raised Limestone Islands	
Steep outer slopes Concave inner basin Sharp karst topography Narrow coastal plains No surface water No or minimal soil	Depending on height may be exposed to storm surge. Exposed to fresh water shortages and drought. Water problems may lead to health hazards.

Source: After Campbell (2006)

effects of these changes on tropical cyclones although they are expected to decline in frequency throughout the region while the number of intense storms may decrease in the north and increase south.

Relevance to Displacement

What are likely to be the implications of these projections for the communities that live on Pacific Islands? A key consideration will be the nature of the physical and natural environment of the islands that they occupy. Table 8.2 summarises the main types of islands found in the region. Large 'continental' or inter-plate type islands, formed by subduction, are found along the boundary between the Australian and Asian (continental) and the Pacific (oceanic) plates. These islands are characterised by high levels of landform and ecological diversity, large river systems with well developed flood plains and river deltas which are often densely populated. However, as previously noted, with the exception of Papua New Guinea (where high population densities are found in the Highlands region), the great majority of inhabitants are found in coastal settlements.

East of the plate boundary there is a large number of smaller islands formed originally as volcanoes over hot spots in the Earth's mantle as the oceanic plate slowly moved away from its origin along the East Pacific Rise. These islands can be found in a range of forms from relatively large volcanic islands (some still being formed) through to smaller high islands reduced in size by subsidence (as the plate moves away from the hot spot) and erosion from wind and wave, through to atolls which are based on coral reefs that have been built as volcanic islands subside below sea level. The atolls are comprised of small islets made from sand, derived from the coral reefs, and deposited on top of them by wave action. There are also some 'raised limestone islands' which are atolls that have been stranded by earlier higher sea levels or uplifted by geological processes.

Climate change is likely to reduce the suitability of some islands, or parts of islands, for human settlement in a number of direct and indirect ways. Working group II of the IPCC AR4 identified several implications of these changes for small island states. These are outlined in the left hand column of Table 8.3. The table also indicates how these changes may in turn negatively impact upon the security of human communities and, in some cases, potentially give rise to migration.

In this chapter we use the notion of community security as a means of considering the ways in which climate change may induce migration.[4] Three important elements of community security include land (locational, cultural) security, habitat (biological, health and safety) security and livelihood (subsistence and commercial) security. Land security refers to communities not only having a physical terrestrial base upon which to build settlements and carry out important livelihood and cultural activities, but one which sustains the visceral link between each individual's kinship group and its land that characterises most Pacific island societies (Campbell 2010a; Pond 1999; Ravuvu 1988). For a great part of the region individual and group identity cannot be separated from the land and vice versa. As noted above, the great majority of land is held under customary forms of ownership (AusAID 2008) and this land cannot be bought, sold or even given away unless sanctioned by traditional forms of land exchange which are relatively rare. This is an important issue when considering migration and relocation within the region – the loss of the link to land happens both

[4]Climate change has often been linked to issues of 'national security'. However we use the term 'security' for want of another term that excludes the state and military aspects often associated with notions of security (e.g. Barnett 2001). We hope that these concerns (relating to larger scale systems of government) can be somewhat overcome by using the modifier community. "Community security" refers here to the ability of a community to have sustainable livelihoods, a place on which to be located and a healthy environment.

Table 8.3 Key effects of climate change on PICs and implications for community security

Sea Level Rise Inundation Coastal erosion Storm surge exacerbated	**Land security** in coastal and atoll locations may be severely reduced and there may be impacts on **livelihood security** through loss of agricultural land and salinisation of soils, plants and water supplies
Water Resources Rainfall uncertainty Increased frequency and magnitude of droughts Reduced quantity and quality of water resources Salinisation	**Livelihood security** may be affected by decreased agricultural productivity and **habitat security** may be adversely affected by water borne diseases
Coral Reefs Reef degradation as a result of increased sea surface temperatures and increased ocean acidity	**Livelihood security** may be compromised by reductions in fisheries and other marine resources dependent upon healthy coral environment. **Land security** may be reduced by increased exposure to high waves and storm surge.
Agriculture Adverse effects from a variety of processes including temperature rise, reduced water availability, salinisation, exposure to tropical cyclones (wind, rain and wave damage)	Reduced agricultural productivity would impinge upon **livelihood security** and where extremely severe may render some locations uninhabitable.
Human Health Changing disease vectors such as malaria, dengue Increased incidence of water borne disease Increased incidence of heat related diseases	Effects on human health are likely to reduce the **habitat security** of island settlement locations and where severe may render some locations uninhabitable.

Source: After Campbell (2012)

for migrants and people at the destination whose land may be used for resettlement. Sea level rise (in conjunction with extreme events, such as storm surge, king tides and high waves) may cause inundation and coastal erosion, and river flooding may cause river bank erosion, processes that physically remove land from its customary owners and reduce, and in extreme cases, potentially remove, land security.

Habitat security refers to having a safe environment in which to live. Climate change effects that result in negative health effects, such as reductions in food and water quality and/or availability, changing disease vectors, and exposure to more frequent or more intense extreme weather events that may cause injury or fatalities are likely to have negative effects upon habitat security. Breakdowns in infrastructure, especially that which is related to sanitation, may also impact upon habitat security. The final aspect of community security is livelihood security which is related to subsistence food production, production of commodities for sale and job availability. In cases where land and habitat security remain intact, livelihood security may be sustained through migration and remittances. However,

Lost land security	Lost livelihood security	Habitat security severely reduced	Livelihood security severely disrupted	Livelihood security significantly but not severely disrupted	Land and livelihood security disrupted periodically
Community relocation	**Community relocation**	**Community relocation or individual migration**	**Individual migration**	**Individual migration**	**Short term migration**
Drivers of reduced land, livelihood and habitat security					
Atoll Submerged	**Salinisation -Water inadequate -Crop failure**	**Changes in disease vectors such as malaria, dengue, ciguatera**	**Severe reductions in - water supply - food production**	**Moderate reductions in - water supply - food production**	**Increased frequency and / or magnitude of climatic extremes**
Coastal sites submerged and / or eroded	**Persistent Drought - water inadequate - crop failure**	**Changes in water - borne disease incidence**	**Severely but not completely reduced land available for settlement or livelihoods**	**Moderately but not completely reduced land available for settlement or livelihoods**	
Delta sites eroded or inundated		**Temperature related illnesses**			
River bank sites eroded					

Fig. 8.2 Forced and induced migration in the context of climate change effects on land, livelihood and/or habitat security

it is important to note that the majority of PIC populations still remain in rural areas and are dependent upon agriculture and fisheries for subsistence and cash livelihoods.

Figure 8.2 seeks to establish the links between climate change effects on community security and migration. The figure distinguishes between involuntary and voluntary migration (after Hugo 2011). The former, often referred to as forced migration, is likely to take place as community relocation as a result of complete or very severe loss of community security. The latter, also referred to as climate change induced migration, is more likely to take place as individual or household mobility in a manner similar to much contemporary Pacific island migration (see section "Migration in Oceania"). In this chapter we distinguish evacuation and resettlement, forms of climate migration, from climate change migration. We see this as reasonable assumption given the failure of countries to significantly reduce greenhouse gas emissions thus ensuring climate change will be a unidirectional process.[5]

Both involuntary and voluntary migration may be seen as forms of climate change adaptation. In particular, voluntary migration may be considered a rational response to climate change enabling continued *in situ* settlement for many affected communities (Barnett and Webber 2009); although Barnett (2012) observes that there are a number of negative implications for communities whose population falls below what may be considered sustainable levels because of migration. These issues are explored in sections "A synthesis of existing case studies on the links between migration and climate" and "Projecting climate change induced and forced migration in the Pacific".

Overview of the Main Current Environmental Issues in the Region

While not characterised by high levels of industrialisation and urbanisation, and the associated environmental problems found in many developing countries, PICs have a number of significant environmental problems apart from climate change. These include a growing waste stream, coastal degradation through marine pollution and coral reef and mangrove destruction, land degradation including deforestation and soil loss, freshwater degradation (both in terms of quality and quantity), problems related to mining, depletion of fisheries and overall loss of biodiversity. Some of these will be worsened by climate change. For example, increased sea surface temperatures and ocean acidification may increase the rate of coral reef decline. In other cases, these environmental problems will increase the impacts of climate change effects such as deforestation resulting in greater runoff and erosion during flood events (which are expected to increase in frequency and intensity). This may in turn result in increasing coastal and lagoon sedimentation and further damage to coral reef ecosystems.

A combination of existing environmental problems and climate change may hasten the transformation of some locales to becoming either less secure or even uninhabitable. For example, coral reef degradation caused by local processes may be worsened by increasing SSTs and acidification and this may result in increased coastal inundation and erosion, reduced coastal protection from normal wave conditions and extreme events such as storm surges while also reducing the

[5]It may be anticipated that climate extremes may result in forced or voluntary evacuation, but following such events resettlement often takes place. Where the need for temporary evacuation occurs with increasing frequency some communities may decide that relocation or migration is necessary.

productivity of coastal fisheries, resulting in reduced habitat and livelihood security. These changes may also result in salinisation of fragile fresh water lenses, especially on coral atolls and reef islands, and it may be that destruction of reservoirs of fresh ground water could be a critical tipping point for environment-induced migration in some parts of the region.

Migration in Oceania

Population movement within and between islands has long been an essential part of life in both the small volcanic and coral islands of the eastern, central and northern Pacific as well as the much larger inter-plate islands of the western Pacific. Almost 20 years ago Tongan sociologist, Epeli Hau'ofa (1993, p. 8), captured the essence of this dimension of everyday life in the region when he observed:

> The world of our ancestors was a large sea full of places to explore, to make homes in, to breed generations of seafarers like themselves. People raised in this environment were at home with the sea. They played in it as soon as they could walk steadily, they worked in it, they fought on it. They developed great skills for navigating their waters, and the spirit to traverse even the few large gaps that separated their island groups.
>
> Theirs was a large world in which peoples and cultures moved and mingled unhindered by boundaries of the kind erected much later by imperial powers. From one island to another they sailed to trade and to marry, thereby expanding social networks for greater flow of wealth. They travelled to visit relatives in a wide variety of natural and cultural surroundings, to quench their thirst for adventure, and even to fight and dominate.

Within what Hau'ofa termed this 'sea of islands' that encompassed the island Pacific as well as Aotearoa (New Zealand) and Hawai'i one way of adjusting to natural disaster and environmental degradation was population movement. In the Pacific countries, where tropical cyclones are annual events, temporary and permanent relocation following destruction of houses and gardens was an adaptation to adverse impacts of climate on environment (Campbell 1977, 1985). Village relocation, within the lands belonging to the group living in the area, was not uncommon as is shown in the next section. The migration of Polynesians over very long distances to New Zealand and to Hawai'i was probably partly a response to rising pressure of population on available resources, just as migration of Europeans to Oceania was partly due to perceptions of opportunities for a better life elsewhere than in the challenging environments of the early industrial city.

During the twentieth and early twenty-first centuries tens of thousands of people have moved from their island homes to countries on the Pacific Rim. Bedford and Hugo (2012, 50) estimated that by 2010 there would have been around 850,000 people of Pacific Island ancestries or ethnicities living in the four 'countries of immigration' on the Pacific rim: New Zealand (350,000), Australia (150,000), USA (300,000) and Canada (50,000). These indicative figures are based on some pro-rata adjustments to the last published census data for 2000 (USA) and 2006 (Australia, Canada and New Zealand). The combined total of these Pacific ethnic/ancestry populations is larger than the Secretariat of the Pacific Community's (SPC) (2010) estimates of the total populations of either Micronesia (547,300) or Polynesia (663,960) in 2010. There are also small Pacific populations in the United Kingdom and Europe as well as in parts of Asia. The overall size of the global diaspora of Pacific peoples, measured in terms of ancestry/ethnic identity, could have been close to one million in 2010.

Migration overseas continues to be a favoured strategy for adjusting to changing circumstances and opportunities at home as well as in other places especially in Polynesia. Migration from islands in Micronesia to the United States, and from Fiji in Melanesia to New Zealand, Australia and North America has also been widespread, although not as extensive as that from Polynesia. Populations in western Melanesia (Papua New Guinea, the Solomon Islands and Vanuatu) have had much less access to residence in Pacific rim countries; their diaspora are very small in relation to the sizes of their populations in the islands (Bedford and Hugo 2012).

In their *World Migration Report 2010*, the IOM (2010, 241) note that "Oceania hosts over 6 million international migrants, who make up 16.8 % of the population in the region. Although only accounting for less than 3 % of the global migrant stock, Oceania is the region with the highest proportion of migrants in its population". While both Australia and New Zealand have long been 'countries of immigration' with active policies to attract skilled migrants in particular from Europe and Asia, much of the international migration that the IOM report for Oceania is intra-regional. The South Pacific migration system (Bedford 1992, 2005, 2008) is dominated by three subsystems of flows: trans-Tasman migration between New Zealand and Australia, migration from PICs to countries on the Pacific rim (especially New Zealand, Australia, the United States and Canada) and a much less significant flow in terms of volume between the different island states and territories.

With regard to international migration from Pacific countries, the IOM (2010, 227) points out that:

> The corridors for emigration from the Pacific Islands are strongly linked to geographical ties. According to Census 2000 data, 50 % of the Pacific Islands' migrants left their countries and remained within the region, with New Zealand being the leading country of 20 % of the total, followed by Australia (17 %) and other Pacific Islands (13 %). Canada and the USA receive 36 % of the Pacific Islands' migrants.

When preparing these estimates of emigration, the IOM drew on a matrix prepared by the University of Sussex's Development Research Centre on Migration, Globalisation and Poverty (DRC 2007), which contains numbers of people enumerated (or estimated) born in particular countries who were living in other countries around 2000–2002. The database contains a snapshot of the distribution of people born in each of the Pacific countries as well as in Australia and New Zealand by their countries of residence in these years. This is a conservative estimate of international population flows within Oceania as well as to North America and to other parts of the world, but it is a very useful broad indicator of the locations of people living outside their countries of birth around 2000.

As Bedford and Hugo (2012, 52–53) have noted, their Pacific-born populations totalled 106,900 (Australia) and 138,400 (New Zealand) at the time of their censuses in 2006 (Table 8.4). The Melanesia-born dominate Australia's Pacific population

Table 8.4 Pacific Island-born populations, New Zealand and Australia, 2006

	New Zealand		Australia		Australia-New Zealand	
Sub-region/country	Number	%	Number	%	Number	%
Melanesia	*41,184*	*29.8*	*75,756*	*71.3*	*116,940*	*47.8*
Fiji Islands	38,679	27.9	48,145	45.3	86,824	35.5
New Caledonia	273	0.2	1,102	1.0	1,375	0.6
Papua New Guinea	1,329	1.0	24,022	22.6	25,351	10.4
Solomon Islands	549	0.4	1,495	1.4	2,044	0.8
Vanuatu	354	0.3	986	0.9	1,340	0.5
Melanesia nfd	0	0.0	6	0.0	6	0.0
Micronesia	*1,203*	*0.9*	*1,059*	*1.0*	*2,262*	*0.9*
Federated States of Micronesia	51	0.0	15	0.0	66	0.0
Guam	24	0.0	78	0.1	102	0.0
Kiribati	846	0.6	395	0.4	1,241	0.5
Marshall islands	21	0.0	33	0.0	54	0.0
Nauru	246	0.2	487	0.5	733	0.3
Northern Mariana Islands	9	0.0	12	0.0	21	0.0
Palau	6	0.0	13	0.0	19	0.0
Micronesia nfd	0	0.0	26	0.0	26	0.0
Polynesia	*96,039*	*69.4*	*29,491*	*27.7*	*125,530*	*51.3*
American Samoa	495	0.4	199	0.2	694	0.3
Cook Islands	14,817	10.7	5,026	4.7	19,843	8.1
French Polynesia	1,053	0.8	337	0.3	1,390	0.6
Niue	4,878	3.5	577	0.5	5,455	2.2
Samoa	51,108	36.9	15,236	14.3	66,344	27.1
Tokelau	1,596	1.2	355	0.3	1,951	0.8
Tonga	20,748	15.0	7,582	7.1	28,330	11.6
Tuvalu	1,248	0.9	116	0.1	1,364	0.6
Wallis and Futuna	12	0.0	18	0.0	30	0.0
Polynesia nfd	84	0.1	45	0.0	129	0.1
Total	*138,426*	*100.0*	*106,306*	*100.0*	*244,732*	*100.0*

Source: Bedford and Hugo (2012, p. 53)

(71 %) with 45 % born in Fiji. There was also a Samoa-born population of over 15,000 in Australia – the third-largest group by country of birth – many of these had moved to Australia following an initial move to New Zealand, entering as New Zealand citizens under the terms of the Trans-Tasman Travel Arrangement that allows for visa-free movement of Australian and New Zealand citizens between their countries. In the case of New Zealand, Samoa is the largest single Pacific country of birth (37 %) for migrants from the Pacific, followed by Fiji (28 %) and Tonga (15 %).

The dominance of countries in Polynesia (69 % in 2006) as sources of Pacific migrants in New Zealand is declining, however, as flows from Fiji and other parts of Melanesia increased during the 2000s. This has occurred partly because of the impact of political instability in that country, partly because of the introduction of a seasonal work scheme in 2007 that has targeted workers from selected Pacific countries, including Vanuatu and the Solomon Islands, and partly because of the increasing numbers of study permits granted to students seeking some of their education in New Zealand. A small number of Pacific-born migrants to New Zealand and Australia in the 2000s have also been seeking new lives because of the impact of environmental change in their island homes, especially people from Tuvalu (Paton 2009; Shen and Gemenne 2011). It is impossible to get precise numbers for migrants entering Australia and New Zealand because of environmental pressures, but both countries are probably receiving some migrants from Kiribati and Tuvalu in their skilled, family and international migration streams that would mention 'environmental factors' if asked to recount the reasons for seeking residence overseas.

When Bedford and Hugo (2012, 54–56) examined approvals for residence in Australia and New Zealand of citizens of Pacific countries between July 2003 and June 2007 they found that there were three times more Pacific people moving to New Zealand than to Australia with the intention of settling (Table 8.5). Melanesia, especially Fiji, is the dominant origin for Pacific Islanders moving to Australia, while the Polynesian countries of Samoa and Tonga remain more important for New Zealand. Distinctively different patterns are in evidence if we consider skilled, family and international (humanitarian) migrants separately. Fiji is the source of over 90 % of skilled migrants in both countries. New Zealand's share of the skilled migrant intake from the Pacific is much less than its share of all settlers from the region indicating perhaps that Australia is a preferred destination for skilled migrants. Family migration is much more important in New Zealand (Table 8.5). Polynesian flows of family migrants are significantly bigger than the skilled inflows in both countries. Even greater shares of migrants in the international/humanitarian streams for both countries enter New Zealand under the special quota systems that apply for Samoa and the Pacific Access Category (PAC) which allows for entry of small numbers each year from Tonga (250), Kiribati (75) and Tuvalu (75) if they can meet certain employment-related conditions (Bedford 2008).

These three PAC countries, along with Samoa, Vanuatu and the Solomon Islands, also receive assistance from the Department of Labour to participate in a seasonal employment scheme that allows for up to 8,000 temporary workers, mainly from the Pacific, to spend up to 7 months (nine for I-Kiribati and Tuvaluans) in New Zealand's horticulture and viticulture industries (Ramasamy et al. 2008). The Recognised Seasonal Employer (RSE) work policy, and an Australian equivalent, the Pacific Seasonal Worker Pilot which has recently become an on-going policy (the Seasonal Work Program), offer considerable scope for those islanders selected for work in New Zealand or Australia to gain work experience in the agricultural sector that might be useful longer-term should they wish to gain a temporary work permit that might allow them to transition to permanent residence. The seasonal work visas are limited-purpose visas and workers must return to their home countries when the visa expires. Such work experience might also provide an opportunity for RSE workers from Tonga, Kiribati and Tuvalu to gain access

Table 8.5 Approvals for residence in Australia and New Zealand, Pacific Citizens July 2003–June 2007

Sub-region	Australia	New Zealand	Total ANZ	% NZ
(a) All Approvals				
Melanesia	*7,835*	*10,369*	*18,204*	*57.0*
Fiji	6,466	10,138	16,604	61.1
Papua New Guinea	1,086	84	1,170	7.7
Micronesia	*77*	*554*	*631*	*94.5*
Kiribati	19	521	540	96.5
Polynesia	*845*	*14,574*	*15,419*	*94.5*
Samoa	369	8,584	8,953	95.9
Tonga	432	5,230	5,662	92.4
Pacific	*8,757*	*25,497*	*34,254*	*74.4*
% Melanesia	89.4	40.7	53.1	
(b) Skilled migrants				
Melanesia	*3,738*	*4,098*	*7,836*	*52.4*
Fiji	3,275	3,992	7,267	54.9
Papua New Guinea	382	20	402	5.0
Micronesia	*28*	*15*	*45*	*33.3*
Kiribati	0	10	10	100.0
Polynesia	*74*	*264*	*338*	*78.1*
Samoa	20	63	83	76.8
Tonga	40	173	213	81.2
Pacific	*3,840*	*4,377*	*8,219*	*53.3*
% Melanesia	97.3	93.6	85.3	
(c) Family categories				
Melanesia	*3,683*	*4,390*	*8,073*	*54.3*
Fiji	2,962	4,281	7,243	59.1
Papua New Guinea	554	54	608	8.9
Micronesia	*40*	*79*	*119*	*66.4*
Kiribati	19	64	83	77.1
Polynesia	*501*	*5,549*	*6,050*	*91.7*
Samoa	130	3,127	3,257	96
Tonga	364	2,194	2,558	85.8
Pacific	*4,224*	*10,018*	*14,242*	*70.3*
% Melanesia	87.2	43.8	56.7	
(d) International/humanitarian categories				
Melanesia	*414*	*1,881*	*2,295*	*82.0*
Fiji	229	1,865	2,094	89.1
Papua New Guinea	150	10	160	6.3
Micronesia	*9*	*460*	*467*	*98.5*
Kiribati	0	447	447	100.0
Polynesia	*270*	*8,761*	*9,031*	*97.0*
Samoa	219	5,394	5,613	96.1
Tonga	28	2,863	2,891	99.0
Pacific	*693*	*11,102*	*11,793*	*94.1*
% Melanesia	59.7	16.9	19.5	

Source: Bedford and hugo (2012, p. 55)

to longer-term residence in New Zealand through the PAC, if an RSE worker is successful in the annual ballot. Bedford and Bedford (2010) have argued that there is the potential, longer term, for greater linkage between the PAC and the RSE work policy in New Zealand, in the interests of increasing the opportunities for eligible Pacific nationals to take up places in the PAC.

Within the islands, Fiji's military government has signalled that it is prepared to consider assisting Kiribati and Tuvalu in the event that climate change-induced environmental change makes it necessary for people to leave their atoll homes (Bedford and Bedford 2010). The governments of New Zealand and Australia have not made any specific plans for resettling people from these countries longer-term; rather there is a preference for dealing with inflows of migrants from the Pacific through their regular migration programmes. At a regional level at the Pacific Islands Forum, the annual meeting of leaders of countries in Oceania, there is frequent reference to the need for a coherent regional approach to climate change, which includes provision for migration as one form of adaptation to environmental change. For the great majority of Pacific residents, this will involve movement within their own countries, rather than emigration to others, and in this context there are some significant challenges which are touched upon in the next section.

A Synthesis of Existing Case Studies on the Links Between Migration and Climate

Pacific Island Countries are exposed to a wide range of natural hazards or potential extreme events. These include geological hazards such as volcanic eruptions, earthquakes, tsunamis (the effects of which may be increased by sea level rise) and landslides (though many are caused by high intensity rainfall events) in addition to a large suite of climatic hazards or forms of climate variability including tropical cyclones, floods, droughts, storm surges and high wave events. As well, disease vectors such as malaria, dengue and ciguatera have climate and or sea surface temperature linkages and water borne health problems are often related to climatic conditions. Table 8.6 and Fig. 8.3 show the incidence of extreme disaster related events in PICs in the first decade of the twenty-first century. As the figure shows

Table 8.6 Disaster events in PICs, 2000–2009

Type of disaster	Number
Tropical cyclones	30
Floods	14
Drought	1
Landslides	4
Volcanic eruptions	7
Earthquakes	4
Tsunami	2

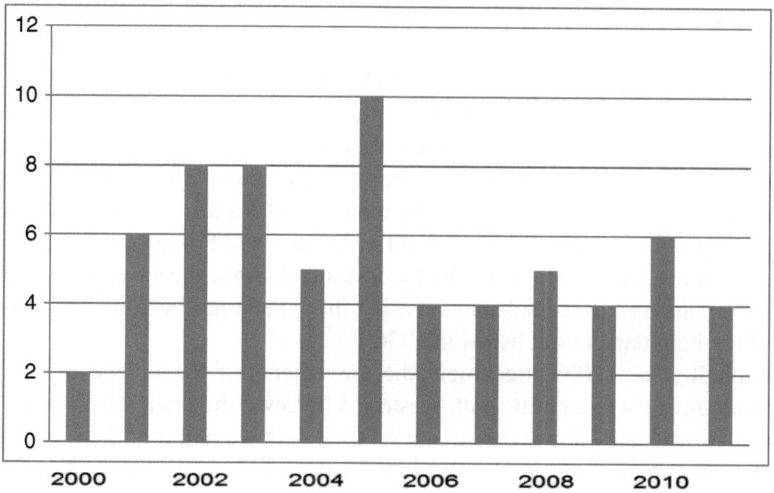

Fig. 8.3 Annual distribution of disaster events in PICs, 2000–2011 (Source of data: Reliefweb)

climatic events (climate variability) dominate in terms of frequency accounting for 45 (73 %) of events (not counting for landslides which may have been triggered by rainfall, flooding or tectonic processes).

Tropical Cyclones, Torrential Rains and Floods

All but a few Pacific countries (the exceptions being Papua New Guinea Highlands, Nauru and Kiribati) are located either fully or partially in areas where tropical cyclones may occur and tropical cyclone events are among the most frequently reported triggers of disasters in the region. Most high intensity rainfalls and flood events are also associated with tropical cyclones though it is not unusual for such events to be caused by low pressure systems that have not reached hurricane or even tropical storm intensity. Generally river flooding is restricted to larger islands and particularly the high continental or inter-plate type islands of the Western Pacific (PNG, Solomon Islands, Vanuatu, New Caledonia and Fiji). The most devastating recorded flood events have occurred on these islands such as in the Ba River, Fiji in 1930 (Yeo et al. 2007), Cyclone Namu in Solomon Islands in 1986 when about 150 died (Britton 1987) and Cyclone Guba in Oro Province, Papua New Guinea, 2007 (IFRC 2008). There is, however, little information available about mobility after these events. Volcanic high islands and atolls are largely porous. In very intense rains small and even ephemeral streams on volcanic high islands can experience flash floods which can cause considerable damage and loss of life.

There have been many tropical cyclone events that have resulted in evacuation of affected communities. There are several instances of communities deciding to relocate on a new site, usually nearby, often within their own customary land boundaries, and nearly always within their national or colonial territorial boundaries, though this has not always guaranteed protection from following events. For example, when the village located at Qaliqali, on Kabara, was devastated by a tropical cyclone in 1936, the community moved to a new site called Naikeleyaga several kilometres along the coast (Campbell 1977). The land for the new village had to be obtained from Tokelau, another village on the island, a procedure that may not have been possible without the direction of Ratu Lala Sukuna, high chief of the Lau province where he also served as District and Provincial Commissioner. In later cyclones, Naikeleyaga was badly devastated. Since the move, the population of Naikeleyaga has grown substantially, but the village has been unable to expand beyond the boundaries established in 1936.

Campbell et al. (2007) document the movements of the village of Biausevu, located alongside a stream in south-western Viti Levu, the main island of Fiji. The village is now located above the stream on the top of a small hill, the top of which, was 'flattened' by bulldozer to make way for the building of houses after Cyclone Oscar in 1983 caused major flooding of the nearby Biausevu Stream. This was the third village relocation by the people of Biausevu in the past century, each move initiated by heavy flooding of the Biausevu stream that the settlements were located adjacent to in order to enable access to fresh water. The latest site is serviced by a water supply sourced inland from the village and at a higher location providing the necessary head to ensure continuity of supply. But, all is not totally satisfactory – the new village has used up all available land on the flattened hilltop and new homes are being built below it, closer to the stream.

Cagilaba (2005) documented the relocation of the most of the members of the village of Solodamu, on the island of Kadavu, Fiji, following several storm surge events in the 1970s. The relocatees moved inland and upslope to avoid further inundation. Part of the new site was on land belonging to another kinship group and was obtained using traditional approaches and a new village was constructed. At the time of Cagilaba's field work, tensions were arising among younger generations over the validity of the land transaction and many of the women in the village described the need to carry water uphill every day as a continuing burden.

Drought and Desertification

While cases of desertification are relatively rare in PICs, droughts are relatively common and cause considerable levels of disruption and often affect large numbers of countries at a time. Typically, the western part of the region is affected most by drought during El Niño events and the central Pacific is most likely to experience drought during La Niña episodes. A number of atolls have chronic freshwater availability and quality problems and these are worsened considerably during drought events although is little in the literature to indicate that droughts have caused migration. The Papua New Guinea Highlands are particularly prone to droughts and associated frosts that destroy the staple sweet potato crops during El Niño conditions that have been recorded for over a century (Allen 1989; Allen et al. 1989).

According to Waddell (1975, 1989) prior to a large government initiated food relief programme in 1972 people migrated temporarily out of the devastated areas to locations down slope and spent time with communities that had not been affected and this was observed by Wohlt (1989) who also noted that several return visits took place in the years that followed. Ties between the communities were maintained in traditional times but these broke down as the result of the relief which rendered them unnecessary in following events. A more recent El Niño event in 1997–1998 resulted in another massive relief programme, downslope migration did not occur, but there is some evidence that highlanders migrated to urban areas as a result of the scarcity and if relief were not supplied the numbers would have been very large (Allen and Bourke 1997).

Sea-Level Rise

Sea-level rise has been identified by numerous observers as likely to be the most disruptive of climate change effects, the greatest risk to PICs and most likely to force migration. This is reflected in international conventions (e.g. UNFCCC), numerous reports by and for multilateral agencies and some scientific writings and popular media discourses of sinking islands such as 'titanic states', 'coalmine canaries' and 'extinction and survival' (Barnett and Campbell 2010 p. 168). Atolls have been identified as being particularly 'vulnerable'. However, there is considerable uncertainty as to what effect sea-level rise will exactly have on PIC coastlines and atolls.

Nevertheless, it might be expected that sea level rise will affect coastal communities, especially low-lying urban areas, in a number of ways and its manifestation is likely to become most notable during extreme events such as tropical cyclones (and associated high waves and storm surges), long-distance waves generated by mid-latitude cyclones and king tides when inundation, erosion, salinisation of water supplies and agricultural resources, and coastal erosion occur.

There are as yet no clear cases of climate change forced sea-level rise causing atolls to become uninhabitable. Two atolls in eastern Papua New Guinea, Tulun (Carteret Islands) and Takuu (Mortlock islands), particularly the former, have become *causes célèbres* as the origins of the 'first climate refugees'. However, the causes of their relative sea-level rise remain controversial and island subsidence as a result of tectonic processes may be as responsible as is eustatic sea-level rise, although the latter may be an exacerbating factor. O'Collins (1990) indicates problems of erosion on the island were observed some 50 years ago (presumably before significant anthropogenic sea level rise). Nevertheless, these case studies (whether caused by climate change or not) do provide us with some useful insights into problems of community relocation as a result of climate change. There have been several attempts over about three decades to relocate part of the Tulun community to Bougainville, but thus far the schemes have struggled to succeed. While church land has been made available for housing and the development of a settlement, local people at the destination are unwilling (or unable) to cede land to the migrants for agriculture and the most recent relocation attempt has been reported to have failed (Campbell 2010a; Vainerere 2009).

A discourse has emerged of sea level rise as the dominant likely cause of climate induced or forced migration – reducing or even eliminating land, livelihood or habitat security, particularly on atolls, the inhabitants of which are portrayed as potential environmental refugees (Kempf 2009). A counter to this discourse has emerged in recent years with a number of observers not only questioning the idea of forced climate migration or community relocation, but also suggesting that considering climate change migration may be counterproductive and potentially damaging for small island dwellers (Mortreux and Barnett 2009; Barnett and O'Neill 2012; Kempf 2009). Mortreux and Barnett (2009), and Barnett and O'Neill (2012) question the ethics of 'climate change migration discourses' as promoting premature consideration of migration, especially community relocation, long before climate change effects on islands are fully understood and likely to be manifested. Such discourses may preclude the consideration or implementation of other (in situ) approaches to adaptation.

We agree with their concerns that notions of climate 'refugees' may create (perhaps self-fulfilling) discourses of vulnerability, that focussing on migration/relocation may foreclose other adaptation options and premature migration/relocation may arise causing major social disruption both at the origin and destination. There are major problems with extremist or alarmist discourses. We also acknowledge that forced migration of communities can be highly problematic (Campbell et al. 2007; Campbell 2010a, b). But, for this very reason alone, it seems that ignoring the possibility of loss of community security, and not engaging in long-term planning for relocation (for both migrants and the 'host' communities) is also highly problematic. There is a need for a more nuanced consideration of migration as a climate change response rather than considering it in dualistic ways as either an inevitability or something that will not happen at all, and one that incorporates both migrating/relocating and 'host' communities (e.g. Farbotko and Lazrus 2012; Gemenne and Shen 2009; Lazrus 2012). Planning for migration, including community relocation, is likely to be something of a balancing act taking these seemingly contradictory considerations into account.

Several studies conducted in 'vulnerable' places such as Tuvalu indicate that people currently do not migrate for environmental reasons (Farbotko and Lazrus 2012; Gemenne and Shen 2009; Lazrus 2012; Mortreux and Barnett 2009) and do not see themselves as potential 'environmental refugees' (McAdam and Loughry 2009). At the UN General Assembly in October, 2012, the ambassador for Palau observed "We assure you, Mr. President, that Palau has no intention to move our people ... Our islands are our home. They are [the] essence of our very being. We will pursue every legitimate recourse available to us before we concede hope." (UN New Centre, October 1, 2012). It remains vital, given the difficulties of large-scale relocation that mitigation of climate change becomes more of an international priority than is currently the case.

It is also important to acknowledge that the effects of sea level rise are likely to not only seriously affect atoll populations. The great majority of PIC people (with the exception of Papua New Guinea) live adjacent, or very near, to the coastline. These communities may also be at significant risk but tend to be ignored in popular accounts of climate impacts because they are backed by higher land. However, unless the elevated locations are customarily owned by the coastal communities, movement inland and upward is likely to be made extremely difficult because of the cultural importance associated to land and its immutable connection to its 'owners'.

Projecting Climate Change Induced and Forced Migration in the Pacific

Despite the debate about the issue of community relocation as a climate change response, there is a history of community relocation in Pacific Islands, especially in relation to natural hazard avoidance, and where appropriate traditional procedures were followed disruption related to land exchange was often limited. The costs of relocation do increase with distance, elevation and with the crossing of boundaries (local land tenure, internal divisions within nations and international borders) (Campbell 2010b). Moreover, if the numbers of migrants are large, the costs of mobility as a form of adaptation are also likely to be high, and the issue of who pays for such a response, which currently falls upon the migrants themselves, will also need to be addressed alongside other *in situ* climate change adaptations, another issue with likely long-term implications (Campbell 2010b).

There have been very few attempts to predict the likely quantum of climate change forced migrants in and from PICs in the years to come. Nicholls et al. (2011), using a modelling approach, estimated that a 0.5–2.0 m rise in sea level would cause the displacement of between 1.1 and 2.2 million people from the islands in the Caribbean, Indian Ocean and Pacific Ocean combined. For the islands of Oceania, detailed data on projected climate change effects remain lacking and there is limited information available about the levels of exposure of PIC communities to the effects of climate change that may significantly reduce their land, livelihood and habitat security. Nevertheless, there is concern in the region, and among Pacific Rim countries that may be destinations for international climate change migrants, about how to respond to the possibility of climate change migration and how to develop appropriate policies.

As has been indicated in section "Migration in Oceania" above the flows of migrants from PICs to the rim countries are not evenly spread with a predominance of Polynesian and Fijian migrants particularly in the South Pacific and relatively large flows of Micronesians to the USA, particularly Hawaii and also Guam, not to mention the USA mainland. However, it is likely that many climate change migrants might originate from the atoll states of Kiribati and Tuvalu which have limited external migration access and from the western Melanesian countries which also have restricted options. The atoll states would also have very limited internal options for relocation and voluntary migration with the likelihood that increased flows of external migrants may be initiated. For the Melanesian countries, where customary land ownership is extremely important, internal options are likely to be culturally unacceptable. For all of these countries, it is not exactly clear what internal or

Table 8.7 Atoll population estimates and projections, 2011–2050

PICs consisting only of atolls			High island countries with atolls	
	Population Estimates[a]			
	2011	2050	Papua New Guinea	
Kiribati	102,697	163,266	Solomon Islands	
Marshall Islands	54,999	61,217	New Caledonia	
			Palau	
Tokelau	1,162	1,148	American Samoa	
Tuvalu	11,206	13,858	Tonga	
			Cook Islands	
Totals	170,064	239,489	French Polynesia	
			Population estimates 2009[b]	c. 55,000
			Population projections 2050[b]	c. 80,000

[a]Source: SPC (2011) Pacific Island Populations – Estimates and Projections of Demographic Indicators for selected years
[b]Source: Campbell (2011)

external migration patterns will emerge. It is also possible that Pacific Islanders will be ‘competing’ with climate change migrants from elsewhere, particularly Southeast Asia, for entry to Pacific rim countries.

The implications of climate change may be particularly serious if the quantum of climate change migrants (forced or voluntary) is large. It is impossible to predict even in broad terms what the numbers of climate change migrants might be, and when they might engage in mobility as a climate change response. It may be that contemporary migration decisions reflect concerns about climate change induced environmental degradation as well as other forms of social, economic and environmental marginality. But, if we consider some possibilities of diminution of community security for the region, it is likely that some forms of migration will emerge as a response to climate change. As Table 8.2 indicates, the population of the region may reach almost 18 million by mid-century. Table 8.7 shows current and projected populations of the four atoll entities (Marshall Islands, Kiribati, Tuvalu and Tokelau) together with other atolls located in countries with high (inter-plate and volcanic) islands. These populations may increase from around 225,000 in 2011 to 320,000 by 2050. Similarly, the majority of PIC inhabitants live near the coast (though data does not exist that indicates the elevation above, or distance from, the sea) and it is possible that the security of their localities may also be compromised. Taking 5 % of the populations of all Pacific countries as being located in coastal areas (excluding Papua New Guinea), a conservative assumption, we would find that by 2050 the number exceeding well over 200,000. These projections, albeit speculative, indicate that there may be half a million people in Oceania that may be exposed to the effects of sea level rise, not to mention other manifestations of climate change. Other groups that may be affected by climate change include the drought prone communities of the Papua New Guinea highlands, the riverbank and delta communities of Melanesia and the concentrations of people in urban areas which tend to be coastal rather than inland. Whether or not some of these people or whole communities become forced or voluntary climate change migrants remains moot, but given the many problems particularly associated with community relocation and the potentially large volume of voluntary migration, it is important to consider mobility as a climate change adaptation in a proactive manner.

References

Allen, B. (1989). Frost and drought through time and space. Part I: The climatological record. *Mountain Research and Development, 9*(3), 252–278.

Allen, B. J., & Bourke, R. M. (1997). *Report of an assessment of the impacts of frost and drought in Papua New Guinea* (p. 22). Canberra: Australian Agency for International Development.

Allen, B. J., Brookfield, H. C., et al. (1989). Frost and drought through time and space. Part II: The written, oral, and proxy records and their meaning. *Mountain Research and Development, 9*(3), 279–305.

Anon. (2012). *Report on site inspection of Natoavatu, Vanua Levu.* 23 December 2011. Unpublished Report for the Kiribati Government.

Australian Agency for International Development (AusAID). (2008). *Making land work: Vol. 1. Reconciling customary land and development in the Pacific*. Canberra: Australian Agency for International Development.

Australian Bureau of Meteorology., & CSIRO. (2011). *Climate change in the pacific: Scientific assessment and new research: Vol. 1. Regional overview*. Aspendale: Pacific Climate Change Science Program.

Barnett, J. (2001). *The meaning of environmental security: Ecological politics and policy in the new security era*. London: Zed.

Barnett, J. (2012). On the risks of engineering mobility to reduce vulnerability to climate change: Insights from a small island state. In K. Hastrup & K. Fog-Olwig (Eds.), *Climate change and human mobility. Global challenges to the social sciences*. Cambridge: Cambridge University Press.

Barnett, J., & Campbell, J. R. (2010). *Climate change and small island states power, knowledge and the South Pacific*. London: Earthscan.

Barnett, J., & O'Neill, S. (2012). Islands, resettlement and adaptation. *Nature Climate Change, 2*(1), 8–10.

Barnett, J., & Webber, M. (2009). *Accommodating migration to promote adaptation to climate change*. Secretariat of the Swedish Commission on Climate Change and Development Policy Brief, Stockholm.

Bedford, R. D. (1967). *Resettlement: Solution to economic and social problems in the Gilbert and Ellice Islands*. Unpublished MA thesis in Geography, University of Auckland.

Bedford, R. D. (1992). International migration in the South Pacific region. In M. M. Kritz, L. L. Lim, & H. Zlotnik (Eds.), *International migration systems: A global approach* (pp. 41–62). Oxford: Clarendon.

Bedford, R. D. (2005). International migration and globalization: The transformation of New Zealand's migration system since the mid-1980s. In R. Patman (Ed.), *Sovereignty under Siege: New Zealand and globalisation* (pp. 129–155). London: Ashgate.

Bedford, R. D. (2008). Pasifika mobility: Pathways, circuits and challenges in the 21st century. In A. Bisley (Ed.), *Pasifika interactions. Pasifika in New Zealand, New Zealand in Pasifika* (pp. 85–134). Wellington: Institute of Policy Studies, Victoria University of Wellington. on line publication: http://ips.ac.nz.

Bedford, R. D., & Bedford, C. E. (2010). International migration and climate change: A post-Copenhagen perspective on options for Kiribati and Tuvalu. In B. Burson (Ed.), *Climate change and migration in the South Pacific Region: Policy perspectives* (pp. 89–134). Wellington: Institute for Policy Studies, Victoria University of Wellington.

Bedford, R. D., & Hugo, G. (2012). *Population movement in the Pacific: A perspective on future prospects*. Wellington: Department of Labour.

Bedford, R. D., Macdonald, B. K., & Munro, D. (1980). Population estimates for Kiribati and Tuvalu: Review and speculation. *Journal of the Polynesian Society, 89*(2), 199–246.

Bedford, R. D., Ho, E. S., Krishnan, V., & Hong, B. (2007). The neighbourhood effect: The Pacific in Aotearoa and Australia. *Asian and Pacific Migration Journal, 16*(2), 251–269.

Britton, N. R. (1987). Disaster in the South Pacific: Impact of tropical cyclone "Namu" on the Solomon Islands, May 1986. *Disasters, 11*(2), 120–133.

Brookfield, H. C. (1972). *Colonialism, development and independence: The case of the Melanesian Islands in the South Pacific*. Cambridge: Cambridge University Press.

Brookfield, H. C., & Hart, D. (1971). *Melanesia. A geographical interpretation of an island world*. London: Methuen.

Butcher, M. (2012). *. . . when the long trick's over': Donald Kennedy in the Pacific*. Kennington: Holland House Publishing.

Cagilaba, V. (2005). *Fight or flight? Resilience and vulnerability in rural Fiji* (p. 123). MSocSci thesis, Department of Geography, University of Waikato, Hamilton.

Campbell, J. R. (1977). Hurricanes in Kabara. In H. C. Brookfield (Ed.), *The hurricane hazard: Natural disaster and small populations* (pp. 149–175). Island Report No. 1, UNESCO/UNFAP Population and Environment Project in the Eastern Islands of Fiji, Canberra.

Campbell, J. R. (1985). *Dealing with disaster. Hurricane response in Fiji*. Suva: Government of Fiji and Honolulu, Pacific Islands Development Program, East-West Center.

Campbell, J. R. (2006). *Traditional disaster reduction in Pacific Island communities* (GNS Science Report 2006/038). Wellington: Institute of Geological and Nuclear Sciences.

Campbell, J. R. (2010a). Climate-induced community relocation in the Pacific: The meaning and importance of land. In J. McAdam (Ed.), *Climate change and displacement: Multidisciplinary perspectives* (pp. 57–79). Oxford: Hart Publishing.

Campbell, J. R. (2010b). Climate change and population movement in Pacific Island Countries. In B. Burson (Ed.), *Climate change and migration in the South Pacific Region: Policy perspectives* (pp. 29–50). Wellington: Institute of Policy Studies.

Campbell, J. R. (2011). Climate change and population displacement. In W. Narsey et al. (Eds.), *Population and development in the Pacific Islands* (pp. 320–341). Suva: UNFPA and USP.

Campbell, J. R., Goldsmith, M., & Koshy, K. (2007). *Community relocation as an option for adaptation to the effects of climate change and climate variability in Pacific Island Countries (PICs)*. Final report for APN project 2005-14-NSY-Campbell. Tokyo: Asia Pacific Network for Global Change Research.

Christensen, J. H., Hewitson, B., Busuioc, A., Chen, A., Gao, X., Held, I., Jones, R., Kolli, R. K., Kwon, W.-T., Laprise, R., Magaña Rueda, V., Mearns, L., Menéndez, C. G., Räisänen, J., Rinke, A., Sarr, A., & Whetton, P. (2007). Regional climate projections. In S. Qin, M. Manning, Z. Chen, M. Marquis, K. B. Averyt, M. Tignor, & H. L. Miller (Eds.), *Climate change 2007: The physical science basis. Contribution of Working Group I to the fourth assessment report of the Intergovernmental Panel on Climate Change Solomon*. Cambridge: Cambridge University Press.

Connell, J., & Lea, J. P. (1993). *Planning the future. Melanesian cities in 2010* (Pacific Policy Paper 11, Pacific 2010). Canberra: National Centre for Development Studies.

Connell, J., & Lea, J. P. (1995). *Urbanisation in Polynesia* (Pacific Policy Paper 14, Pacific 2010). Canberra: National Centre for Development Studies.

Connell, J., & Lea, J. P. (1998). *Island Rowns. Managing urbanization in Micronesia* (Occasional Paper No. 40, Center for Pacific Island Studies, University of Hawai'i) Honolulu: University of Hawai'i.

Development Research Centre on Migration, Globalisation and Poverty (DRC). (2007). *Global migrant origin database*. Brighton: University of Sussex. www.migrationdrc.org. Accessed Mar 2011.

Farbotko, C., & Lazrus, H. (2012). The first climate refugees? Contesting global narratives of climate change in Tuvalu. *Global Environmental Change, 22*(2), 382–390.

Gemenne, F., & Shen, S. (2009). *Tuvalu and New Zealand case study report: United Nations University EACH-FOR Environmental Change and Forced Migration Scenarios*. Bonn: United Nations University – Institute for Environment and Human Security (UNU-EHS).

Hau'ofa, E. (1993). Our sea of islands. In E. Waddell (Ed.), *A new Oceania: Rediscovering our sea of islands*. Suva: SSED, University of the South Pacific.

Hugo, G. (2011). Future demographic change and its interactions with migration and climate change. *Global Environmental Change, 21S*, S21–S33.

International Federation of Red Cross and Red Crescent Societies (IFRC). (2008c, July 31). *Papua New Guinea: Cyclone Guba* (Operations Update No.4). http://www.reliefweb.int/rw/RWFiles2008.nsf/FilesByRWDocUnidFilename/YSAR-7H4LHZ-full_report.pdf/$File/full_report.pdf. Accessed 22 Sept 2008.

International Organization for Migration (IOM). (21010). *World migration report 2010. The future of migration: Building capacities for change*. Geneva: International Organization for Migration.

Jupp, J. (2007). *From White Australia to Wommera: The story of Australian immigration*. Cambridge: Cambridge University Press.

Kempf, W. (2009). A sea of environmental refugees? Oceania in an age of climate change. In E. Hermann, K. Klenke, & M. Dickhardt (Eds.), *Form, Marcht, Differenze; Motive und Felder ethnologischen Forschens* (pp. 191–205). Göttingeng: Universitätsverlag Göttingeng.

Knudson, K. E. (1965). *Titiana. A Gilbertese community in the Solomon Islands*. Eugene: Department of Anthropology, University of Oregon.

Knudson, K. E. (1977). Sydney Island, Titiana and Kamaleai: Southern Gilbertese in the Phoenix and Solomon Islands. In M. D. Lieber (Ed.), *Exiles and migrants in Oceania* (pp. 195–241). Honolulu: University of Hawai'i Press.

Kock, K. F. (1973). *Logs in the current of the sea. Neli Lifuka's story of Kioa and the Vaitupu colonists*. Canberra: Australian National University Press.

Lambert, B. (1975). Makin and the outside world. In V. Carroll (Ed.), *Pacific atoll populations* (pp. 212–285). Honolulu: University of Hawai'i Press.

Lazrus, H. (2012). Sea change: Island communities and climate change. *Annual Review of Anthropology, 41*, 285–301.

Macdonald, B. K. (1982). *Cinderellas of the Empire. Towards a history of Kiribati and Tuvalu*. Canberra: ANU Press.

Maude, H. E. (1938). *Colonization of the Phoenix Islands by the surplus population of the Gilbert and Ellice Islands*. Confidential Report to the Resident Commissioner, Gilbert and Ellice Islands Colony. Suva: Government Press.

Maude, H. E. (1952). The colonization of the Phoenix Islands. *Journal of the Polynesian Society, 61*(1&2), 62–89.

McAdam, J., & Loughry, M. (2009, June 30). Kiribati – Relocation and adaptation. *Inside Story*. http://inside.org

McKinnon, M. (1996). *Immigrants and citizens: New Zealanders and Asian immigration in historical context*. Wellington: Institute of Policy Studies, Victoria University of Wellington.

Mimura, N., Nurse, L., McLean, R. F., Agard, J., Briguglio, L., Lefale, P., Payet, R., & Sem, G. (2007). Small islands. In M. L. Parry, O. F. Canziani, J. P. Palutikof, P. J. van der Linden, & C. E. Hanson (Eds.), *Climate change 2007: Impacts, adaptation and vulnerability. Contribution of Working Group II to the fourth assessment report of the Intergovernmental Panel on Climate Change* (pp. 687–716). Cambridge: Cambridge University Press.

Mortreux, C., & Barnett, J. (2009). Climate change, migration and adaptation in Funafuti, Tuvalu. *Global Environmental Change, 19*(1), 105–112.

Nicholls, R. J., Marinova, M., Lowe, J. A., Brown, S., Vellinga, P., de Gusmao, D., Hinkel, J., & Tol, R. S. J. (2011). Sea-level rise and its possible impacts given a 'beyond 4°C world' in the twenty-first century. *Philosophical Transactions of the Royal Society, 369*, 1–21.

O'Collins, M. (1990). Carteret islanders at the Atolls Resettlement Scheme: A response to land loss and population growth. In J. Pernetta & P. Hughes (Eds.), *Implications of expected climate changes in the South Pacific Region: An overview* (pp. 247–269). Nairobi: United Nations Environment Programme.

Paton, K. L. (2009). *At home or abroad: Tuvaluans shaping a Tuvaluan future*. Unpublished Masters of Development Studies thesis, Victoria University of Wellington.
Pincus, J., & Hugo, G. (2012). *A Greater Australia: Population, policies and governance*. Melbourne: The Committee for Economic Development of Australia (CEDA).
Pond, W. (1999). *The land with all woods and waters* (Rangahaua Whanui series). Wellington: Waitangi Tribunal.
Ramasamy, S., Krishnan, V., Bedford, R. D., & Bedford, C. E. (2008). The recognised seasonal employer policy: Seeking the elusive triple wins for development through international migration. *Pacific Economic Bulletin, 23*(3), 176–186.
Ravuvu, A. D. (1988). *The Fijian ethos*. Suva: Institute of Pacific Studies.
Secretariat of the Pacific Community (SPC). (2010). *Population data sheet 2010*. www.spc.int/spd/
Secretariat of the Pacific Community (SPC). (2011). *Pacific Island populations – Estimates and projections of demographic indicators for selected years*. Noumea: Secretariat of the Pacific Community.
Shen, S., & Gemenne, F. (2011). Contrasted views on environmental change and migration: The case of Tuvaluan migration to New Zealand. *International Migration, 49*, e224–e242.
Silverman, M. G. (1971). *Disconcerting issue: Meaning and struggle in a resettled Pacific Community*. Chicago: University of Chicago Press.
Silverman, M. G. (1977). Making sense: A study of a Banaban meeting. In M. D. Lieber (Ed.), *Exiles and migrants in Oceania* (pp. 121–160). Honolulu: University of Hawai'i Press.
Spoonley, P., & Bedford, R. D. (2012). *Welcome to our world? Immigration and the reshaping of New Zealand*. Palmerston North: Dunmore Press.
Vainerere, T. (2009). *Climate change refugees return to Bougainville atolls*. Port Moresby: New Guinea Post-Courier.
Waddell, E. (1975). How the Enga cope with frost: Responses to climatic perturbations in the Central Highlands of New Guinea. *Human Ecology, 3*(4), 249–273.
Waddell, E. (1989). Observations on the 1972 frosts and subsequent relief programme among the Enga of the Western Highlands. *Mountain Research and Development, 9*(3), 210–223.
Williams, M., & Macdonald, B. K. (1985). *The phosphateers. A history of the British Phosphate Commissioners and the Christmas Island Phosphate Commission*. Melbourne: Melbourne University Press.
Wohlt, P. (1989). Migration from Yumbisa, 1972–1975. *Mountain Research and Development, 9*(3), 224–234.
Yeo, S. W., Blong, R. J., et al. (2007). Flooding in Fiji: Findings from a 100-year historical series. *Hydrological Science Journal, 52*(2), 1004–1015.

The Changing Hindu Kush Himalayas: Environmental Change and Migration

9

Soumyadeep Banerjee, Richard Black, Dominic Kniveton, and Michael Kollmair

This chapter is focused on the relationship between environmental change and migration in mountain areas, and specifically the Hindu Kush Himalayan region of Asia, which contains the world's highest and most iconic mountains. Mountain populations have long adjusted to living in fragile and marginal environments, and partially as a result, mountains have for many decades been zones of out-migration, as people seek alternative and more reliable livelihoods. At the same time, mountain regions are often seen as particularly vulnerable to climate change, as well as to a range of environmental shocks and hazards. As such, they represent a particularly interesting case study to set alongside the broader regional overviews contained elsewhere in this volume – a special case of how major environmental change could have dramatic migration consequences in the future.

The chapter is organized into three main parts. First, we review available knowledge about climate and other environmental change in the Hindu Kush Himalayan region, based on a range of scientific reports and projections. A key element of this review is uncertainty about the nature of future change, albeit it is clear that temperatures in the region are rising, likely above global averages. Second, we review evidence on migration in the region, including labour migration, internal displacement and flows of refugees. Then, we systematically review evidence from

S. Banerjee (✉)
International Centre for Integrated Mountain Development (ICIMOD), Kathmandu, Nepal

University of Sussex, Brighton, UK
e-mail: sbanerjee@icimod.org

R. Black • D. Kniveton
University of Sussex, Brighton, UK

M. Kollmair
Independent consultant, Munich, Germany

E. Piguet and F. Laczko (eds.), *People on the Move in a Changing Climate*, Global Migration Issues 2, DOI 10.1007/978-94-007-6985-4_9,

case studies that link these two phenomena – environmental change and migration – together, including looking specifically at the extent to which migrants remittances help to build resilience against environmental change. Finally, in the conclusion, we examine the nature of causal linkages, and potential patterns of future change.

What Do We Know About Climate Change in the Himalayas?

Mountains, with their steep relief, high precipitation, and quickly changing climatic patterns, are particularly sensitive to environmental change, with significant influence on human wellbeing (Körner and Ohsawa 2005). There is high confidence that future anthropogenically enduced changes in the climate including enhanced heat waves, glacial retreat, and permafrost degradation are likely to lead to increasing slope instabilities, movements of mass, and glacial lake outburst floods (IPCC 2012). While it is also anticipated that changes in heavy precipitation will affect landslides in some regions (IPCC 2012). Although, the direct impacts of climate change will be most marked at high elevations, they will have a greater impact at lower elevations, the 'cascading' of effects from high to low altitude areas, for example, including increased runoff at high altitude leading to floods and increased sand deposition on agricultural land at lower altitudes (Tse-ring et al. 2010).

However the recent furore over the validity of claims in the Intergovernmental Panel on Climate Change (IPCC) fourth assessment report (AR4) that all glaciers in the Himalayas could disappear by 2035 (Cruz et al. 2007) has highlighted the dearth of data on the environment in mountain regions, and the space that this creates for claims about the future impacts of climate change. In this case, it was claimed that the disappearance of glaciers within 25 years would severely affect the water supplies in the region. While highly likely if glaciers did melt on this timescale, the claim was later retracted when doubts were cast over the evidence base of the claim.

At its core, the Hindu Kush Himalayan (HKH) region – in common with other mountain regions – suffers from a lack of data on the state of the environment. In turn, the cause of this scarcity is primarily due to problems of access. An example of the lack of data on the environment is that available on glacial mass balances. Understanding glacial mass balances is vital for assessing the impact of climate change particularly on water resources in mountain regions and surrounding areas. Yet according to Singh et al. (2011), data on mass balances only exists for roughly 10 out 50,000 glaciers in the HKH region. It should therefore be understood that any scenarios of future climate change in the region are currently built on a poor understanding of baseline conditions and the processes that effect change.

Given uncertainty in terms of unknown future greenhouse gas and aerosol emissions, as well as this lack of data and highly heterogeneous conditions in the region, it is difficult to develop scenarios of climate change. For example at their most detailed, global climate models create climate outputs at a resolution of a 100 km by 100 km. While regional climate models create scenarios at approximately 25 km by 25 km, this is still of limited use in mountainous regions where climate varies over distances of kilometres. In this sense the climate change scenarios developed for the region should be viewed as speculative and illustrative of only a range of changes that may occur rather than representative of the full range of potential futures. It should be borne in mind when considering future climate scenarios that models vary between each other in the ways in which processes are parameterised so that scenarios developed from only one model may contain systematic biases and thus not be representative of a large field of climate possibilities.

Despite this lack of confidence in forecasting, the Hindu Kush Himalayan region is still widely believed to be likely to be one of the planet's hot spots of future climate change impacts (Maplecroft 2011). As with other mountain environments there exists a fine equilibrium between snow, ice, and water that effects biodiversity and ecosystem services, such as the regulation of water resources. This equilibrium is particularly sensitive to small changes in temperature and precipitation. The sub-sections below cover expected changes in temperature and precipitation, as well as a range of other potential environmental changes.

Temperature Change

Mountain regions have experienced above average warming in the twentieth century (IPCC 2007a, b; Nogues-Bravo et al. 2007), although available studies are limited to isolated parts of the HKH region, and may not be representative of elsewhere. The warming rate for a major part of the region is higher than 0.01 °C per year, ranging from 0.01 to 0.03 °C per year in the Western and Eastern Himalayas. However, warming is higher in the central Himalayas and the whole of the Tibetan plateau at 0.03–0.07 °C per year (Shrestha 2009). In turn, progressive warming at higher altitudes is three times greater than the global average (Eriksson et al. 2009; Xu et al. 2009). The warming trend in the three elevation zones – <1,000 masl, 1,000–4,000 masl, and >4,000 masl – over the past one and a half decades has been higher than the global average. This trend is greater at >4,000 masl zone compared to the other zones (Singh et al. 2011).

Looking to the future, IPCC assessment reports attempt to integrate the numerous studies undertaken to create scenarios of future climate change. Largely global in outlook, the last report (AR4) indicated an increased likelihood for a continuation of the current warming in the HKH region with increases in temperature likely to be greater at the higher altitudes and in winter. Spatially, some regional climate modelling work has indicated that the maximum warming (projected to the end of the century) will likely be in the Western Himalayan region (Kulkarni et al. 2011), although this was carried out with a single climate model. By comparison a study using three global climate models suggested that for the Eastern Himalayas the rate of future winter warming up to the 2080s is projected to vary between 3.6 and 5.3 °C, while the increase in summer temperature is projected to vary between 2.8 and 3.8 °C (Tse-ring et al. 2010).

Changes in Precipitation

In contrast to studies of temperature change, most of the studies of precipitation in the Hindu Kush Himalayan region (Shrestha et al. 2000; Shrestha 2009; Dimri and Dash 2011) report a lack of notable trends. The lack of long-term observations even on spatial scales of a few tens of kilometres forms the major obstacle to assess the long term pattern of precipitation. Regardless of the reports of an increased frequency of heavy rainfall events and that of moderate rainfall events in the region had decreased, the available data is insufficient to support these claims (Singh et al. 2011).

As with many parts of the world, projections of changes in precipitation in the HKH are also highly uncertain with some models indicating increases and others decreases. More recent studies have confirmed this variation in precipitation projections into the future between models, with some projecting 10–20 % increases in precipitation and others a 10–20 % decrease (Immerzeel 2008). However, it should be recognised that many models produce scenarios of increased precipitation and an accelerated hydrological cycle (Singh et al. 2011).

Once again, there are regional variations within the HKH region. For example, winter precipitation averaged across the Eastern Himalayas is likely to increase by 23–35 %, and summer monsoon precipitation by 17–28 % (Tse-ring et al. 2010). This compares to 20–40 % over the entire HKH region at the end of the century (Kulkarni et al. 2011).

Noticeably in terms of water availability, these figures indicate a much greater loss through evapotranspiration due to warmer temperatures than increase in water from the increase in rainfall in the summer period (Singh et al. 2011)

Changes in Snow Cover and Glaciers

Whilst in many parts of the world, the key determinant of water availability is precipitation, in the HKH region, large volumes of water were stored in the form of ice and snow that are released gradually as water over a long period during the dry seasons (Messerli and Ives 1997; Chettri et al. 2011). The estimated snow cover area in the HKH region for the 2000–2010 period was estimated to be 0.76 million square kilometer (Gurung et al. 2011a, b). During this period, the snow cover of this region was more or less stable, or only decreased slightly (Gurung et al. 2011a).

Notwithstanding the lack of information about glacier mass balances noted above, there is some evidence on glacier retreat. First, about 70 % of the glaciers that have been studied in the Hindu Kush Himalayan region have retreated 15 m per year or less in recent years (Singh et al. 2011). A substantial decrease in the total area of glaciers in Bhutan and Nepal accompanied by fragmentation of glacier was reported by several studies (Bajracharya et al. 2010, 2011; Bajracharya and Shrestha 2011). The rate of retreat for the Gangotri Glacier in the Indian Himalayas over the last three decades is more than three times the rate during the preceding 200 years

(Srivastava 2003). In the last half century, 82.2 % of the glaciers in western China have retreated (Ding et al. 2006; Kang et al. 2010). One exception is the Karakoram, where some high altitude glaciers are advancing (Hewitt 2005). Glacier retreat had also been reported for other parts of the region (Kulkarni et al. 2010; Nie et al. 2010).

However, recent studies (Immerzeel et al. 2010; Bolch et al. 2011) reported that rates of glacial retreat in this region were less than that suggested by the AR4 (Cogley et al. 2010; Miller et al. 2013). The rate of glacier melting in the Himalayas varies across the region. The shrinkage generally decreased from the Himalayas to the continental interior. The eastern Pamir is characterised by the least glacial retreat, area reduction and positive mass balance. The Himalayas, excluding Karakoram, experienced the greatest reduction in glacial length and area and the most negative mass balance (Yao et al. 2012). In the Himalayas, the glaciers are of significance as they are an important source of water for springs and rivers, particularly during the dry season. The contribution of snow to the runoff of major rivers varies between Eastern and Western Himalayas. It is about 10 % for the former and more than 60 % for the latter (Vohra 1981).

Changes in River Flow

Notwithstanding the uncertainties noted so far, a number of studies have attempted to model the impact of climate change on river flows (see Singh et al. 2011). For example, one of these studies indicated that the mean upstream water supply was likely to decrease between the two time slices 2000–2007 and 2046–2065 by −8.4 % for the Upper Indus, −17.6 % for the Ganges, −19.6 % for the Brahmaputra, and −5.6 % for the Yangtze (Immerzeel et al. 2010). It should be noted that these decreases were less than would be expected from the loss of snow cover and glacial retreat alone, as they were compensated to varying extents by an increase in upstream rainfall depicted in the model runs used (Immerzeel et al. 2010).

Biodiversity

Mountain systems support about half of the world's biological diversity and nearly half of the world's biodiversity hotspots (Myers et al. 2000; Hassan et al. 2005). Of the world's total acreage of land-based protected areas, some 27.6 % are situated in mountains (Kollmair et al. 2005). With rising temperatures, upward shifts of vegetation belts to higher elevations and northward advances in the geographical ranges of species in the northern hemisphere were expected (Nogues-Bravo et al. 2007). Some mountain species were likely to be losers including large territorial animals, late successional plant species, species with small and restricted populations, and species confined to summits (Körner 2009).

However, these processes should not be regarded as entirely negative – they may also present new opportunities. Because temperatures decrease with altitude, mountain species are in the privileged position of being able to migrate upwards to cooler areas, whereas lowland species usually have no other option than to adapt to higher temperatures, which is much more difficult (Körner 2009). Some of the observed impacts of climate change on biodiversity in Eastern Himalayas were loss and fragmentation of habitat, reduction in forest biodiversity, degradation of wetland and riverine island ecosystems, decline in forage and fodder resources, reduction in agrobiodiversity, increase in forest fires, rise in exotic and invasive species, soil fertility degradation, changes in land use pattern, more growth/biomass production in forests and variable productivity in agriculture (Tse-ring et al. 2010). Thus, mountains could serve as refuges for species that can no longer survive in the lowlands (Singh et al. 2010).

Natural Hazards and Disasters

Mountains are typically exposed to multiple hazards (Kohler and Maselli 2009). According to Guha-Sapir et al. (2011), hydrological disasters – flood, mass movement and drought – accounted for 48 % of the total annual disasters. Other forms of disasters included storms (23 %), earthquakes and landslides (14 %), epidemics (8 %), extreme temperature events (6 %), wildfires (1 %). In Nepal, the number of floods days and consecutive days of flood events had been on the rise (Shrestha et al. 2003). Continued glacier recession is and will increase the incident of glacial lake outburst floods (GLOFs). Globally in the past century, GLOFs have caused disasters in many regions of the world (Rosenzweig et al. 2007), including the Himalayas (Vuichard and Zimmermann 1987; Xin et al. 2008; Bajracharya and Mool 2009; Osti and Egashira 2009: as all cited in IPCC 2012). Some increasing trends have been observed in the number of GLOF events in the Himalayas, in the latter half of the last century, but these findings suffer from incomplete documentation (Richardson and Reynolds 2000; IPCC 2012). However, currently about 204 glacial lakes have been listed as having the potential to breach in the Himalayan region (Ives et al. 2010).

The Relationship Between Environmental Change and Human Wellbeing

If current knowledge and data of the HKH environment is insufficient to pinpoint realistic detailed scenarios of future change for the region, by extension it is also beyond current capabilities to create meaningful scenarios of change for the diverse populations of the HKH. Certainly mountain communities are sensitive to changes in ecosystem services and these impacts are likely to be manifest in human wellbeing.

One of the key sensitivities of livelihoods is exposure to stresses and shocks in terms of the availability and changing state of water. It is possible to imagine a likely scenario of continued impacts from hydrospheric change of increased ground instability in permafrost regions; increased rock avalanches and erosion of soils; reduced water supplies and hydropower potential as well as changes in the seasonality of flows in basins supplied by meltwater from snow and ice; reductions in water quality and freshwater species due to the adverse impacts of temperature increases on wetlands, freshwater lakes, and rivers (Singh et al. 2011).

As with much of the globe, the rule of thumb predictions indicate that drought-affected areas are likely to increase in extent with climate change, leading to substantial increases in the need for water for irrigation, while the increased variability in precipitation, increase in precipitation intensity and seasonal shifts in runoff are likely to negatively impact water supply, water quality, and flood risk.

An Assessment of the Main Migration Flows to and from the Region

As with environmental and climate change, a lack of disaggregated mountain specific migration data and literature for this region remains a major challenge to assess the migration flows to and from the Hindu Kush Himalayan region. The available data and literature are often based on national administrative boundaries, which do not necessarily correspond with the boundaries of ecosystems such as mountains. This section provides an overview of the main migration flows in the region.

Internal Migration

Evidence suggests that most of the migrants from the Hindu Kush Himalayan region remain within their country of origin. There are, generally, no restrictions on internal migration and fewer resources required for moving within the country of origin compared to that of migrating abroad. A study by the Ministry of Agriculture in Bhutan (2006) reported that 47 % of the sampled rural households had one or more migrants. Most of these rural migrants had moved to an urban area, primarily to those in western Bhutan. Only 6 % of rural migrants had moved to another village (Ministry of Agriculture in Bhutan 2006). Based on the Census of India in 2001, over three-quarters of all migrant workers from the Uttarakhand province of India had moved to other parts of the country, with concentrations in cities like Delhi, Mumbai, Lucknow, Chandigarh and Ambala. Others had moved to urban areas within the province such as Dehradun, Nainital and the district headquarters (Mamgain 2004).

Pakistani cities such as Muzaffarabad, Peshawar, Rawalpindi, Lahore, and Karachi were major destinations of migrants from the mountain areas. Migration from mountain areas of northern Pakistan to nearby urban areas in the plains was a traditional livelihood strategy among rural households (Nadeem et al. 2009; Schutte and Kreutzmann 2011). According to the Census of Nepal in 2001, around 13.2 % of the total native born population were classified as internal migrants. The rural-to-rural (68.2 %) and rural-to-urban stream (25.5 %) were the major streams of internal migration. The same Census figures showed that 11 % of the internal migrants had moved for employment related reasons (Subedi 2009). The rural areas of Mid- and Far-West Nepal were major areas of net out-migration as the migrants moved to the Kathmandu valley and other urban areas, and lowlands of Terai (Lokshin et al. 2007).

International Migration

There are several well established international migration streams in the Hindu Kush Himalayan region. In a departure from the general pattern in this region, rural migrant households in Afghanistan were twice as likely to send members abroad rather than to destinations within the country. A study by Ghobadi et al. (2005) found that compared to the 22 % of internal migrants, around 43 % of the Afghan migrants had moved abroad to destinations in Iran, Pakistan, Middle East, Europe and North America.[1] The Afghan migrants, besides being employed in low skilled non-primary sector professions such as transport, construction, services, lifting work, trade, and chemical industries, were also hired in large numbers by landlords in Iran and Pakistan during the summer and autumn seasons to do heavy agricultural work (Olimova 2005; Olimova and Olimov 2007).

The flow of migrant workers from Nepal to the Middle East-, South East-, and East Asia had been widely documented (Kollmair et al. 2006; Shrestha 2008; Adhikari and Hobley 2011). According to the Census of Nepal of 2001, about 3.3 % of the total native population of the country were international migrants; and over two-thirds of these had migrated for employment (Subedi 2009). In spite of the growing popularity of destinations such as Malaysia, Qatar, Saudi Arabia, United Arab Emirates, South Korea, and Kuwait among economically better off Nepalese migrants, India still remained the most important destination for unskilled and marginal population of rural Nepal (Bhattrai 2007; Adhikari and Hobley 2011). Adhikari et al. (2006) estimated that approximately 1.5 million Nepalese had found employment in India. There is a lot of variation between available estimates since there was no way to ascertain the actual number of Nepalese migrants in India because of the open border between the two countries. The India-Nepal Treaty of Friendship of 1950 created an open border between the two countries, allowing visa and passport free entry and access to employment without a work permit. Any citizen of India or Nepal can migrate to the other country and stay for as long as desired (Subedi 1991; Adhikari et al. 2008).

[1] The destination was unknown for 32 % of the migrants (Ghobadi et al. 2005).

The countries around the Persian Gulf have provided unprecedented economic opportunities for migrant workers from mountain regions of Pakistan, where they were predominantly occupied in semi to low skilled professions in transportation, construction, cargo services, brick kilns, chemical enterprises, and oilfields (Olimova and Olimov 2007; Hunzai 2010). Based on data from the Bureau of Emigration and Overseas Employment, Arif (2010) estimated that about a quarter of the international labour migrants originated from the Khyber Pakhtunkhwa province in Northern Pakistan. Given the labour market demand in the Middle East, most Pakistani migrant workers to the region are male (BOE 2005).

Olimova and Olimov (2007) observed a direct relationship between the type of migration and altitudinal profile of the origin communities in Afghanistan and Pakistan. Migrants from mid-altitude regions were seasonal workers with relatively sophisticated professional skills, including a significant number of seasonal employees in trade and commerce. Migration from high-altitude areas was dominated by younger semi-skilled or unskilled individuals, who were usually employed in low skilled occupations in destination that required heavy manual work. Only few were involved in trade and commerce. On an average, there were more female migrant workers from this region than from other regions.

Internal Displacement

The cases of conflict induced internal displacement have been reported from all across the Hindu Kush Himalayan region. As of 2011, 448,000 people remained internally displaced in Afghanistan because of armed conflict, human rights abuses, and other forms of generalised violence. (UNHCR IDP data 2012 cited in IDMC 2012a). The majority of this displacement has taken place in the southern, south-eastern, eastern and western regions of Afghanistan (IDMC 2012a). In neighbouring Pakistan, some 850,000 people were displaced due to the military operations against the insurgents in Khyber-Pakhtunkhwa and Federally Administered Tribal Areas (FATA) (FDMA 2011 cited in IDMC 2012c). In Nepal, although the signing of the Comprehensive Peace Accord (CPA) between the Government and the Unified Communist Party of Nepal (Maoist) in 2006 ended a decade old conflict, by the end of 2011, about 50,000 people still remained displaced and were unable or unwilling to return to their places of origin (RCHCO 2011 cited in IDMC 2012b). Inter-ethnic violence in various parts of Northeast India such as western Assam, along the border between Assam and Meghalaya, and in Tripura has also displaced scores of people. As of 2011, conservative estimates suggest that more that 76,000 people were still displaced (IDMC 2011).

Presently, there is no coordinated mechanism in the Hindu Kush Himalayan region to collect and compile data on mobility due to natural disasters. The available evidence is based on post-disaster rapid assessments or sporadic case studies. A rapid assessment by IOM in 11 districts of the Khyber-Pakhtunkhwa province of northern Pakistan affected by the 2010 floods found that over 257,000 households had been temporarily displaced (IOM 2010). The 2005 earthquake in Northern Pakistan resulted in the displacement 2.8–3.5 million people (DFID 2006, ERRA 2005 cited in Yasir 2009). Banerjee et al. (2011) reported permanent as well as temporary displacement in flood affected communities of the Koshi sub-basin of East Nepal, Eastern Brahmaputra sub-basin in Assam province of India, and Upper Indus sub-basin in Chitral district of Pakistan.

Refugees

For the past three decades, most refugees in the region originated from Afghanistan. At the end of 2009, almost 2.9 million Afghans were still refugees. Almost 96 % of Afghan refugees were based in Pakistan and the Islamic Republic of Iran (UNHCR 2010). Since 2006, security concerns and poor prospects of economic and social improvement had caused a decline in the rate of returns to Afghanistan (Altai Consulting 2009). It is also estimated that around two million Myanmarese refugees are in Thailand (Brees 2008). The exact estimate of Myanmarese refugees in other neighbouring countries – Bangladesh, India, China and Malaysia – is unknown (Brees 2008; Shukla 2008).

Over the years, though, the refugee population in Pakistan had decreased due to voluntary repatriation, registration and resettlement. In 2009, Pakistan was still the country with largest number of refugees, around 1.7 million. Most of these refugees were from neighbouring Afghanistan (UNHCR 2010). India hosted some 110,000–150,000 refugees from the Tibetan Autonomous Region of China, about 75,000 ethnic Chin refugees from Myanmar, around 36,000 Chakma and Hajong refugees from the Chittagong Hill Tracts of Bangladesh, between 15,000 and 30,000 ethnic Nepalese from Bhutan, and over 31,000 Afghans (HRLN 2007; United States Committee on Refugees and Immigrants 2008).

Case Studies on Links Between Migration and Environmental Change

The multi-causal nature of migration is widely acknowledged (Castles and Miller 1993; Boyle et al. 1998; Foresight: Migration and Global Environmental Change 2011). Indeed, even in the Hindu Kush Himalayan region, where the livelihoods of many households are dependent on ecosystem services, a disruption of lives and livelihoods due to environmental shocks and stresses may not be a sufficient motivation to migrate, as the environmental driver of migration does not function in isolation from non-environmental drivers and intervening obstacles. For this reason, this last part of the chapter considers first environmental drivers of migration in the Hindu Kush Himalayan region, but then turns to other drivers that interact with environmental drivers in the context of climate change.

Environmental Drivers of Migration

There is a relative lack of specific empirical evidence on the role of environmental drivers of in the Hindu Kush Himalayan region. This sub-section reviews all available empirical evidence from this region. Some of the reviewed case studies (Massey et al. 2007; Shrestha and Bhandari 2007; Banerjee et al. 2011; Bohra-Mishra and Massey 2011) specifically focus on the relationship between environmental change and migration. In other case studies (Ghobadi et al. 2005; IUCN 2005), the effects of environmental variables are mentioned as a passing reference within a wider study.

Drought affected households in Afghanistan are more likely to have migrant members than those unaffected by drought (Ghobadi et al. 2005). During long winters when the valleys are covered with snow, migrations to urban centres in the plains had been a traditional strategy among the lower income rural migrants of northern Pakistan (IUCN 2005). To represent environmental change in Chitwan valley of south-east Nepal, Massey et al. (2007) used variables such as neighbourhood population density, perceptions of changing agricultural productivity, share of neighbourhood land covered by flora, time required to collect fodder, and time required to collect firewood. This study found that environmental change had a greater chance of influencing local (within Chitwan valley) rather than long-distance (outside Chitwan Valley) mobility. The likelihood of moving within the Chitwan valley was greater if a decline in agricultural productivity was perceived, the share of the neighbourhood covered in flora declined, or time required gathering firewood increased. For long distance mobility, only a perceived decline of agricultural productivity was significant but the effect was considerably less powerful (Massey et al. 2007).

Yet in a separate study in the same area, Shrestha and Bhandari (2007) conceptualised the notion of environmental security in terms of changes in time required to collect firewood from forest or common land. The results from Chitwan valley in Nepal showed that increases in environmental insecurity raised the likelihood of labour migration, regardless of destination (Shrestha and Bhandari 2007). This contradictory result from two case studies (Massey et al. 2007; Shrestha and Bhandari 2007) set in the same area – Chitwan valley in Nepal – may be due in part to differences in the definitions of migration adopted by them. Massey et al. (2007) focused on local (i.e. within Chitwan valley) versus long-distance mobility (i.e. outside Chitwan valley). Whereas, Shrestha and Bhandari (2007) considered domestic (i.e. within Nepal) versus international mobility (i.e. outside Nepal) (Bohra-Mishra and Massey 2011).

In order to resolve this contradiction, Bohra-Mishra and Massey (2011) distinguished three streams of migration – within Chitwan, outside Chitwan but within Nepal, and those outside Nepal – in the same database that had been used by the earlier studies. This study identified five environmental indicators: change in time required to collect fodder, change in time required to collect firewood, change in agricultural productivity, change in quality of drinking water, and population density (Bohra-Mishra and Massey 2011). The likelihood of undertaking a local move had a strong and consistent relationship with neighbourhood density, rise in time required to collect fodder and firewood, and declining agricultural productivity. The effects of environmental drivers were more prevalent on local migration of women than that for men. Environmental deterioration had little influence on migration outside the Chitwan district, either to other districts in Nepal or overseas. However, the likelihood of male migration to other districts of Nepal or international destinations increased with rise in time to collect firewood (Bohra-Mishra and Massey 2011).

A recent study by Banerjee et al. (2011) across the Hindu Kush Himalayan region has assessed patterns of labour migration in rural communities that were exposed to water hazards. The environmental variables used in this study were the type of water hazard (i.e. rapid or slow onset water hazard), natural hazard proneness of livelihoods prior to migration, and impact of water hazard on owned agricultural land. This study found that the likelihood that household members would migrate for work was higher in communities exposed to rapid onset water hazard (i.e. flood and flash flood) than those exposed to slow onset water hazard (i.e. drought). Among the communities exposed to slow onset water hazards, the likelihood of labour migration was higher in communities affected by very severe hazards rather than those affected by less severe ones. In flood or flash flood affected communities, a comparison between the households where agricultural land had been damaged by the hazard to those where it had not been, showed that members of the former were less likely to migrate for work. The likelihood of labour migration from households that depended on natural hazard prone livelihoods (e.g. farming or animal husbandry) in the past was lower when exposed to slow onset water hazards (Banerjee et al. 2011).

Economic Drivers of Migration

Livelihood opportunities in the rural areas of the Hindu Kush Himalayan region are generally restricted to primary sector occupations. Factors such as market volatility, environmental shocks and stress, land degradation, and lack of basic infrastructure undermined agricultural growth, its labour absorption potential, income generating capacity and role in food security. At the same time, the introduction of modern agricultural technology and natural increase in the rural population has created a surplus of rural labour (Bohle and Adhikari 1998; Liang and Ma 2004; Huo et al. 2006; Olimova and Olimov 2007). There is a growing awareness of prospects beyond the mountains through education, communication, and social networks (Opel 2005; Hoermann et al. 2010). In such circumstances, labour migration has become one of the most important livelihood strategies across this region: Afghanistan (Ghobadi et al. 2005; Opel 2005), Bhutan (Ministry of Finance in Bhutan 2005); China (Liang and Ma 2004; Zhu and He 2010); Nepal (Seddon et al. 2002; Sharma 2008, 2011); and Pakistan (Nadeem et al. 2009; Arif 2010).

Economic opportunities within the Hindu Kush Himalayas also attract migrants from within the region and neighbouring lowlands (Liang and Ma 2004;

Nepal 2007; Brusle 2008). For instance, tourism induced growth in the service sector in Yunnan attracted 'floating migrants' (those without the *hukou* or household registration status) from within the province and elsewhere (Liang and Ma 2004).

Social Drivers of Migration

Social networks based on familial links or affiliation to a social group have a strong influence on the migration decision and choice of destination. These networks support migration by extending loans, assisting in logistics, arranging jobs and accommodation, and providing emotional support to the migrant or family left behind. The influence of social networks on migration had been documented in Afghanistan (Opel 2005; Ghobadi et al. 2005); Bhutan (Walcott 2009); China (Liang and Ma 2004); India (Mamgain 2004); Nepal (Seddon et al. 2002; Thieme 2006; Sharma 2008); and Pakistan (Nadeem et al. 2009).

Education has emerged as another important social determinant of migration. The lack of adequate education facilities was the most commonly cited reason for leaving rural homes in Bhutan (Ministry of Finance in Bhutan 2005; Ministry of Agriculture in Bhutan 2006). Access to better education facilities for their children was one of the factors considered by the migrants in Ladakh district of India (Goodall 2004) and Far West Nepal (Poertner et al. 2011).

Demographic Drivers of Migration

The demographic factors such as household composition, age, and gender influence migration process. In Afghanistan, likelihood of migration has been found to be higher in households with more persons of working age (Ghobadi et al. 2005). Banerjee et al. (2011) report that a rise in the number of males in the working age group in a household increased the likelihood of the household to send one of its members to work somewhere else. In Nepal, Massey et al. (2007) reported that both local mobility and long-distance migration was age-selective in nature. The same study found that the likelihood of local mobility and long-distance migration declined with rising age. This effect was more pronounced for local rather than long distance migration.

In Nepal, Shrestha and Bhandari (2007) found that the presence of both men and women was important for international migration but only the availability of men positively contributed to internal migration. Also, households that engaged men in firewood collection were less likely to send members to work somewhere else in Nepal or abroad. In another study from Nepal, Massey et al. (2007) reported that an increase in the time required to gather firewood raised the odds of male migration but had no effect on female migration. Similarly, an increase in the time needed to collect fodder raised the odds of female long-distance mobility but had no effect on the odds of male long-distance mobility.

Political Drivers of Migration

Some policies, either explicitly or implicitly, seek to control migration, and may have an independent effect on whether people move or not. The Government of India-sponsored Mahatma Gandhi National Rural Employment Guarantee Act provides a legal guarantee for 100 days of wage employment in a financial year to every rural household (Ministry of Rural Development in India 2008). Jain (2010) found that this programme has reduced the need for seasonal migration to some extent in the province of Uttarakhand, mainly among unskilled or less educated persons.

In turn, other policies seek to facilitate migration. The India-Nepal Treaty of Friendship of 1950 created an open border between the two countries, which includes visa and passport-free entry and access to employment without the necessity of a work permit. Citizens of either country can migrate to the other country and stay as long as desired (Subedi 1991; Adhikari et al. 2008). For the poor, even the acquisition of official documents such as a passport is frequently an insurmountable hurdle (Hoermann and Kollmair 2008). The open border permits any national identification documents, such as electoral identity cards or driving license, to gain entry.

In practice, the political drivers of migration go well beyond this. For example, the Hindu Kush Himalayan region has also witnessed conflicts in various forms. These conflicts continue to uproot many across this region, and hinder the return of refugees and internally displaced persons (IDPs), including Tibetan refugees in India and Nepal (Baral 2003; Mahajan et al. 2008); Bhutanese refugees in Nepal (Mazumdar 2005); Afghan refugees in Iran and Pakistan (AREU 2006; UNHCR 2010); Chin refugees from Myanmar in India (Mahajan et al. 2008); and IDP populations in Pakistan (IDMC 2012c). Meanwhile, there are a number of policies which, whilst not specifically targeted at migration, may have an effect on whether it occurs or not – including broader development policies, and policies towards disasters and displacement.

Intervening Obstacles

With progress in communication, electrification, and transportation networks, marginal mountain communities have become connected to the main market economies of the region (Ediger and Huafang 2006; Massey et al. 2007; Olimova and Olimov 2007). According to Du et al. (2004), the low population density and high transportation costs of rural mountainous regions of China has been a major challenge to the growth of industries in the interior rural areas, which implies that migration may be an important component of the structural change occurring there. Yet the creation of supportive infrastructure, particularly roads and communication facilities, could facilitate out-migration as well as in-migration (Bhandari 2004; Massey et al. 2007). Conversely, the probability of migration has been shown to be lower among households in Afghanistan that resided in large communities with more irrigated land and services such as markets, public transportation or health facilities (Ghobadi et al. 2005).

Some households may not be able to meet the financial cost of migration. Yet, migration for work may be a necessary livelihood choice for them. If the expected income in destination is higher than the actual income at the place origin, some households borrowed loans to finance the migration (see Opel 2005; Ghobadi et al. 2005; Jain 2010; Nadeem et al. 2009).

Migrant Remittances: Closing the Circle Between Migration and Environmental Change

In general, the environmental migration discourse has been overtly focused on the impacts of environmental variability and change on migration (Barnett and Webber 2009). In contrast, the feedback from migration in context of adaptation such as the role of financial and social remittances, and the influence of social networks, has received little empirical research attention (Banerjee et al. 2012), including in mountain areas. Thus only two case studies (Suleri and Savage 2006; Banerjee et al. 2011) from the Hindu Kush Himalayan region have explicitly discussed the role of remittances in context of adaptation, though other examples cited in this sub-section have discussed the role of remittances in mountain households but not in context of environmental change. These findings can still be used to illustrate the potential role that migration outcomes can have on adaptation to environmental variability and change.

Financial remittances are a significant source of cash income for many mountain households (Blaikie et al. 2002; Olimova and Olimov 2007; Subedi 2009), and the percentage contribution of remittances to recipient household's income is significant. They often supplement household income from other sources such as agriculture, livestock, wage labour, salary or business (Kreutzmann 1993). Suleri and Savage (2006) reported that 96 % of the studied households in Northern Pakistan claimed that remittances were their primary source of income, and half declared it to be their only source of income. Remittances contributed to recipient household's wellbeing as these were widely used to purchase food, repay loan, afford healthcare and education, buy consumer goods and construct or repair houses across the Hindu Kush Himalayas: Afghanistan (Opel 2005; Ghobadi et al. 2005); Bhutan (Ministry of Agriculture in Bhutan 2006); India (Jain 2010); Nepal (Lokshin et al. 2007; Adhikari and Hobley 2011); and Pakistan (Arif 2010). To a lesser extent they have been invested in farming, business ventures and savings (Ghobadi et al. 2005; Banerjee et al. 2011).

Remittances have also been used to procure food and other basic needs during or in aftermath of a disaster and re-establish livelihoods and rebuild lost assets (Suleri and Savage 2006; Banerjee et al. 2011). In some cases, remittances have been used for disaster preparedness such as strengthening of housing quality or procurement of boat in flood affected communities and buying irrigation equipment in drought affected settlements (Banerjee et al. 2011). The impacts of remittance are not limited to recipient households. Suleri and Savage (2006) reported in the aftermath of the 2001 earthquake, remittance cash generated demand for goods and services in the local economy.

Besides remittance cash, migrants also bring back social remittances – ideas, behaviours, identities, social capital, knowledge, and skills – from destination to origin communities (Levitt 1998; Bailey 2010). Globally, their role in promoting innovation, entrepreneurship, community and family formation, and political integration has been widely documented (Levitt 1998; Levitt and Lamba-Nieves 2011). However, knowledge gaps still exist in terms of their role in building adaptation to climate change in general, and specifically the role of social remittances in the Hindu Kush Himalayan region.

Conclusion

Mountain regions are often perceived as particularly vulnerable to climate change, as well as to a range of environmental shocks and hazards. Mountain communities are sensitive to these stressors, and their impacts are likely to be manifest in human wellbeing through effects on local livelihoods and infrastructure. Though there is uncertainty about the nature of future change, it is clear that temperatures in the region are rising, likely above global averages. In addition, there is a dearth of data on the state of environment in Hindu Kush Himalayan region, and the space that this creates for claims about the future impacts of climate change. Nonetheless, despite this lack of confidence in forecasting, the Hindu Kush-Himalayan region is still widely considered as one of the planet's hot spots of future climate change impacts.

Mountain people have long adjusted to living in fragile and marginal environments, and partially as a result, mountains have for many decades been zones of out-migration, as people seek alternative and more reliable livelihoods. There are several well-established internal and international migration streams in the Hindu Kush Himalayan region. Evidence, though sporadic, from across this region indicates that environmental stressors do influence migration decisions but not in isolation from non-environmental drivers and intervening obstacles. The impact of the environment is largely manifest through the other determinants of migration. Remittances have a significant role in the relationship between environmental change and migration, as it contributes to the well being of recipient households in normal time as well as during or in aftermath of a disaster. In the process, it is has a key role in reducing vulnerability and building resilience to environmental stresses and shocks.

At the start of this chapter, we highlighted some of the specificities of mountain areas and particular vulnerability to environmental change and especially environmental shocks and hazards. This, and the extent of downstream effects of changes that occur in mountains make it particularly important for regional policy-makers to work together to address both environmental risks and shocks themselves, and associated migration flows. Migration is long-established as a consequence of the marginal conditions within many mountain zones, but also represents an important potential form of adaptation to such conditions.

Yet the consequences of migration in terms of adaptation to climate change and environmental shocks, through the role of financial and social remittances, and influence of social networks, have received little empirical research attention in these regions. In this context, Banerjee et al. (2012) have identified a number of knowledge gaps within the wider environmental migration discourse, including the role of migrant remittances in structural and non-structural adaptation measures to reduce household vulnerability to environmental hazards; the circumstances – social, political or economic – that are most propitious for this kind of spending; the extent to which benefits from migration spread beyond migrant households to the wider community; the additional risks for migrants themselves and for family members left behind in origin communities; the sustainability of remittance flows at the household level; and effect of migration on resilience and vulnerability. Future research in the Hindu Kush Himalayan region could usefully focus on any or all of questions.

Acknowledgement The authors would like to thank Etienne Piguet (University of Neuchatel), Frank Laczko (IOM), Nand Kishor Agrawal (ICIMOD) and Neera Shrestha Pradhan (ICIMOD) for their insightful feedback at various stages during this research. The authors would also like to thank the anonymous reviewers for their helpful comments. This research was funded by the Ministry of Foreign Affairs (MFA) of the Government of Norway as a part of ICIMOD's Himalayan Climate Change Adaptation Programme (HICAP). This chapter draws upon the DR9 Drivers of Migration in Mountain Regions of the Developing World: A Review, which was submitted to the Foresight: Migration and Global Environment Change project.

References

Adhikari, J., & Hobley, M. (2011). *Everyone is leaving – Who will sow our fields? The Effects of Migration from Khotang District to the Gulf and Malaysia*. Kathmandu: Swiss Agency for Development and Corporation.

Adhikari, J., Gurung, G., Bhadra, C., Seddon, D., Niraula, B. P., Gurung, P., & Gurung, K. (2006). *Nepali women and foreign labour migration*. Kathmandu: UNIFEM and Nepal Institute for Development Studies.

Adhikari, K., Suwal, B., & Sharma, M. (2008). *Foreign labour migration and remittance in Nepal: Policy, institution and services*. Washington, DC: International Food Policy Research Institute.

Altai Consulting. (2009). *Study on cross border population movements between Afghanistan and Pakistan. Report commissioned by the Office of the United Nations High Commissioner for Refugees*. Kabul: Altai Consulting.

AREU. (2006). *Afghans in Karachi: Migration. Settlement and social networks* (Case studies series). Kabul: Afghanistan Research and Evaluation Unit (AREU).

Arif, G. M. (2010). *Economic and social impacts of remittances on households: The case of Pakistani migrants working in Saudi Arabia*. Geneva: International Organisation for Migration.

Bailey, A. J. (2010). Population geographies, gender, and the migration-development nexus. *Progress in Human Geography, 34*(3), 375–386.

Bajracharya, S. R., & Mool, P. (2009). Glaciers, glacial lakes and glacial lake outburst floods in the Mount Everest region, Nepal. *Annals of Glaciology, 50*(53), 81–86.

Bajracharya, S. R., & Shrestha, B. (2011). *The status of glaciers in the Hindu Kush-Himalayan region*. Kathmandu: International Centre for Integrated Mountain Development.

Bajracharya, S. R., Maharjan, S. B., & Shrestha, F. (2010). *Glaciers shrinking in Nepal Himalaya*. www.intechopen.com/articles/show/title/glaciers-shrinking-in-nepal-himalaya. Accessed 12 Mar 2012.

Bajracharya, S. R., Maharjan, S. B., & Shrestha, F. (2011). Glaciers shrinking in Nepal Himalaya. In J. Blanco & H. Kheradmand (Eds.), *Climate change: Geophysical foundations and ecological effects* (pp. 445–458). Rijeka: InTech.

Banerjee, S., Gerlitz, J. Y., & Hoermann, B. (2011). *Labour migration as a response strategy to water hazards in the Hindu Kush-Himalayas*. Kathmandu: International Centre for Integrated Mountain Development.

Banerjee, S., Black, R., & Kniveton, D. (2012). *Migration as an effective mode of adaptation to climate change*. Policy paper for the European Commission. Brussels: European Commission.

Baral, L. (2003). Nepal: Forced migration, challenges and prospects. *Nepali Journal of Contemporary Studies, III*(1), 51–66.

Barnett, J., & Webber, M. (2009). *Accommodating migration to promote adaptation to climate change*. Policy brief prepared for the Secretariat of the Swedish Commission on Climate Change and Development and the World Bank World Development Report 2010 Team. Retrieved from http://www.ccdcommission.org/Filer/documents/Accommodating %20Migration.pdf

Bhandari, P. (2004). Relative deprivation and migration in an agricultural setting of Nepal. *Population and Environment, 25*(5), 475–499.

Bhattrai, R. (2007). *Open borders, closed citizenships: Nepali labor migrants in Delhi*. Paper prepared for Conference on International migration, multi-local livelihoods and human security: Perspectives from Europe, Asia and Africa, Institute of Social Studies, 30–31 August 2007, The Netherlands.

Blaikie, P. M., Cameron, J., & Seddon, D. (2002). Understanding 20 years of change in west-central Nepal: Continuity and change in lives and ideas. *World Development, 30*, 1255–1270.

BOE. (2005). *International migration statistics*. Islamabad: Government of Pakistan, Bureau of Emigration and Overseas Employment. www.beoe.gov.pk. Accessed 16 Sept 2009.

Bohle, H. G., & Adhikari, J. (1998). Rural livelihoods at risk: How Nepalese farmers cope with food insecurity. *Mountain Research and Development, 18*, 321–332.

Bohra-Mishra, P., & Massey, D. S. (2011). Environmental degradation and out-migration: New evidence from Nepal. In E. Piguet, A. Pécoud, & P. de Guchteneire (Eds.), *Migration and climate change* (pp. 74–101). Cambridge: Cambridge University Press.

Bolch, T., Pieczonka, T., & Benn, D. I. (2011). Multi-decadal mass loss of glaciers in the Everest area (Nepal Himalaya) derived from stereo imagery. *The Cryosphere, 5*, 349–358.

Boyle, P., Halfracree, K., & Robinson, V. (1998). *Exploring contemporary migration*. Harlow: Pearson Education.

Brees, I. (2008). Forced displacement of Burmese people. *Forced Migration Review, 30*, 4–5.

Brusle, T. (2008). Choosing a destination and work. *Mountain Research and Development, 28*(3), 240–247.

Castles, S., & Miller, M. J. (1993). *The age of migration: International population movements in the modern world*. Basingstoke: Macmillan.

Chettri, N., Sharma, E., Shrestha, A. B., Zhoali, Y., Hua, Q., & Bajracharya, B. (2011). Real world protection for the 'third pole' and its people. In F. Huettmann (Ed.), *Protection of three poles* (pp. 113–133). Tokyo: Springer.

Cogley, J. G., Kargel, J. S., Kaser, G., & van der Veen, C. J. (2010). Tracking the source of glacier misinformation. *Science, 327*(5965), 522.

Cruz, R. V., Harasawa, H., Lal, M., Wu, S., Anokhin, Y., Punsalmaa, B., Honda, Y., Jafari, M., Li, C., & Huu Ninh, N. (2007). Asia. In M. L. Parry, O. F. Canziani, J. P. Palutikof, P. J. van der Linden, & C. E. Hanson (Eds.), *Climate change 2007: Impacts, adaptation and vulnerability. Contribution of Working Group II to the fourth assessment report of the Intergovernmental Panel on Climate Change* (pp. 469–506). Cambridge: Cambridge University Press.

Department for International Development (DFID). (2006). *South Asia earthquake* (Situation Report No. 44). Available at: http://www.reliefweb.int/rw/RWFiles2005.nsf/FilesByRWDocUNIDFileName/HMYT-6HRLDP-dfid-southasia-02nov.pdf/$File/dfid-southasia-02nov.pdf. Accessed 12 Feb 2012.

Dimri, A. P., & Dash, S. K. (2011). Wintertime climate trends in the western Himalayas. *Climate Change*. doi:10.1007/A10584-011-0201-y.

Ding, Y., Liu, S., Li, J., & Shangguan, D. (2006). The retreat of glaciers in response to recent climate warming in western China. *Annals of Glaciology, 43*(1), 97–105.

Du, Y., Park, A., & Wang, S. (2004). *Is migration helping China's poor?* Paper prepared for the Conference on Poverty, Inequality, Labour Market and Welfare Reform in China. Canberra: Australia National University.

Ediger, L., & Huafang, C. (2006). Upland China in transition. *Mountain Research and Development, 26*(3), 220–226.

Eriksson, M., Jianchu, X., Shrestha, A. B., Vaidya, R. A., Nepal, S., & Sandstrom, K. (2009). *The changing Himalayas: Impact of climate change on water resources and livelihoods in the greater Himalayas*. Kathmandu: ICIMOD.

Foresight: Migration and Global Environmental Change. (2011). *Final project report*. London: The Government Office for Science.

Ghobadi, N., Koettl, J., & Vakis, R. (2005). *Moving out of poverty: Migration insights from Rural Afghanistan* (Working Paper Series). Kabul: Afghanistan Research and Evaluation Unit (AREU).

Goodall, S. (2004). Rural-to-urban migration and urbanization in Leh, Ladakh: A case study of three nomadic pastoral communities. *Mountain Research and Development, 24*(3), 220–227.

Guha-Sapir, D., Vos, F., Below, R., & Ponserre, S. (2011). *Annual disaster statistical review 2010: The numbers and trends*. Brussels: CRED.

Gurung, D. R., Giriraj, A., Aung, K. S., Shrestha, B., & Kulkarni, A. V. (2011a). *Snow-cover mapping and monitoring in the Hindu Kush-Himalayas*. Nepal: ICIMOD.

Gurung, D. R., Kulkarni, A. V., Giriraj, A., Aung, K. S., Shrestha, B., & Srinivasan, J. (2011b). Change in seasonal snow cover in Hindu Kush–Himalayan Region. *The Cryosphere Discuss, 5*(2), 755–777.

Hassan, R., Scholes, R., & Ash, N. (Eds.). (2005). *Ecosystems and human well-being: Current state and trends* (Findings of the Condition and Trends Working Group of the Millennium Ecosystem Assessment, Vol. 1). Washington, DC: Island Press.

Hewitt, K. (2005). The Karakoram anomaly? Glacier expansion and the 'elevation effect', Karakoram Himalaya. *Mountain Research and Development, 25*(4), 332–340.

Hoermann, B., & Kollmair, M. (2008). *Labour migration and remittances in the Hindu Kush-Himalayan region*. Kathmandu: ICIMOD.

Hoermann, B., Banerjee, S., & Kollmair, M. (2010). *Labour migration for development in the western Hindu Kush-Himalayas*. Kathmandu: ICIMOD.

HRLN. (2007). *Report of refugee populations in India*. New Delhi: Human Rights Law Network.

Hunzai, K. (2010). *Labour migration and remittances in the mountains of Pakistan*. Kathmandu: ICIMOD.

Huo, W., Deng, G., Wang, P., & Yang, X. (2006). The pull of inter-provincial migration of rural labor force and its influence on policy making. *Ecological Economy, 2*, 25–31.

IDMC. (2011). *North-East India: People displaced by ethnic violence "forgotten"* (Press Release). Geneva: Internal Displacement Monitoring Centre.

IDMC. (2012a). *Afghanistan: Durable solutions far from reach amid escalating conflict*. Geneva: Internal Displacement Monitoring Centre.

IDMC. (2012b). *Nepal: Unresolved property issues and IDP policy hiatus undermine search for durable solutions*. Geneva: Internal Displacement Monitoring Centre.

IDMC. (2012c). *Pakistan: Displacement caused by conflict and natural disasters, achievements and challenges*. Geneva: Internal Displacement Monitoring Centre.

Immerzeel, W. (2008). Historical trends in future predictions of climate variability in the Brahmaputra basin. *International Journal of Climatology, 28*(2), 243–254.

Immerzeel, W. W., van Beek, L. P. H., & Bierkens, M. F. P. (2010). Climate change will affect the Asian water towers. *Science, 328*(5984), 1382–1385.

IOM. (2010). *IOM appeal in support of Pakistan Initial floods emergency response plan*. Geneva: International Organisation for Migration.

IPCC. (2007a). *Fourth assessment report*. Geneva: Intergovernmental Panel on Climate Change Secretariat.

IPCC. (2007b). *Climate change 2007: Summary for policymakers*. Fourth assessment report of the Intergovernmental Panel on Climate Change. Cambridge: Cambridge University Press.

IPCC. (2012). *Managing the risks of extreme events and disasters to advance climate change adaptation. A special report of Working Groups I and II of the Intergovernmental Panel on Climate Change*. Cambridge/New York: Cambridge University Press. 582 pp.

IUCN. (2005). *Project proposal on baseline study regarding the socio-economic condition including socio-economic survey of 10 selected villages in Shishi valley of Chitral with special focus on chilghoza pine and its contribution to livelihoods; situation of disadvantaged groups including women and their involvement in Chilghoza business; mapping of resource base (conditions and practices)*. Islamabad: IUCN.

Ives, J. D., Shrestha, R. B., & Mool, P. K. (2010). *Formation of glacial lakes in the Hindu Kush-Himalayas and GLOF risk assessment*. Kathmandu: ICIMOD.

Jain, A. (2010). *Labour migration and remittances in Uttarakhand*. Kathmandu: International Centre for Integrated Mountain Development.

Kang, S., You, Q., Flügel, W., Pepin, N., Yan, Y., Xu, Y., & Huang, J. (2010). Review of climate and cryospheric change in the Tibetan Plateau. *Environmental Research Letters, 5*(1). doi:10.1088/1748-9326/5/1/015101.

Kohler, T., & Maselli, D. (Eds.). (2009). *Mountains and climate change – From understanding to action*. Berne: Geographica Bernensia and Swiss Agency for Development and Cooperation.

Kollmair, M., Gurung, G. S., Hurni, K., & Maselli, D. (2005). Mountains: Special places to be protected? An analysis of worldwide nature conservation efforts in mountains. *International Journal of Biodiversity Science and Management, 1*, 181–189.

Kollmair, M., Manandhar, S., Subedi, B., & Thieme, S. (2006). New figures for old stories: Migration and remittances in Nepal. *Migration Letters, 3*(2), 151–160.

Körner, C. (2009). Conservation of mountain biodiversity in the context of climate change. In: *Proceedings of the international mountain biodiversity conference*, Kathmandu, 16–18 November 2008. Kathmandu: ICIMOD.

Körner, C., & Ohsawa, M. (2005). Mountain systems. Chapter 24. In R. Hasser, R. Scholes, & N. Ash (Eds.), *Millennium ecosystem assessment, 2005. Current state and trends: Findings of the condition and trends working group* (Ecosystems and human well-being, Vol. 1). Washington, DC: Island Press.

Kreutzmann, H. (1993). Challenge and response in the Karakoram: Socioeconomic transformation in Hunza, Northern Areas, Pakistan. *Mountain Research and Development, 13*(1), 19–39.

Kulkarni, A. V., Rathore, B. P., & Singh, S. K. (2010). Distribution of seasonal snow cover in central and western Himalaya. *Annals of Glaciology, 51*(54), 123–128.

Kulkarni, A., Patwardhan, S., Kumar, K. K., Ashok, K., & Krishnan, R. (2011, August 18–19). *Projected climate change over Hindukush-Himalayan region using high Resolution Regional Climate Model PRECIS*. Paper presented at the Authors' Workshop for the Regional Report on Climate Change in the Hindu Kush–Himalayas: The State of Current Knowledge. Kathmandu: ICIMOD.

Levitt, P. (1998). Social remittances: Migration-driven local-level forms of cultural diffusion. *International Migration Review, 32*(4), 926–948.

Levitt, P., & Lamba-Nieves, D. (2011). Social remittances revisited. *Journal of Ethnic and Migration Studies, 37*(1), 1–22.

Liang, Z., & Ma, Z. (2004). China's floating population: New evidence from the 2000 census. *Population and Development Review, 30*(3), 467–488.

Lokshin, M., Bontch-Osmolovski, M., & Glinskaya, E. (2007). *Work-related migration and poverty in Nepal* (World Bank Policy Research Working Paper 4231). Washington, DC: World Bank.

Mahajan, D., Sharma, D., Vivekanandan, J., & Talwar, S. (2008). *Climate change induced migration and its security implications in India's neighbourhood* (Project Report No. 2008RS10). New Delhi: TERI.

Mamgain, R. P. (2004). *Employment, migration and livelihoods in the hill economy of Uttaranchal*. PhD thesis submitted to Jawaharlal Nehru University, Centre for the Study of Regional Development, School of Social Sciences. New Delhi: Jawaharlal Nehru University.

Maplecroft. (2011). *Climate change risk atlas 2011*. http://maplecroft.com/search/?q=climate+change+vulnerability+ranking. Accessed 10 July 2012.

Massey, D., Axinn, W., & Ghimire, D. (2007). *Environmental change and out-migration: Evidence from Nepal* (Population Studies Center Research Report 07-615). Ann Arbor: Population Studies Center.

Mazumdar, A. (2005). Bhutan's military action against Indian insurgents. *Asian Survey, 45*(4), 566–580.

Messerli, B., & Ives, J. D. (Eds.). (1997). *Mountains of the world: A global priority*. London: The Parthenon Publication Group.

Miller, J., Rees, G., Warnaars, T., Young, G., Collins, D., & Shrestha, A. (2013). What is the evidence for glacial shrinkage across the Himalayas? CEE Review 10–008. Collaboration for Environmental Evidence.

Ministry of Agriculture. (2006). *Rural-urban migration in Bhutan*. Thimphu: Royal Government of Bhutan.

Ministry of Finance. (2005). *Bhutan National Human Development Report 2005*. Thimphu: Royal Government of Bhutan.

Ministry of Rural Development. (2008). *The National Rural Employment Guarantee Act 2005 (NREGA)*. Operational guidelines (3rd ed.). New Delhi: Ministry of Rural Development.

Myers, N., Mittermeier, R. A., Mittermeier, C. G., da Fonseca, G. A. B., & Kent, J. (2000). Biodiversity hotspots for conservation priorities. *Nature, 403*, 853–858.

Nadeem, S., Elahi, I., Hadi, A., & Uddin, I. (2009). *Traditional knowledge and local institutions support adaptation to water-induced hazards in Chitral, Pakistan*. Kathmandu: International Centre for Integrated Mountain Development.

Nepal, S. (2007). Tourism and rural settlements: Nepal's Annapurna region. *Annals of Tourism Research, 34*(4), 855–875.

Nie, Y., Zhang, Y., Liu, L., & Zhang, J. (2010). Glacial change in the vicinity of Mt. Qomolangmc (Everest), central high Himalayas since 1976. *Journal of Geographical Sciences, 20*(5), 667–686.

Nogues-Bravo, D., Araujo, M. B., Erra, M. P., & Martinez-Rica, J. P. (2007). Exposure of global mountain systems to climate warming during the 21st century. *Global Environmental Change, 17*, 420–428.

Olimova, S. (2005). *Impact of external migration on development of mountainous regions: Tajikistan, Kyrgyzstan, Afghanistan, and Pakistan*. Paper prepared for the International Workshop on Strategies for Development and Food Security in Mountainous Areas of Central Asia, 6–10 June 2005, Dushanbe, Tajikistan.

Olimova, S., & Olimov, M. (2007). Labor migration from mountainous areas in the Central Asian Region: Good or evil? *Mountain Research and Development, 27*(2), 104–108.

Opel, A. (2005). *Bound for the city: A study of Rural to Urban labour migration in Afghanistan* (Working Paper Series). Kabul: Afghanistan Research and Evaluation Unit (AREU).

Osti, R., & Egashira, S. (2009). Hydrodynamic characteristics of the Tam Pokhari glacial lake outburst flood in the Mt. Everest region, Nepal. *Hydrological Processes, 23*(20), 2943–2955.

Poertner, E., Junginger, M., & Müller-Böker, U. (2011). Migration in far west Nepal: Intergenerational linkages between internal and international migration of rural-to-urban migrants. *Critical Asian Studies, 43*(1), 23–47.

Richardson, S. D., & Reynolds, J. M. (2000). An overview of glacial hazards in the Himalayas. *Quaternary International, 65*, 31–47.

Rosenzweig, C., Casassa, G., Karoly, D. J., Imeson, A., Liu, C., Menzel, A., Rawlins, S., Root, T. L., Seguin, B., & Tryjanowski, P. (2007). Assessment of observed changes and responses in natural and managed systems. In M. L. Parry, O. F. Canziani, J. P. Palutikof, P. J. Van Der Linden, & C. E. Hanson (Eds.), *Climate change 2007. Impacts, adaptation and vulnerability. Contribution of Working Group II to the fourth assessment report of the Intergovernmental Panel on Climate Change* (pp. 79–131). Cambridge: Cambridge University Press.

Schutte, S., & Kreutzmann, H. (2011). Linking relief and development in Pakistan-administered Kashmir. *Mountain Research and Development, 31*(1), 5–15.

Seddon, D., Gurung, G., & Adhikari, J. (2002). Foreign labour migration and the remittance economy of Nepal. *Himalayan Research Bulletin, XVIII*(2), 3–10.

Sharma, J. R. (2008). Practices of male labor migration from the hills of Nepal to India in development discourses: Which pathology? *Gender, Technology and Development, 12*, 303–323.

Sharma, J. R. (2011). *The impact of environmental change on labour migration from Nepal to the Gulf States*. Foresight case study. London: The Government Office for Science.

Shrestha, B. (2008). Contribution of foreign employment to Nepalese Economy. *Economic Review, Occasional Paper, 20*, 1–15. Kathmandu: Nepal Rastra Bank.

Shrestha, A. B. (2009). Climate change in the Hindu Kush-Himalayas and its impacts on water and hazards. *APMN (Asia Pacific Mountain Network) Bulletin, 9*, 1–5.

Shrestha, S. S., & Bhandari, P. (2007). Environmental security and labor migration in Nepal. *Population and Environment, 29*, 25–38.

Shrestha, A. B., Wake, C. P., Dibb, J. E., & Mayewski, P. A. (2000). Precipitation fluctuations in the Nepal Himalaya and its vicinity and relationship with some large scale climatological parameters. *International Journal of Climatology, 20*(3), 317–327.

Shrestha, K. L., Shrestha, M. L., Shakya, N. M., Ghimire, M. L., & Sapkota, B. K. (2003). Climate change and water resources of Nepal. In A. Muhammed & F. J. Hansen (Eds.), *Climate change and water resources in South Asia: Proceedings of year end workshop, Kathmandu, Nepal, 7–9 January 2003, Kathmandu, Nepal*. Islamabad: Asianics Agro Development International.

Shukla, K. (2008). The international community's responsibility to protect. *Forced Migration Review, 30*, 7–9.

Singh, S. P., Singh, V., & Skutsch, M. (2010). Rapid warming in the Himalayas: Ecosystem responses and development options. *Climate Change and Development, 2*, 1–13.

Singh, S. P., Bassignana-Khadka, I., Karky, B. S., & Sharma, E. (2011). *Climate change in the Hindu Kush Himalayas: The state of current knowledge*. Kathmandu: ICIMOD.

Srivastava, D. (2003). Recession of Gangotri glacier. In D. Srivastava, K. R. Gupta, & S. Mukerji (Eds.), *Proceedings of a workshop on Gangotri glacier, Lucknow, 26–28 March 2003* (Special Publication No. 80). New Delhi: Geological Survey of India.

Subedi, B. (1991). International migration in Nepal: Towards an analytical framework. *Contribution to Nepalese Studies, 18*(1), 83–102.

Subedi, M. K. (2009). *The impact of labor migration and remittances on household income and welfare in Nepal*. Paper prepared for Pacific Conference for Development Economics, March 14, 2009, San Francisco State University.

Suleri, A. Q., & Savage, K. (2006). *Remittances in crises: A case study from Pakistan* (An Humanitarian Policy Group Background Paper). London: Overseas Development Institute.

Thieme, S. (2006). *Social networks and migration: Far west Nepalese labour migrants in Delhi*. Muenster: LIT Publishing House.

Tse-ring, K., Sharma, E., Chettri, N., & Shrestha, A. (2010). *Climate change vulnerability of mountain ecosystems in the Eastern Himalayas: Climate change impact and vulnerability in the Eastern Himalayas – Synthesis report*. Kathmandu: ICIMOD.

UNHCR. (2010). *Statistical yearbook 2009: Trends in displacement, protection and solutions*. Geneva: United Nations High Commissioner for Refugees.

United States Committee for Refugees and Immigrants. (2008). *World Refugee Survey 2008 – India*. Available at: http://www.unhcr.org/refworld/docid/485f50d82.html. Accessed 19 Apr 2012.

Vohra, C. P. (1981). Himalayan glaciers. In J. S. Lal & A. D. Moddie (Eds.), *The Himalaya: Aspect of change* (pp. 138–151). New Delhi: Oxford University Press.

Vuichard, D., & Zimmermann, M. (1987). The 1985 catastrophic drainage of a moraine-dammed lake, Khumbu Himal, Nepal: Cause and consequences. *Mountain Research and Development, 7*(2), 91–110.

Walcott, S. (2009). Urbanization in Bhutan. *Geographical Review, 99*(1), 81–93.

Xin, W., Shiyin, L., Wanqin, G., & Junli, X. (2008). Assessment and simulation of glacier lake outburst floods for Longbasaba and Pida Lakes, China. *Mountain Research and Development, 28*(3), 310–317.

Xu, J., Grumbine, R. E., Shrestha, A., Eriksson, M., Yang, X., Wang, Y. U. N., & Wilkes, A. (2009). The melting Himalayas: Cascading effects of climate change on water, biodiversity, and livelihoods. *Conservation Biodiversity, 23*(3), 520–530.

Yao, T., Thompson, L., Yang, W., Yu, W., Gao, Y., Guo, X., Yang, X., Duan, K., Zhao, H., Xu, B., Pu, J., Lu1, A., Xiang, Y., Kattel, D. B., & Joswiak, D. (2012). Different glacier status with atmospheric circulations in Tibetan Plateau and surroundings. *NClimate* 1580. www.nature.com/natureclimatechange. Accessed 20 July 2012.

Yasir, A. (2009). *The political economy of disaster vulnerability: A case study of Pakistan earthquake 2005*. London: London School of Economics & Political Science.

Zhu, H., & He, S. (2010). Land pressure and adaptation in the mountainous region of northern China: An empirical analysis of 21 small watersheds. *Journal of Geographical Science, 20*(6), 913–922.

Regional Policy Perspectives

10

Karoline Popp

Introduction

This volume has chosen to focus on the region as its scale of analysis for the interactions between human mobility and the environment. According to existing evidence, and as corroborated by the preceding chapters, internal mobility and cross-border migration among neighbouring countries are the predominant migratory patterns where environmental factors are at play. In keeping with the chosen scale of analysis, this chapter will present an overview of selected policy and cooperation efforts among governments on the issue of environmental migration[1] at regional level. It thereby aims to complement existing accounts of the normative and policy context surrounding environmental migration, principally from the perspective of international law (e.g. Zetter 2009; McAdam 2011a, 2012), international institutions (e.g. Gemmene 2011; Koser 2011; Warner 2010, 2011; McNamara 2007) and national policy (e.g. Martin 2009).

Calls for regional approaches to environmental migration have been heard in a number of inter-state fora as well as the research community.[2] A priori, regional

The author would like to acknowledge and thank Sarah Lynn Harris and Daniel Salmon for their research support for this chapter. The author is also grateful to Alex Flavell and Naomi Shiferaw for their comments and inputs on earlier drafts of the chapter.

[1]The chapter will utilize the term "environmental migration" to summarize the various aspects of the relationship between environmental events and processes and human mobility addressed in this volume (see also section "Methodology" below).

[2]See for instance the Global Forum on Migration and Development (summary of discussions and background paper of roundtable 3.2 in 2010); paragraph 14(f) of the 2010 Cancun Agreements adopted at the 16th Conference of the Parties of the UN Framework Convention on Climate

K. Popp (✉)
Migration Policy Officer, International Organization for Migration, Geneva, Switzerland
e-mail: kpopp@iom.int

E. Piguet and F. Laczko (eds.), *People on the Move in a Changing Climate*,
Global Migration Issues 2, DOI 10.1007/978-94-007-6985-4_10,

collaboration on environmental migration would seem attractive to policymakers for several reasons: firstly, most migration is regional, including environmental migration. Briefly, pre-existing migration routes and networks and the typically low levels of resources available to those affected by environmental degradation and natural disasters combine to favour nearby destinations over distant ones. As a result, regional partners are frequently the most relevant in addressing migration issues. Secondly, regions are likely to face similar environmental phenomena and hazards, as well as similar effects of climate change. These shared challenges represent another incentive for cooperation. Thirdly, regional cooperation has become the status quo in most parts of the world. Regional political and economic integration processes have expanded significantly over the past few decades and today virtually all countries belong to a regional organization, if not to several. Kälin and Schrepfer argue that "regional and sub-regional organizations are often more coherent in terms of interests of member states and thus more likely to reach consensus on issues" (2012:76). More specifically, migration governance has witnessed a marked shift to the regional level, including by way of dedicated regional consultative processes on migration, or the inclusion of migration within the scope of activity of regional economic and political bodies.[3] Existing institutional structures, albeit at different levels of maturity, thus offer a natural space to extend collaboration to new and different challenges, such as environmental migration.

An examination of policy developments on environmental migration at regional level begs a preliminary question: what is the track record of regional interstate cooperation on migration, on the one hand, and on environmental/climate change policy, on the other hand? While an exhaustive answer to this question is beyond the scope of this chapter, it quickly becomes apparent that multilateral action on both issues, taken separately, is relatively underdeveloped, in particular when compared to other domains such as trade or development.[4] From a global governance perspective, migration on the one hand and environment/climate change on the other rank high on the list of concerns, but equally high among the most intractable issues on the international policy agenda: political sensitivities are palpable, positions and interests continue to diverge widely, consensus is inchoate at best, and the powers of regional and international institutions to impose policies or laws are limited.

It will therefore come as little surprise that strong and coherent policy action on a *combination* of the two, at regional or global levels, remains a rare occurrence.

Change; the IOM International Dialogue on Migration "Climate change, environmental degradation and migration" (March 2011, Chair's Summary); the UN High Commissioner for Refugees' expert roundtable "Climate change and displacement: identifying gaps and responses" (February 2011, Summary of Deliberations); Warner (2011), Zetter (2011), Kälin and Schrepfer (2012), Asian Development Bank (2012).

[3]See IOM (2010b, 2011) and Popp (2012).

[4]For a more detailed treatment of this question, see IOM (2010a, 2011) and Betts (2011).

An additional problem that arises is that of institutional(ized) blindspots: where policy silos – such as development, security, or environment – are deeply entrenched, a cross-cutting issue such as migration is liable to be dealt with in a partial manner, or not at all. This problem is considerably more acute where environmental migration is concerned. Blindspots are also reflected in and perpetuated by the segregation between policy communities: for example, most policy processes on migration would not typically include environmental experts, or vice versa. McAdam sums up the dilemma, in stating that "the governance of climate-change related movement (like global migration more broadly) suffers from significant fragmentation, both vertically – with actors at the international, regional and local levels – and horizontally – with the phenomenon addressed in part or, more rarely, under the auspices of a range of other 'policy categories' and associated institutions... despite (or because of) the plethora of existing, as well as potential, governance mechanisms, processes and institutions, no coherent multilateral governance framework exists for this purpose" (2012:213).

Having established that, as a policy matter, approaches to environmental migration are likely to be fraught with fragmentation, it would be narrow-minded to analyse environmental migration exclusively within the logic of migration policy. Instead, other policy domains have contributed significantly to shaping the debate. This chapter posits that, at this point in time, environmental migration needs a "carrier": rather than a stand-alone policy item, it tends to be embedded in broader discussions, principally in the fields of migration, climate/environment, security, and human rights. Secondly, given the novelty of the subject and the sensitivities it tends to stir, this chapter seeks to demonstrate how the issue of environmental migration is "entering the DNA" (Somerville 2011:13) of regional policy and institutions. Thus, the second strand of this chapter's argument runs as follows: while there is little evidence of concrete regional action, environmental migration has begun to permeate policy awareness, mainly through exhortatory statements of intent and soft policy tools. Following a brief discussion of this chapter's scope and methodology, it will go on to examine four principal "tracks" through which environmental migration has entered regional policymaking and cooperation: the migration track, the climate track, the security track, and the human rights track.

Methodology

This chapter proceeds via a review of primary documentation and secondary literature to assess if, how and to what extent the issue of environmental migration has entered regional policy and cooperation processes. In determining the scope of the analysis, the chapter limits itself to processes and initiatives which are state-led or at least have a demonstrated policy component and significant degree of state involvement. The overview covers some of the most important policy developments at regional level, but lays no claim to being exhaustive. Pure research initiatives and those consisting of only international governmental or non-governmental organizations are excluded.

Other than these limitations, the chapter deliberately casts its net wide in order to capture the variety of ways in which governments and other regional actors have engaged with the topic: firstly, the analysis of primary documentation carried out for this chapter encompassed key documents of regional processes, institutions and initiatives, such as declarations, policy papers, workplans or conference reports, irrespective of their binding or non-binding nature. The principal criterion for selection was the evidence of an *explicit link* between discussions of human mobility and environmental/climatic factors. Secondly, the chapter considers formal and informal settings, one-off events and on-going processes, and a variety of institutional set-ups: for instance, it examines the work of existing regional institutions or regional economic and political integration processes, as well as the activities of informal regional consultative processes on migration (RCPs) and spontaneous partnerships between states and other actors at a regional level and unrelated to any pre-existing institutions. Thirdly, the chapter uses a flexible geographical definition of "region", focussing mainly on regional and sub-regional settings as determined by the contours of relevant regional institutions. Thematically, as discussed above, the relationship between environmental factors and the movement of people has been approached from a range of policy angles. As a result the chapter admits to its analysis initiatives dealing with migration at large, displacement more narrowly, climate change, natural disaster and broader environmental issues. Finally, a note on terminology: to the extent possible the following analysis cites terms as used in the original documents. Otherwise, and as a shorthand, the chapter employs the term "environmental migration" to denote the complex linkages between climate change, environmental factors and human mobility. For the purpose of the chapter, the term thus encapsulates the full range of temporary, circular and permanent movements of people as a result of slow- or sudden-onset changes in their natural environment, such as environmental degradation or natural disasters, which can take the form of forced migration or represent an adaptive strategy to changing environments.[5]

A number of recurrent themes emerged from the research. Consequently, the chapter is organized along some of the main "tracks" that have transported the issue of environmental migration into regional policy settings – migration, climate, security, and human rights. Naturally, more or different "tracks" could have been devised. Equally, not every policy initiative under examination in this chapter corresponds unambiguously to one category or the other and some, arguably, could fit into several. The below overview should thus be read as one attempt at representing the principal discursive strategies and policy domains through which the issue of environmental migration has begun to gain prominence at the regional level.

[5] A critical analysis of the terminology challenges related to "environmental migration" is beyond the scope of this chapter. See IOM (2009, 2012) and Foresight (2011).

The Migration Track

The **African Union** (AU) pioneered the inclusion of environmental considerations in regional migration policy, specifically in its Migration Policy Framework for Africa[6] adopted in 2006 by the AU Executive Council in Banjul: the document recognizes environmental factors, both degradation and disasters, among the drivers of "mass migration and forced displacement" in Africa, including internal displacement, refugee movements, rural-urban migration, and cross-border migration in its analysis. Its principal recommendations concern the incorporation of environmental considerations in national and regional migration policies, enhanced research and data collection on the nexus between migration and the environment, and measures to prevent environmental degradation and natural disasters.

Unlike most other documents analysed in this chapter, the AU Framework also pays attention to the environmental impact of migration and displacement, such as pressure on natural resources and urban infrastructure and services as well as environmental degradation caused by large numbers of displaced persons. In its recommendations, it calls on States to "counter environmental degradation caused by the protracted presence of displaced persons by implementing relevant and targeted environmental protection programmes" and to consider environmental factors in resettlement decisions.[7]

To what extent the goals stated on paper are implemented in practice is an open question. No systematic evaluation of the level of operationalization of the AU Migration Policy Framework seems to exist and according to one author "it remains difficult to assess at this stage whether the AU and REC [regional economic commissions] initiatives have led to noticeable changes in the member states' legal frameworks" (Klavert 2011:iv). The same author provides an insight into the controversy and resistance that continues to surround multilateral policy proposals on migration: though not legally binding on states, the framework took 2 years to negotiate and an initial idea for an implementation mechanism was abandoned in the process, thereby giving states even greater flexibility over which elements to implement and which to leave aside (2011:4). As a result, "the AU strategic framework... remains largely unimplemented", according to Adepoju (2010:8) who adds that weak institutional capacity at the level of the AU Secretariat and at national level in most countries constitute further impediments to implementation. As concerns the incidence of climate change on migration more specifically, Adepoju concludes that "African countries lack the capacity to develop efficient early-warning systems, to monitor the effects of climate change on migration, and to factor these effects into migration policies" (2010:9).

---

[6]http://www.iom.int/jahia/webdav/shared/shared/mainsite/microsites/rcps/igad/au_migration_policy_framework_africa.pdf (Accessed 15 July 2013).

[7]Ibid, pages 24, 25 and 30.

In Europe, the issue was first tabled in the **Council of Europe**'s Parliamentary Assembly in 2006 in a motion on "the problem of environmental refugees" presented by the Turkish representative.[8] It appears, however, that the motion – which examines the inadequacies of the existing refugee and asylum regime in assisting those moving for environmental reasons and calls for awareness-raising, research and dialogue, the elaboration of definitions and policy guidelines, and sustainable development programming for areas at risk – only gained limited support. As a result, it was not discussed in the Assembly, nor did it reach the Council of Europe's Committee of Ministers for their consideration. In 2009, the Committee on Migration, Refugees and Population of the Parliamentary Assembly launched a new initiative: On the basis of a report on "Environmentally induced migration and displacement: a 21st century challenge" the Parliamentary Assembly passed resolution 1655[9] in which it "recognizes that natural disasters and environmental degradation will increasingly determine the nature of human mobility as well as its humanitarian and security dimensions" and makes far-reaching calls on the Council of Europe's member states as well as on the European Union. Besides further research, clarification of definitions, measures to reduce vulnerability and allocation of adequate resources, the resolution also calls for elaboration of "national legislation that would recognize environmental migrants and their protection needs not only through the principle of *non-refoulement* under Articles 2 and 3 of the European Convention on Human Rights (ETS No. 5), but also through subsidiary protection" (paragraph 24.5). The same Assembly debate which adopted resolution 1655 also issued recommendation 1862[10] to the Committee of Ministers, inviting them to inter alia "consider adding a new protocol to the European Convention on Human Rights (ETS No. 5), concerning the right to a healthy and safe environment". It also proposes the Committee of Ministers work with the United Nations (UN) and other international partners in pursuing an extension of the Guiding Principles on Internal Displacement to include cases of displacement through gradual environmental degradation and to develop similar guidelines concerning the rights of those compelled to cross borders due to environmental factors. The reply subsequently adopted by the Committee of Ministers[11] put a break on the ambitious proposals: while conceding the need for further research, the Committee expressed hesitation regarding an additional protocol to the European Convention on Human Rights, pointing out that the Convention already offers a certain degree of protection in

[8]Council of Europe **Doc. 11084** 23 October 2006 **The problem of environmental refugees** http://assembly.coe.int/ASP/Doc/XrefViewHTML.asp?FileID=11438&Language=EN (Accessed 15 July 2013).

[9]http://assembly.coe.int/Main.asp?link=/Documents/AdoptedText/ta09/ERES1655.htm (Accessed 11 May 2012).

[10]http://assembly.coe.int/main.asp?Link=/documents/adoptedtext/ta09/erec1862.htm (Accessed 11 May 2012).

[11]**CM/AS(2009) Rec1862 final 15** July 2009. https://wcd.coe.int/ViewDoc.jsp?Ref=CM/AS(2009)Rec1862&Language=lanEnglish&Ver=final&Site=CM&BackColorInternet=DBDCF2&BackColorIntranet=FDC864&BackColorLogged=FDC864 (Accessed 11 May 2012).

relation to environmental issues and that the European Court of Human Rights' case law will continue to evolve in this area (see also Kälin and Schrepfer 2012:47). No further initiatives on environmental migration have been tabled in the context of the Council of Europe since.

The **European Union** (EU) is commonly regarded as among the most advanced regional integration processes, in terms of its institutional development, its binding force on member states and the wide range of thematic issues covered. Relatively speaking, however, European migration policy has lagged behind other policy areas advanced by EU institutions. In so far as external migration policy – as opposed to the liberalisation of intra-EU mobility – is concerned, it has largely remained a matter of national sovereignty and inter-governmental rather than supranational governance, and has only recently become an item on the list of EU policy priorities (see Cholewinski et al. 2007; Margheritis and Maldonado 2007; Somerville 2011; Jordan 2011; Pascouau and McLoughlin 2012). By contrast, environmental policy is far more established as a EU policy domain, with EU institutions having considerably greater competencies to influence national policies as compared to migration issues (Somerville 2011).

Paradoxically, the more advanced a regional institution, the more difficult it can be to develop comprehensive and integrated policy approaches to cross-cutting issues such as environmental migration. With respect to the EU, some commentators therefore raise the point that regional bodies may not only reproduce, but sometimes even "accentuate" (Jordan 2011:7) the patterns discussed above that have proven an impediment to policymaking on migration, and environmental migration more specifically, at national levels: where policymaking is fragmented across numerous, well-entrenched sectors and institutions and where multiple competencies are at stake, "... the institutional and cognitive barriers to better coordination in a complex, multi-level system such as the EU are extremely daunting" (Jordan 2011:10).

As a preface to an overview of the EU's policy developments in this area, it is important to note that the EU has almost exclusively treated environmental migration as a matter *external* to the EU region, not an intra-EU phenomenon.[12] It is typically situated in the context of the EU's external migration policy, which is discussed in this section, as well as to a lesser extent in development, environment policy (see section "The climate track" below) and security (see section "The security track" below). While the issue was still absent from the 2005 EU Global Approach on Migration,[13] a communication by the European Commission (EC) in 2008 on "Strengthening the Global Approach on Migration" recommends that "the EU should also formulate a policy in response to recent developments *such as the increasing impact of climate change on migratory movements*" (emphasis

[12]Kälin and Schrepfer (2012) discuss one possible if unlikely exception as residing in the "solidarity clause" of the Lisbon treaty which stipulates mutual assistance among EU member states in the event of natural or man-made natural disaster.

[13]http://register.consilium.europa.eu/pdf/en/05/st15/st15744.en05.pdf (Accessed 6 July 2012).

added) and "explore the relationship between climate change and migration and gain a better understanding of the number of people affected now and in the future".[14] Notably, these recommendations are found in a section on migration and development.

A year later, in 2009, similar calls are taken up in the Stockholm Programme passed by the European Council.[15] Section 6.1.2. on migration and development states that "the connection between climate change, migration and development needs to be further explored, and the European Council therefore invites the Commission to present an analysis of the effects of climate change on international migration, including its potential effects on immigration to the Union." In the corresponding Action Plan Implementing the Stockholm Programme of 2010[16] the EC commits to issuing a communication on the topic by 2011. In explaining this commitment, Somerville (2011:7) argues that "the context for producing a communication was not a high-powered debate; although the issue did come up in the [Swedish] Presidency conference in Malmö in 2009, the debate there indicated that the issue was not a priority in any form except to avoid any connection to asylum policy". In May 2011 the EC released a "Communication on Migration", which only contains a single reference to migration and climate change[17] and promises a Commission Staff Working Paper on migration and climate change for November 2011. The Commission Staff Working Paper was subsequently postponed and eventually released in April 2013 as a Commission Staff Working Document on "Climate change, environmental degradation and migration" accompanying a Communication by the Commission on an EU Strategy on Adaptation to Climate Change. The significance of this working document resides in the fact that it expresses broad agreement on a subject among all (relevant) EC directorates consulted in the drafting process. Unlike an EC Communication, however, it does not make legislative and policy proposals. In its analysis the Commission Staff Working Document relies extensively on the 2011 Foresight report on "Migration and Global Environmental Change" and balances development, protection and adaptation considerations in looking at displacement by natural disasters, slow-onset process and migration as an adaptation strategy. The document is noteworthy in its focus on creating coherence between environmentally induced migration (sic) and different existing EU policy or funding instruments, such as the Global Approach to Migration and Mobility (see below), directives for subsidiary and temporary protection, the Global Climate Change Alliance, the Civil Protection Mechanism,

[14] http://eur-lex.europa.eu/LexUriServ/LexUriServ.do?uri=COM:2008:0611:FIN:EN:PDF (Accessed 6 July 2012), pages 7 and 8.

[15] http://register.consilium.europa.eu/pdf/en/09/st14/st14449.en09.pdf (Accessed 11 May 2012).

[16] http://eur-lex.europa.eu/LexUriServ/LexUriServ.do?uri=CELEX:52010DC0171:EN:NOT (Accessed 11 May 2012).

[17] "Special attention should be given to the relationship between migration and climate change." (page 15) http://ec.europa.eu/home-affairs/news/intro/docs/1_EN_ACT_part1_v11.pdf (Accessed 11 May 2012).

development cooperation and disaster risk reduction and a future EU Disaster Risk Management framework. With respect to the Development Cooperation Instrument, for instance, the paper suggests that "the growing relevance of this topic for the EU's relations with some non-EU countries may lead to the topic being integrated as a priority in future programming."[18] While the comprehensiveness is encouraging, it remains somewhat unclear how such coherence across the different branches of the Commission's work could be implemented concretely, what factors would persuade the different directorates involved to do so, and what resource implications this may have. Although acknowledging "the need to strengthen policy coherence at the EU level", the paper limits itself to general general recommendations in the areas of research, dialogue and cooperation.

In the meantime, environmental migration received further visibility in the revised EU Global Approach to Migration and Mobility, issued in 2011. With reference to the Stockholm Programme, the document stipulates that "addressing environmentally induced migration, also by means of adaptation to the adverse effects of climate change, should be considered part of the Global Approach."[19] A Commission Staff Working Paper on Migration and Development accompanying the Global Approach makes numerous references to climate change and migration, including the "environmental consequences of migration and asylum policy". As such, the revised Global Approach continues to frame migration – and particularly "forced displacement" – related to climate change and the environment in an adaptation/development discourse.[20]

Finally, the EC Thematic Programme on Cooperation with third countries in the areas of Migration and Asylum (strategy for 2011–2013)[21] comes closest to clear evidence of concrete action – and financial commitments – by the EU to the issue of environmental migration. Climate change is referenced as a driver of migration throughout the strategy document (particularly with respect to the Pacific region) and "migration and climate change" is one of 12 thematic priorities for funding.

Somerville rather pessimistically sums up that "policy measures on environmental migrants have barely been proposed and certainly not been implemented at the European level. Moreover, in the short term at least, it appears likely that there is limited interested in developing policy on the issue" (2011:7). Institutional obstacles and political sensitivities notwithstanding, it seems that the EU's weight

[18]See Commission Staff Working Document "Climate change, environmental degradation and migration": http://ec.europa.eu/clima/policies/adaptation/what/docs/swd_2013_138_en.pdf (Accessed 2 May 2013) pages 33–34.

[19]See corresponding EC Communication: http://eur-lex.europa.eu/LexUriServ/LexUriServ.do?uri=COM:2011:0743:FIN:EN:PDF (Accessed 15 July 2013), pages 6 and 7.

[20]http://ec.europa.eu/home-affairs/news/intro/docs/2_EN_autre_document_travail_service_part1_v3.pdf (Accessed 11 May 2012). See in particular page 18 ("More focused attention should be paid to the debate on the connection between forced displacement and development, in particular the links between migration, climate change and environmental degradation."), page 2, and others.

[21]http://ec.europa.eu/europeaid/what/migration-asylum/documents/strategy_2011-2013_en.pdf (Accessed 11 May 2012).

as a global development actor make the migration and development discourse and programming of the EU the most likely area for future policy advancement on environmental migration.

Turning to regional cooperation processes dedicated specifically to migration, only a few regional consultative processes on migration (RCPs) have taken up the issue of environmental migration, and those who have, have done so only recently and in a limited fashion. In Africa, a workshop on Inter-State and Intra-Regional Cooperation on Migration Management held by the **Intergovernmental Authority on Development** (IGAD), the African Union Commission and IOM in 2008 made several passing references to environmental factors.[22] Recommendations emanating from the meeting which was attended by government officials from IGAD member states of Djibouti, Ethiopia, Kenya, Somalia, Sudan and Uganda as well as Egypt, Libya, Niger, Tunisia and Yemen in representation of transit and destination states, included "linking migration management to economic, productive, *environmental*, conflict management, governance and social issues to address both push and pull factors of migration" as well as "incorporating migration management issues in all IGAD regional strategies as well as those of its member States [food security, *environment* and natural resources, HIV and AIDS, etc.]" (emphasis added). At the same meeting, IGAD member states launched the IGAD-RCP which remains the only RCP that makes mention of "migration and environment" in its founding document – even if only as one of 14 "related issues of common interest and concern".[23] Given this fact, the relative youth of this RCP and the many structural governance deficits plaguing the region, it may come as no surprise that the item has not reappeared in any subsequent IGAD-RCP discussions.

There may, however, be room for future developments considering the strong emphasis of the IGAD regional body on the challenges posed by recurrent drought and its consequences for agriculture, livestock herding and pastoralism, including cross-border movements by pastoralist herders. The first meeting of senior officials of the IGAD-RCP in 2010 identified "the migration challenges of nomadic communities" as an issue of concern and recommended to put in place mechanisms to engage such communities.[24] Furthermore, the AU Migration Policy Framework (see above)

---

[22] http://www.igad.int/index.php?option=com_docman&task=doc_details&gid=21&Itemid=144 (Accessed 11 May 2012).

[23] Ibid. For other RCPs, see Hansen (2010) and www.iom.int/rcps.

[24] http://www.iom.int/jahia/webdav/shared/shared/mainsite/microsites/rcps/igad/2010-Meeting-Report-on-Migration-Management.pdf (Accessed 11 May 2012). For more about pastoralism and climate change adaptation in the IGAD region, see also the final report of Security in Mobility Initiative by the United Nations Office for the Coordination of Humanitarian Affairs (UN-OCHA), United Nations Environment Programme (UNEP), International Organization for Migration (IOM) and the Institute for Security Studies (ISS) which recommends the development of "a regional migration framework that seeks to address the security imperatives of the pastoralist communities focusing on the gaps in the cross-border protection of pastoralists" http://reliefweb.int/report/ethiopia/security-mobility-advocating-safe-movement-climate-change-adaptation-strategy (Accessed 16 July 2012).

serves as a conceptual basis for the IGAD-RCP which has committed to building capacity and furthering implementation of the framework among its member states. On the basis of the AU model, IGAD-RCP member states will also be deliberating a proposed IGAD regional migration policy framework, which contemplates climate change, environmental degradation, natural disasters, nomadic pastoralism and the adaptation potential of migration extensively.[25]

The South American RCP, the **South American Conference on Migration** (SACM), had requested the International Organization for Migration (IOM) to prepare a discussion paper on migration and climate change which featured at a dedicated session during the SACM's tenth conference in Bolivia in 2010. Subsequently, the conference adopted the South American Human Development Plan for Migration which in passing recognizes "environmental factors" as among the possible causes of migration flows. The conference's final "Declaration of Migration Principles and Overall Guidelines",[26] by contrast, makes no mention of the issue. Finally, a Record of Agreements and Commitments[27] calls on the presidency-pro-tempore, together with the SACM technical secretariat, to build strategic alliances in order to generate a better understanding of environmental migration and to promote the exchange of good practices and the creation of a common position in this regard. It further calls on SACM member states to participate in the relevant roundtable session at the Global Forum on Migration and Development in Puerto Vallarta in 2010. While the SACM has a tradition of exhortatory, yet non-binding statements, its influence on regional policymaking should not be underestimated: particularly the close overlap and growing harmonization between SACM and other regional institutions, such as the Common Market of the South (MERCOSUR) and the more recent Union of South America Nations, suggest that policy discussions taking place in the context of SACM may infiltrate institutions which possess greater scope for implementation.

In Asia, RCPs tended to be firmly focussed on labour migration on the one hand (mainly through the Colombo Process) and on human trafficking and smuggling on the other (the principal domain of the Bali Process on People Smuggling, Trafficking in Persons and Related Transnational Crime). At the Fourth Ministerial Consultations of the **Colombo Process** in Dhaka, Bangladesh in 2011, the final declaration urged participating states "to further explore the possible nexus between environmental degradation and climate change on one hand and human mobility on the other, and its likely implications on labour migration".[28] Khadria (2011:16)

[25]As this framework has not been officially adopted or at the time of writing, no firm conclusions can be drawn as to its final scope or eventual implementation.

[26]http://www.iom.int/jahia/webdav/shared/shared/mainsite/microsites/rcps/sacm/Final-Declaration-2010-EN.pdf (Accessed 11 May 2012).

[27]http://www.iom.int/jahia/webdav/shared/shared/mainsite/microsites/rcps/sacm/Final-Declaration-2010-ES.pdf (Accessed 16 July 2012), section 5.

[28]Dhaka Declaration of the Colombo Process Member Countries, adopted on 21 April 2011. The cited text is found under the heading "Emergency Response and Emerging Issues".

comments that "it is the national, rather than the multilateral or bilateral, framework that is likely to be strengthened, and environmental migration would be dealt with only under 'emergency measures'." He also posits that the Colombo Process may provide a mechanism to organize evacuations (of migrant workers, presumably, although this remains unclear in the paper) across borders in case of natural disasters and that adaptive seasonal migration, facilitated by existing "transnational ties" among Colombo Process countries, would offer a form of protection against the impacts of environmental change. This, however, remains a matter of speculation: the Colombo Process was founded with a view to lending a common policy voice to Asian labour-sending countries vis-à-vis destination regions. Only to a much lesser extent has it conceived of itself as a policy forum to address intra-regional migration flows, i.e. among Colombo Process member countries. Similarly, the Asian Development Bank suggests the Colombo Process as a venue to negotiate an Asian regional convention on migration (ADB 2012:56), following in the footsteps of the Cartagena Declaration in Latin America and the Kampala Convention for Africa (see section "The human rights track" below). Considering, however, the low levels of ratification of some key *existing* international conventions concerning the movement of people, such as 1990 International Convention on Protection of the Rights of All Migrant Workers and Members of their Families or the 1951 Convention relating to the Status of Refugees, among South and South East Asian states, it is highly questionable how realistic such a proposal is.[29]

Lastly, some of the regional migration policies in the (South) Pacific may be worth mentioning in this context, if only to clarify their nature: frequently misconstrued in the literature as addressing the threats of climate change to Pacific Island populations, New Zealand's Pacific Access Category as well as Australia's more recent Pacific Seasonal Workers Pilot Scheme[30] are in fact classic admission categories for the purpose of permanent settlement (in the New Zealand example) or temporary labour migration (in the Australian case). Eligibility is determined on the basis of education, age and employability, and not in any way in relation to vulnerability to environmental factors. (For more on the Pacific region, see the next section).

[29]Of the 11 Colombo Process States only three have ratified the International Convention on Protection of the Rights of All Migrant Workers and Members of their Families (Bangladesh, Philippines, Sri Lanka) and one has signed (Indonesia): http://treaties.un.org/Pages/ViewDetails.aspx?src=TREATY&mtdsg_no=IV-13&chapter=4&lang=en. As far as the 1951 Refugee Convention is concerned, it has been ratified only by Afghanistan, China and Philippines: http://treaties.un.org/Pages/ViewDetailsII.aspx?&src=UNTSONLINE&mtdsg_no=V~2&chapter=5&Temp=mtdsg2&lang=en#Participants (as of 11 May 2012).

[30]See McAdam (2011b):121 and ADB (2012):58. The Australian programme is to be launched on a 3-year trial basis in mid-2012.

The Climate Track

An alternative route to environmental migration is via climate, or more broadly, environmental policy. Environment and migration, admittedly, have been rather distant policy cousins. Nonetheless, with increasing awareness among policy and research communities of the far-reaching social, economic and other consequences of climate change, policy developments have started taking account also of migratory dimensions of climate change. Generally, migration has been approached as a negative outcome of climate change in the event that mitigation or adaptation efforts should fail. Much less has been made of the adaptation potential of migration.

At the level of the **European Union**, the EC Green Paper on "Adapting to Climate Change in Europe – Options for EU Action"[31] of 2007 relates climate change to conflict and displacement. The paper identifies an explicit role for migration policy in mitigating the impacts of climate change and recommends that adaptation be integrated into EU external action including "in enhancing the EU's capacity to prevent and deal with conflicts such as border disputes and tensions over access to natural resources and natural disasters accentuated by climate change as well as their potential consequences such as *forced migration and internal displacements of persons. EU migration policy should also take the impacts of climate change into account, in particular in migration management.*" (emphases added). Two years later, an EC White Paper on "Adapting to climate change: towards a European framework for action"[32] (2009) pays rather less attention to the topic of migration, but again recognizes the importance of addressing climate change and its consequences, including migratory consequences, from a range of policy angles: "Failure to adapt could have security implications. The EU is therefore strengthening its analysis and early warning systems and integrating climate change into existing tools such as conflict prevention mechanisms and security sector reform. The effects of climate change on *migratory flows* should also be considered in the broader EU reflection on security, development and *migration* policies." (emphases added). In the European context, the discourse on climate change thus seems to veer into security considerations (see also section "The security track" below). The same trend is also evidenced in the 2011 Council conclusions on EU Climate Diplomacy[33] which identify the "security implications" of climate change as a "threat multiplier" which may intensify "migratory pressures", among other consequences. In fact, environment and climate change are recognized as a separate policy field for the European External Action Service (EEAS) which specifies on its website that "environment and climate change issues are increasingly acknowledged by the international and diplomatic communities in their own right but also due

[31] http://eur-lex.europa.eu/LexUriServ/site/en/com/2007/com2007_0354en01.pdf (Accessed 6 July 2012), see pages 21 and 22.

[32] http://eur-lex.europa.eu/LexUriServ/LexUriServ.do?uri=COM:2009:0147:FIN:EN:PDF (Accessed 15 July 2013), see page 15.

[33] http://ec.europa.eu/clima/events/0052/council_conclusions_en.pdf (Accessed 6 July 2012).

to the growing interdependencies with other policy fields such as trade, security, conflict prevention and *migration*." (emphasis added).[34]

The **Pacific Islands**, as may be expected, have a significant track record in collective regional engagement on climate change. It is therefore perhaps surprising that there has been comparatively little joint action on the migratory consequences of climate change which for at least some islands states may spell the inevitable permanent departure of their populations. The 2008 Niue Declaration on Climate Change encourages "the Pacific's Development Partners to increase their technical and financial support for climate change action on adaptation, mitigation and, if necessary, *relocation*,..." (emphasis added). Similarly, the 2008 Communiqué of the 39th Pacific Islands Forum on advancement of the Pacific Plan highlights the displacement of populations due to climate change and natural disasters as a "security concern" in connection with land issues, although the section on climate change in the same document does not make reference to migration or displacement consequences.[35] A stronger statement is found in the 2011 Communiqué of the 42nd Pacific Islands Forum in which, under the subheading climate change, "leaders stressed the critical and urgent need for adaptation finance to enable Forum Island Countries to respond to the adaptation needs of its people, *in particular those already suffering, are displaced or are being displaced* as a result of the detrimental impacts of climate change." (emphasis added).[36]

These were significant advances, as previous regional policy developments – notably the Pacific Plan for Strengthening Regional Cooperation and Integration (first endorsed in 2005, revised in 2007) and the 2006–2015 Pacific Islands Framework for Action on Climate Change – make no mention of migration, displacement and least of all of relocation in relation to environmental factors. One reason for the relative silence on the movement of people provoked by climate change is the fact that Pacific Island states do not necessarily have a coherent position on the subject (McAdam 2011b:111): while some states, such as Kiribati and the Maldives, have openly discussed the inevitability of relocation (including the much-mediatized proposal by the Government of the Maldives to purchase land for relocation of its citizens[37]), others, such as the Federated States of Micronesia and Tuvalu, have persistently resisted references to relocation in international agreements, concerned that such concessions would reduce the international community's sense of urgency in the fight against climate change (McAdam 2011b:111). Meanwhile, Kiribati is actively pursuing gradual "merit-based" migration options for its citizens in

[34] http://eeas.europa.eu/environment/index_en.htm (Accessed 6 July 2012).

[35] http://www.forumsec.org/resources/uploads/attachments/documents/2008%20Forum%20Communique,%20Alofi,%20Niue,%2019-20%20Aug.pdf (Accessed 17 July 2012).

[36] http://www.forumsec.org/resources/uploads/attachments/documents/2011%20Forum%20Communique,%20Auckland,%20New%20Zealand%207-8%20Sep1.pdf (Accessed 15 July 2013).

[37] See for example http://www.guardian.co.uk/environment/2008/nov/10/maldives-climate-change (Accessed 11 July 2012), also Martin (2010:10), Khadria (2011:7), and Leckie et al. (2012).

the region, specifically to Australia and New Zealand. By increasing the size of i-Kiribati communities abroad, the government hopes to ease the process of relocation and integration of their compatriots if and when the whole population is forced to move (McAdam 2011b:121, Martin 2010:10, Mclellan 2012).

A similar reluctance to confront the issue of migration, relocation and potential permanent displacement of populations is reflected in the work of the **Alliance of Small Island States** (AOSIS), which is not limited to one geographical area but rather unites countries which share similar geographical features. While climate change, sea level rise and natural disasters are a high priority, references to any migration or displacement consequences or relocation plans have been relatively muted. A joint declaration on climate change adopted at an AOSIS summit in 2009 focuses largely on necessary mitigation measures, though also expresses an "urgent need to consider and address the security implications and the human dimensions of climate change, *including where necessary, initiatives for preparing communities for relocation*" (emphasis added).[38] Yet, AOSIS statements made in the context of and following the 2011 UN climate change negotiations in Durban, South Africa, while underlining the threat climate change poses to the "survival" of their nations, do not tackle questions of displacement, migration and relocation.

Another example of a region not in the geographical sense, but "in spirit", the **Climate Vulnerable Forum** should not be neglected for its contribution in bringing not only climate change, but also its migration consequences, to the attention of a range of countries not necessarily known for being at the forefront of multilateral efforts on such subjects. Adopted at the first meeting of the Forum, the Male Declaration (2009)[39] issued a call to the state parties to the UN Framework Convention on Climate Change (UNFCCC) "to also consider and address the health, human rights and security implications of climate change, including the need to prepare communities for *relocation*, to protect *persons displaced across borders due to climate change-related impacts*, and the need to create a legal framework to protect the human rights of those left *stateless* as a result of climate change" (emphases added). This was echoed in the 2010 Ambo Declaration[40] which brought together a number of countries beyond the original group, including, notably, Australia, Brazil, China, Cuba, Japan and New Zealand. The language used in the 2010 declaration is somewhat weaker and makes no explicit mention of statelessness and relocation: "Support consideration of the development and implementation of strategies and actions directed at *protecting people displaced within or across borders* as a result of adverse effects arising from climate change extreme events"

[38]http://www.unohrlls.org/en/orphan/720/ (Accessed 15 July 2013).

[39]http://daraint.org/wp-content/uploads/2010/12/Declaration-of-the-CVF-FINAL2.pdf (Accessed 16 July 2012). States present at the conference were Maldives, Kiribati, Bangladesh, Barbados, Bhutan, Ghana, Kenya, Nepal, Rwanda, Tanzania and Vietnam.

[40]Adopted at the Tarawa Climate Change Conference on 10 November 2010 by Australia, Brazil, China, Cuba, Fiji, Japan, Kiribati, Maldives Marshall Islands, New Zealand, Solomon Islands and Tonga. http://en.wikipedia.org/wiki/Tarawa_Climate_Change_Conference (Accessed 17 July 2012).

(emphasis added). At its most recent gathering in Dhaka, Bangladesh in 2011,[41] the Climate Vulnerable Forum's position on the issue took a significant turn in that it acknowledged the role of migration as a mechanism for adaptation to climate change. The text indicates that it was influenced by relevant language in the Cancun Agreements adopted at the 16th Conference of the Parties of the UNFCCC, particularly paragraph 14(f). The salient passages of the Dhaka Declaration thus read: "Aware that climate change induced displacement of people is a major concern and their relocation puts enormous pressure on infrastructures and service facilities; and furthermore, large-scale displacement has the potential to transform into security concerns [. . .] Recognising that migration is a viable adaptation strategy to ensure that populations are not compelled to reside in high risk and affected areas, and to manage risks during displacement; and furthermore a planned strategy in the long-term to offer displaced populations with enhanced options for a dignified and diversified livelihood". The declaration further calls for the "immediate implementation" of paragraph 14(f), although the text that follows leaves the intentions of the signatories somewhat ambiguous when it states "that migration is a viable adaptation strategy to address human displacement induced by climate change". The same paragraph also emphasises the need for "planned relocation".

The **Council of Arab Ministers** responsible for the environment have yet a different take on environmental migration, where it appears within the parameters of disaster risk reduction and sustainable development in the 2010 Arab Strategy for Disaster Risk Reduction.[42] The framework recognizes the displacement consequences of natural and human-induced hazards and concludes that "in combination with the current demographic trends most parts of the region will experience severe migration pressures as a result of which the most vulnerable groups, especially women, are likely to be the most affected". Internally displaced persons are considered as a vulnerable group in the Strategy's key priorities for action.

Finally, two minor examples from the Americas include initiatives by the **Central American Parliament** and the **Caribbean Development Bank**. The former adopted the Political Declaration of Managua on Climate Change in 2010 which considers "migration due to climate change", among other issues, and pledges support to "the protection and recognition of the rights and needs of migrants, both internal and external, forced to leave their communities due to climatic causes." In a 2008 position paper, the Caribbean Development Bank 2008 makes passing reference to the vulnerability of "the poorest people" to "geographical displacement" in the context of the adverse effects climate change is likely to have for poverty reduction efforts in the region.

---

[41] http://daraint.org/2011/11/14/2748/climate-vulnerable-forum-declaration-adopted/ (Accessed 15 July 2013).

[42] http://www.unisdr.org/files/17934_asdrrfinalenglishjanuary2011.pdf (Accessed 17 May 2012), see pages 9 and 11 in particular.

The Security Track

A third entry door to the issue of environmental migration is security: it has been a common feature of policy discourse to see climate change, migration, and ultimately environmental migration "lumped... together with a broader 'threat' of a more insecure world" (Somerville 2011:14). Following initial and rather uncritical temptations to conjure up menacing scenarios, a growing body of literature[43] has contributed to a more nuanced understanding of interactions between climate change, livelihoods, migration and conflict.

At a regional level, the **Organisation for Security and Co-operation in Europe** (OSCE) raised the issue at its 15th Ministerial Council in 2007 at which it adopted the Madrid Declaration on Environment and Security. Conclusion 2 states that "environmental degradation, including both natural and man-made disasters, *and their possible impact on migratory pressures*, could be a potential additional contributor to conflict. Climate change may magnify these environmental challenges." (emphasis added). The OSCE is also part of the "Environment and Security Initiative (EnvSec), a partnership created in 2002 between the OSCE, the North Atlantic Treaty Organization and several UN agencies "to contribute to the reduction of environment and security risks through strengthened cooperation among and within countries in four regions: Central Asia, Eastern Europe, Southern Caucasus, and South-Eastern Europe."[44] While the introduction to EnvSec on its website makes reference to "competition over declining natural resources such as forests, fresh water, fisheries or fertile soils, exacerbated by climate change impacts and *migratory movements of population* [as] a threat to stability and peace" (emphasis added), there is little evidence that this has been a priority in the group's work.[45] More recently, the Economic and Environmental Activities section of the OSCE launched a project on the "Security Implications of Climate Change", set to run from 2009 to 2012, with the objective to create regional scenarios on the impact of climate change on security. A scoping study on the topic was commissioned and published in 2010.[46] The report clearly identifies migration and population displacement as a security risk, while also making curiously loose, if not to say inaccurate, use of the term "refugee" (including referring to "environmental refugees", "internal

[43]See for example UNEP (2011) and McLeman (2011).

[44]http://www.envsec.org/index.php?option=com_content&view=article&id=60&Itemid=60&lang=en and http://www.osce.org/eea/43651 (Accessed 17 May 2012).

[45]See, however, the statement of the EnvSec Chair at the 2009 OSCE conference on Security Implications of Climate Change in the OSCE Region which makes numerous references to the challenges posed by transboundary migration and internal displacement in the context of climate change.

[46]"Shifting Bases, Shifting Perils: A Scoping Study on Security Implications of Climate Change in the OSCE region" available at http://www.osce.org/eea/78356 (accessed 17 May 2012) In particular, see pages 8 and 32.

refugee challenges" and "Katrina refugees" in the United States).[47] The Southern Mediterranean is highlighted as a region of particular concern, and the report conjectures that of the "50 million African environmental refugees [sic]" most will "want to relocate to Europe", although it remains unclear how the report arrives at this conclusion. This type of analysis is echoed in other parts of the study. Being an externally produced document, the report should not be seen as representing a formal OSCE position on the subject, but nevertheless is likely to have influenced the approach taken in the project.

While the OSCE may be institutionally obliged to give priority to security dimensions in its analyses, the EU has approached the issue from a range of angles, as described earlier in the chapter. In 2008, in the same year as the impact of climate change on migration was put forward for consideration in the EU Global Approach to Migration (see section "The migration track"), the then High Representative Javier Solana and the European Commission submitted a report to the **European Council** on "Climate Change and International Security" which became known as the Solana Report.[48] Again, environmentally-induced migration is identified as a potential future threat. The report warns of increased migration pressures towards Europe and links with instability and conflict, highlighting Africa and South Asia as regions of concern in this regard. Possible areas for actions recommended by the report include "considering environmentally-triggered additional migratory stress in the further development of a comprehensive European migration policy, in liaison with all relevant international bodies." Somerville (2011:6) notes that the initiative was led by the foreign affairs apparatus of the EU, principally by Solana and Benita Ferrero-Waldner, the then European Commissioner for external relations, not home affairs as might have been expected. The Directorate of Home Affairs has since claimed a significant, if not the predominant role in the EC on this issue, although rather by linking climate change with migration and development and to the Global Approach, than with security concerns. As cited in section "The climate track" above, more recently, the EEAS has identified a role for itself with regards to climate change and security. A Joint Reflection Paper by the EEAS and the EC accompanying the conclusions on EU Climate Diplomacy of 2011 explicitly references migration, stating that "while climate change alone does not cause conflict, it is leading to increased competition for scarce resources, further weaken fragile governments and *exacerbates migratory pressures*." (emphasis added).[49]

[47]The relevant literature on this subject concurs that the term "refugee" as defined in the 1951 Convention relating to the Status of Refugees cannot be applied to situations in which persons move due to environmental factors. For instance, see Zetter (2009) and McAdam (2012) or the Working Paper of the informal group on migration/displacement and climate change of the Inter-Agency Standing Committee (2008) http://unfccc.int/resource/docs/2008/smsn/igo/022.pdf (Accessed 17 July 2012).

[48]http://www.consilium.europa.eu/uedocs/cms_data/docs/pressdata/en/reports/99387.pdf (Accessed 17 May 2012).

[49]http://eeas.europa.eu/environment/docs/2011_joint_paper_euclimate_diplomacy_en.pdf (Accessed 6 July 2012).

As a final example in this section, in 2010 the Madariaga-College of Europe Foundation and the Folke Bernadotte Academy, in cooperation with the General Secretariat of the Council of the EU, launched a multistakeholder initiative entitled "**Africa, Climate Change, Environment and Security** (ACCES) Dialogue Process".[50] This initiative has its origins in the above-mentioned Solana report and related events. Through a series of consultations and working groups, which included participation by EU institutions and numerous international and regional bodies and UN organizations, the initiative elaborated a discussion paper covering five areas of vulnerability related to climate change (water security, food security, energy security, natural hazards and migration). The paper was presented at the first ACCES Dialogue Forum in October 2010, held as part of the Seventh African Development Forum and which brought together representatives of relevant ministries from national governments in Africa, the UN, EU and AU, and others. The "Climate change and security in Africa: Vulnerability report" issued following the forum cautions against drawing simplistic links between climate change, migration and security and makes a number of recommendations relating to the need for better data, international cooperation, and the inclusion of migration in national adaptation plans. The ACCES Dialogue Process is continuing its work by focussing on research and dialogue in a limited number of African countries/regions. Its stated objective is to enhance local resilience and collaboration between African and international partners to tackle the security consequences of climate change "from a development and human security perspective". The policy impact of this initiative remains uncertain at this stage.

The Human Rights Track

The applicability of regional legal instruments, drawn mainly from the realm of refugee and human rights protection, to environmental migration has been widely discussed and shall only receive brief treatment in this chapter.[51] The common denominator to note is that, by and large, these are pre-existing, binding and non-binding instruments which may have direct or indirect relevance for population movements induced by natural disasters, environmental degradation and climate change.

As regards regional human rights instruments, the **African Charter of Human and Peoples' Rights** includes a "right to a general satisfactory environment favourable to their development" (Article 24). According to Zetter (2011:18), this provision could be used to argue for a duty on states to relocate populations whose right to a satisfactory environment is violated. More recently, the 2009 **AU Convention for the Protection and Assistance of Internally Displaced Persons**

[50]http://www.envirosecurity.org/acces/docs/brochure.pdf and http://www.envirosecurity.org/acces/docs/ACCES_Programme_Outline.pdf (Accessed 17 May 2012).

[51]For more detail, see Norwegian Refugee Council (2009), Zetter (2009 and 2011), McAdam (2012), Asian Development Bank (2012), Kälin and Schrepfer (2012), and Leckie et al. (2012).

in Africa (known as the "**Kampala Convention**") takes a significant step in linking displacement and climate change, stipulating that "States Parties shall take measures to protect and assist persons who have been internally displaced due to natural or human made disasters, including climate change" (under article 5: obligations of State Parties relating to protection and assistance). The convention entered into force in December 2012 with the necessary 15 ratifications, yet remains a long way from implementation at domestic level, making it difficult to assess its impact at this point.[52]

Other regional instruments commonly referenced in this context are the 1984 **Cartagena Declaration** of the Organisation of American States, a non-binding instrument, and the 1969 **Organization of African Unity Convention Governing specific aspects of Refugee Problems in Africa**. Both provide a more expansive refugee definition than the 1951 Convention relating to the Status of Refugees, including threats emanating from "serious disturbances of public order". This provision could potentially be stretched to include natural disasters. Given the date of both conventions, however, it is highly questionable whether climate change or even environmental factors were on the drafters' minds and can be considered part of the conventions' intention (see Zetter 2009:414; Kälin and Schrepfer 2012:26, 34).

Also on the African continent, the Lomé Declaration on Protection Challenges to Climate Change in West Africa of 2009,[53] initiated by the UN and the regional body ECOWAS (**Economic Community of West African States**), makes ambitious calls for a new legal instrument to protect those displaced by climate change and, unusually, highlights migrants as a group vulnerable to climate change: "recognizing the urgency to encourage the recognition of the protection needs of populations which are not taken into due account by the existing legal instruments (International Refugee Law, the United Nations Guiding Principles on Internal Displacement, the Refugee Convention of the Organization of African Union)" the Declaration recommends "the establishment of measures to protect the various categories of populations affected by climate change, including *migrants*, and especially, women, youth, children, disabled people and other vulnerable groups, in order to preserve the full enjoyment of their fundamental human rights" as well as "the drafting of a new legal instrument aimed at ensuring protection for *climate change displaced persons, residing outside their country of origin*" (emphases added). It seems, however, that nearly 3 year after the adoption of the Declaration no further steps have been taken to transform its exhortations into action, let alone into legal instruments (see Kälin and Schrepfer 2012:48).

Lastly, there has been some debate surrounding the relevance of **EU protection instruments**, including the Temporary Protection Directive (Council Directive

[52]By May 2013 the number of ratifications had increased to 17. See http://www.internal-displacement.org/kampala-convention (Accessed 23 May 2013). See also Koser (2011) and IOM (2012).

[53]http://www.iag-agi.org/bdf/en/corpus_document/fiche-document-914.html (Accessed 17 May 2012).

2001/55/EC), the Qualification Directive (Council Directive 2004/83/EC) and Regional Protection Programmes. Neither directive makes reference to climate change or natural disasters as grounds for protection; in fact, insights into the negotiation records for the Temporary Protection Directive suggest that while this element was raised, it was later dropped (see Kolmannskog and Myrstad 2009). Arguably, there may be some scope for temporary protection in the case of mass influx of persons, presumably following a major natural disaster. Any such decision, however, is at the discretion of the European Council on a case-by-case basis and to date the Temporary Protection Directive has never been used. Similarly, although the inclusion of environmental disasters was reportedly raised by the EC in discussions on subsidiary protection as early as 1999, it was not included in the proposal for the Qualification Directive on subsidiary protection which resulted in narrow wording making the directive's application to displacement situations caused by environmental factors or natural disasters unlikely (see Kolmannskog and Myrstad 2009; Kälin and Schrepfer 2012). More recently, the European Parliament (EP) has begun to examine available European policy and legal instruments for incidences of "environmentally induced displacement". A study commissioned by the EP's Committee on Civil Liberties, Justice and Home Affairs reviews the Qualification and Temporary Protection Directive, the European Charter of Fundamental Rights, the Lisbon Treaty, the Global Approach to Migration and Mobility (see section "The migration track" above), Regional Protection Programmes, Mobility Partnerships, and European resettlement policies, as well as relevant provisions in the legislation of EU member states.[54] Calling for a "human rights based approach", the paper finds that currently, there is "no distinct instrument applicable to 'environmentally displaced individuals'" at EU level (2011:49), but identifies potential scope in the Lisbon Treaty to revise existing asylum and immigration policy to account for "environmentally displaced individuals" and in the Global Approach to promote protection and resilience of climate change outside the EU. Overall, however, the paper mainly clarifies the existing legal and policy landscape at EU level and it is unclear to what extent concrete follow-up action on the part of the EP can be expected.

Conclusion

The preceding overview, while not an exhaustive list, provides a summary of the principal policy developments on environmental migration at the regional level over the course of the past decade. It suggests that environmental migration is not an isolated policy item, but has emerged via other discourses and policy priorities – the main ones selected in this chapter being migration, environment, security, and human rights. Furthermore, the chapter argues that there is evidence that the issue

[54] http://www.europarl.europa.eu/committees/de/studiesdownload.html?languageDocument=EN&file=60931 (Accessed 12 July 2012).

of environmental migration has started to permeate policy awareness, though to varying degrees, while implementation of concrete policies remains rare. Regional policy and cooperation on environmental migration have often remained at the level of **informal, non-binding dialogue** (such as those held by the Colombo Process, the SACM and IGAD-RCP, by the Central American Parliament or as part of the ACCES forum); have in some instances translated into **"soft" regional policy** (including the AU Migration Policy Framework, the EU Global Approach to Migration and Mobility, or the Arab Strategy for Disaster Risk Reduction); and in a few cases have acquired **legal force** (the main example being the Kampala Convention).

A few tentative conclusions may also be drawn regarding the conduciveness of different institutional arrangements to taking up this issue: contrary to expectations, RCPs have so far played a rather limited role in putting environmental migration on regional policy agendas. Their informal, non-binding nature and relative flexibility in agenda-setting would seem well-suited for a potentially controversial topic (see Hansen 2010); but a combination of limited interest, limited capacity and limited participation – many RCPs are based on a well-established circle of officials specialized in a certain area of migration policy, such as labour migration or counter-trafficking, and rarely involve officials from other policy domains – may explain why RCPs have been less prominent as a forum to raise awareness of environmental migration. However, at a global meeting of RCPs in 2011, one working group was dedicated to the nexus between migration, climate change and the environment and may have helped in bringing the issue to the attention of RCPs.[55] More formal regional institutions, by contrast, have made comparatively greater strides, although some have approached environmental migration more narrowly or from several angles at once, in particular where the institutional architecture is sufficiently elaborate, such that different policy sectors within the same regional body may take interest in environmental migration.

In sum, regional policy action on environmental migration remains **incipient, indirect, informal and often incoherent**: incipient, because most initiatives are barely half a decade old and are in many instances not fully mature, making assessments of levels of implementation or effectiveness very difficult. Indirect, as the examples show that the linkages between migration and the environment are entering the policy discourse via other issues which may perpetuate segmented policy approaches: whether environmental migration is addressed in a migration, environment, security or human rights context will influence how it is conceptualised and the kind of policy responses that would be contemplated. With very few exceptions, policy and cooperation remains informal and does not produce outcomes that would be binding on states or lead to legislative changes. As a result, it is doubtful whether any regional cooperation mechanisms could be relied upon in the event that a large number of people move in response to environmental factors.

[55] http://www.iom.int/cms/en/sites/iom/home/what-we-do/regional-processes-1/global-rcp-meetings/2011-global-rcp-consultation.html (Accessed 15 July 2013).

Lastly, the existence of numerous initiatives in parallel and from different policy angles, sometimes within the same region, spells the risk of incoherence.

One of the most pressing questions for the future is how existing and nascent regional migration arrangements will respond to environmental migration. A number of regions have moved towards varying degrees of free movement of people – Europe, West Africa and South America being among the most advanced examples – and others are exploring similar schemes.[56] The Asian Development Bank, for instance, has recommended using current migration channels to absorb environmental migration, which would serve to meet labour demand in Asian countries and allow for adaptation and risk management though mobility, remittances and other beneficial socio-economic effects of migration (2012:55, see also Khadria 2011). Whether regional mobility regimes will adapt to the migration of individuals in search of environmental safety and livelihoods, or whether regional mechanisms will falter and shut down in the face of what could be significant population movements, remains, for the moment, a matter of speculation.

References

Adepoju, A. (2010). *The future of migration policies in Africa*. Background Paper IOM World Migration Report: The future of migration – Building capacities for change. Available electronically only at http://publications.iom.int/bookstore/index.php?main_page=product_info&cPath=37&products_id=652. Accessed 15 July 2013.

Asian Development Bank. (2012). *Asian development bank addressing climate change and migration in Asia and the Pacific. Mandaluyong City, Philippines: Asian Development Bank*. http://www.adb.org/publications/addressing-climate-change-and-migration-asia-and-pacific

Betts, A. (Ed.). (2011). *Global migration governance*. Oxford: Oxford University Press.

Cholewinski, R., Perruchoud, R., & MacDonald, E. (Eds.). (2007). *International migration law – Developing paradigms and key challenges*. The Hague: IOM and T.C.M. Asser Press.

Foresight: Migration and Global Environmental Change. (2011). *Final project report*. London: The Government Office for Science.

Gemmene, F. (2011). How they became the human face of climate change. Research and policy interactions in the birth of the "environmental migration" concept. In É. Piguet, A. Pécoud, & P. De Guchteneire (Eds.), *Migration and climate change*. Cambridge: Cambridge University Press.

Hansen, R. (2010). *An assessment of regional consultative processes on migration* (IOM Migration Research Series No. 38). Geneva: International Organization for Migration.

IOM. (2009). *Migration, environment and climate change: Assessing the evidence*. Geneva: International Organization for Migration. Available at http://publications.iom.int/bookstore/index.php?main_page=product_info&cPath=1&products_id=539. Accessed 15 July 2013.

IOM. (2010a). *World migration report: The future of migration – Building capacities for change*. Geneva: IOM.

IOM. (2010b). *Free movement of persons in regional integration processes*. Report of the International Dialogue on Migration. Red Book. No. 13. Geneva: International Organization for Migration.

[56] See IOM (2010b).

IOM. (2011). *World migration report: Communicating effectively about migration*. Geneva: International Organization for Migration. Available at http://publications.iom.int/bookstore/index.php?main_page=product_info&cPath=37&products_id=752. Accessed 15 July 2013

IOM. (2012). *Climate change, environmental degradation and migration*. Report of the International Dialogue on Migration. Red Book No.18. Geneva: International Organization for Migration. www.iom.int/idmclimatechange/lang/en

Jordan. (2011). *Policy Development Review No.18 commissioned as part of the UK Government's Foresight Project, Migration and Global Environmental Change*. Crown copyright 2011 Foresight: London. Available electronically only at http://www.bis.gov.uk/assets/foresight/docs/migration/policy-development/11-1152-pd18-european-union-coordinating-environment-and-migration.pdf. Accessed 15 July 2013.

Kälin, W., & Schrepfer, N. (2012). *Protecting people crossing borders in the context of climate change: Normative gaps and possible approaches* (UNHCR Legal and Protection Policy Research Series). Geneva: UNHCR. Available electronically only at http://www.unhcr.org/4f33f1729.html. Accessed 15 July 2013.

Khadria, B. (2011). *State of Science Review No.20 commissioned as part of the UK Government's Foresight Project, Migration and Global Environmental Change*. Crown copyright 2011 Foresight: London. Available electronically only at http://www.bis.gov.uk/assets/foresight/docs/migration/science-reviews/11-1139-sr20-impact-on-migration-system-if-colombo-process-succeeded.pdf. Accessed 15 July 2013.

Klavert, H. (2011). *African Union framework for migration: Current issues and questions for the future* (ECDPM Discussion Paper No. 108). Maastricht: ECDPM. Available electronically at www.ecdpm.org/dp108. Accessed 15 July 2013.

Kolmannskog, V., & Myrstad, F. (2009). Environmental displacement in European asylum law. *European Journal of Migration and Law, 11*(2009), 313–326.

Koser, K. (2011). Climate change and internal displacement: Challenges to the normative framework. In É. Piguet, A. Pécoud, & P. De Guchteneire (Eds.), *Migration and climate change*. Cambridge: Cambridge University Press.

Leckie, S., et al. (Eds.). (2012). *Climate change and displacement reader*. New York: Routledge.

Margheritis, A., & Maldonado, M. (2007). Regional integration and (hauled) migration policy: What does the European experience teach us? *Journal of European Public Policy, 14*(1), 152–166.

Martin, S. F. (2009). Managing environmental migration. In IOM (Ed.), *Migration, environment and climate change: Assessing the evidence*. Geneva: IOM.

Martin, S. F. (2010). *Climate change and international migration*. Background Paper IOM World Migration Report. Available electronically only at http://publications.iom.int/bookstore/index.php?main_page=product_info&cPath=37&products_id=634. Accessed 15 July 2013.

McAdam, J. (2011a). Environmental migration. In Betts (Ed.), *Global migration governance*. Oxford: Oxford University Press.

McAdam, J. (2011b). Refusing 'refuge' in the Pacific: (De)constructing climate induced displacement in international law. In É. Piguet, A. Pécoud, & P. De Guchteneire (Eds.), *Migration and climate change*. Cambridge: Cambridge University Press.

McAdam, J. (2012). *Climate change, forced migration and international law*. Oxford: Oxford University Press.

Mclellan, N. (2012). *Kiribati's policy for "migration with dignity"*. Devpolicyblog from the Development Policy Centre (2012, January 12). Available at http://devpolicy.org/kiribati_migration_climate_change/. Accessed 15 July 2013.

McLeman, R. (2011). Climate change, migration and critical international security considerations (IOM Migration Research Series No. 42). Geneva: International Organization for Migration. Available electronically at http://publications.iom.int/bookstore/free/MRS42.pdf. Accessed 15 July 2013.

McNamara, K. E. (2007). Conceptualizing discourses on environmental refugees at the United Nations. *Population and Environment, 29*(1), 12–24.

Norwegian Refugee Council. (2009). *A thematic report from the Norwegian Refugee Council*. Available electronically at http://www.nrcfadder.no/arch/img.aspx?file_id=9913616. Accessed 15 July 2013.

Pascouau, Y., & McLoughlin, S. (2012, January 24). *EU single permit directive: A small step forward in EU migration policy* (EPC Policy Brief). Available electronically at http://www.epc.eu/documents/uploads/pub_1398_eu_single_permit_directive.pdf. Accessed 15 July 2013.

Popp, K. (2012). Regional processes, law and institutional developments on migration. In B. Opeskin, R. Perruchoud, & J. Redpath-Cross (Eds.), *Foundations of international migration law*. Cambridge: Cambridge University Press.

Somerville, W. (2011). *Policy Development Review No.17 commissioned as part of the UK Government's Foresight Project, Migration and Global Environmental Change*. Crown copyright 2011 Foresight: London. Available electronically only at http://www.bis.gov.uk/assets/foresight/docs/migration/policy-development/11-1151-pd17-environmental-migration-governance-european-union.pdf. Accessed 15 July 2013.

United Nations Environment Programme. (2011). *Livelihood security: Climate change, conflict and migration in the Sahel*. Geneva: United Nations Environment Programme. Available at http://www.unep.org/publications/contents/pub_details_search.asp?ID=6231. Accessed 15 July 2013.

Warner. (2010). Background paper produced for the German Marshall Fund of the United States Transatlantic Study Team on Climate Change & Migration. Available electronically only at http://www.ehs.unu.edu/article/read/gmf. Accessed 15 July 2013.

Warner, K. (2011). *Climate change-induced displacement: Adaptation policy in the context of the UNFCCC climate negotiations* (UNHCR Legal and Protection Policy Research Series). Geneva: UNHCR. Available electronically only at http://www.unhcr.org/4df9cc309.pdf. Accessed 15 July 2013.

Zetter, R. (2009). The role of legal and normative frameworks for the protection of environmentally displaced people. In IOM (Ed.), *Migration, environment and climate change: Assessing the evidence*. Geneva: IOM.

Zetter, R. (2011). *Protecting environmentally displaced people: Developing the capacity of legal and normative frameworks*. Research Report. Available electronically at http://www.rsc.ox.ac.uk/pdfs/workshop-conference-research-reports/Zetter-%20EnvDispRep%2015022011.pdf. Accessed 15 July 2013.

MIX
Papier aus verantwortungsvollen Quellen
Paper from responsible sources
FSC® C105338

If you have any concerns about our products,
you can contact us on
ProductSafety@springernature.com

In case Publisher is established outside the EU,
the EU authorized representative is:
Springer Nature Customer Service Center GmbH
Europaplatz 3, 69115 Heidelberg, Germany

Printed by Libri Plureos GmbH
in Hamburg, Germany